Lecture Notes in Computer Science   10718

Commenced Publication in 1973
Founding and Former Series Editors:
Gerhard Goos, Juris Hartmanis, and Jan van Leeuwen

More information about this series at http://www.springer.com/series/7411

Antonio Fernández Anta · Tomasz Jurdzinski
Miguel A. Mosteiro · Yanyong Zhang (Eds.)

Algorithms for Sensor Systems

13th International Symposium on Algorithms and Experiments
for Wireless Sensor Networks, ALGOSENSORS 2017
Vienna, Austria, September 7–8, 2017
Revised Selected Papers

 Springer

Editors

Antonio Fernández Anta
IMDEA Networks Institute
Leganés
Spain

Tomasz Jurdzinski
University of Wrocław
Wroclaw
Poland

Miguel A. Mosteiro
Pace University
New York
USA

Yanyong Zhang
Rutgers University
North Brunswick, NJ
USA

ISSN 0302-9743   ISSN 1611-3349   (electronic)
Lecture Notes in Computer Science
ISBN 978-3-319-72750-9   ISBN 978-3-319-72751-6   (eBook)
https://doi.org/10.1007/978-3-319-72751-6

Library of Congress Control Number: 2017962890

LNCS Sublibrary: SL5 – Computer Communication Networks and Telecommunications

Printed on acid-free paper

This Springer imprint is published by Springer Nature
The registered company is Springer International Publishing AG
The registered company address is: Gewerbestrasse 11, 6330 Cham, Switzerland

Preface

ALGOSENSORS, the International Symposium on Algorithms and Experiments for Wireless Sensor Networks, is an international conference dedicated to algorithmic aspects of networks of restricted devices. The 13th edition of ALGOSENSORS was held during September 7–8, 2017, in Vienna, Austria, as a part of the ALGO 2017 event.

While ALGOSENSORS was created to focus on sensor networks, in recent years it has broadened its scope to topics around the common theme of wireless networks of computational entities. For example, networks of (static or mobile) sensors, cyber-physical systems, mobile robots, Internet of Things (IoT) devices, and drones. Aspects explored include optimization, security and privacy, energy management, localization, coordination and pattern formation, data collection and aggregation, and fault tolerance. This year, ALGOSENSORS had two tracks: Algorithms and Theory, and Experiments and Applications.

For ALGOSENSORS 2017, we received 30 submissions from more than 20 countries. These manuscripts were rigorously reviewed by our Program Committee of 25 members and some external reviewers. Each submission had three or more reviews. As a result, 17 papers were accepted and presented at the conference. The program was completed with the keynote presentation by Jie Gao (Stony Brook University), to whom we are very grateful for delivering an excellent talk. This volume contains the technical details of the papers presented at the conference.

This year, for the first time, one paper was selected to receive the ALGOSENSORS Best Paper Award in the Algorithms and Theory Track. The paper selected was "Parameterized Algorithms for Power-Efficient Connected Symmetric Wireless Sensor Networks" by Matthias Bentert, René van Bevern, André Nichterlein, and Rolf Niedermeier. Congratulations to the authors.

We want to thank all the Program Committee members, and their external reviewers, for their efforts in selecting the best papers. The strong final program of ALGO-SENSORS 2017 is a reflection of their excellent work. We also want to thank the Organizing Committee of ALGO 2017 for the great coordination of the ALGO event, and facilitating the hosting of ALGOSENSORS 2017 in such a nice environment. We are particularly grateful to the Organizing Committee chair, Stefan Szeider, who was always there to help. Finally, we want to thank the Steering Committee of ALGO-SENSORS for trusting us with the task of driving ALGOSENSORS 2017, and in particular the Steering Committee chair, Sotiris Nikoletseas, for all the help and guidance he provided.

November 2017

Miguel A. Mosteiro
Antonio Fernández Anta
Tomasz Jurdzinski
Yanyong Zhang

Organization

Program Committee

Ashwin Ashok	Georgia State University, USA
Evangelos Bampas	LIF, Aix-Marseille University and CNRS, France
Amotz Bar-Noy	City University of New York, USA
Fernando Boavida	University of Coimbra, Portugal
Jacek Cichon	Wroclaw University of Technology, Poland
Gianluca de Marco	University of Salerno, Italy
Robert Elsasser	University of Salzburg, Austria
Martin Farach-Colton	Rutgers University, USA
Antonio Fernández Anta	IMDEA Networks Institute, Spain
Ben Firner	NVIDIA, USA
Leszek Gasieniec	University of Liverpool, UK
James Gross	RWTH Aachen University, Germany
Tomasz Jurdzinski	University of Wroclaw, Poland
Evangelos Kranakis	Carleton University, Canada
Joseph S. B. Mitchell	Stony Brook University, USA
Miguel A. Mosteiro	Pace University, USA
Jorge Ortiz	IBM, USA
Hui Pan	The Hong Kong University of Science and Technology, SAR China
Gopal Pandurangan	University of Houston, USA
Dror Rawitz	Bar-Ilan University, Israel
Guiling Wang	New Jersey Institute of Technology, USA
Dongxiao Yu	Huazhong University of Science and Technology, China
Lan Zhang	Tsinghua University, China
Yanyong Zhang	Rutgers University, USA
Rong Zheng	McMaster University, Canada

Additional Reviewers

Augustine, John	Molla, Anisur Rahaman
Chatterjee, Soumyottam	Montangero, Manuela
Chlebus, Bogdan	Rescigno, Adele
Even, Guy	Robinson, Peter
Karousatou, Christina	Scquizzato, Michele

Contents

Collaborative Delivery by Energy-Sharing Low-Power Mobile Robots

Evangelos Bampas[1], Shantanu Das[1(✉)], Dariusz Dereniowski[2],
and Christina Karousatou[1]

[1] LIF, Aix-Marseille University and CNRS, Marseille, France
{evangelos.bampas,shantanu.das,christina.karousatou}@lif.univ-mrs.fr
[2] Faculty of Electronics, Telecommunications and Informatics,
Gdańsk University of Technology, Gdańsk, Poland
deren@eti.pg.edu.pl

Abstract. We study two variants of delivery problems for mobile robots sharing energy. Each mobile robot can store at any given moment at most two units of energy, and whenever two robots are at the same location, they can transfer energy between each other, respecting the maximum capacity. The robots operate in a simple graph and initially each robot has two units of energy. A single edge traversal by an robot reduces its energy by one unit and the robot can only perform such move initially having at least one unit of energy. There are two distinguished nodes s and t in the graph and the goal for the robots is to deliver the *package* initially present on s to the node t. The package can be passed from one robot to another when they are colocated. In the first problem we study, the robots are initially placed at some given nodes of the graph and the question is whether the delivery is feasible. We prove that this problem is NP-complete. In the second problem, the initial positions of the robots are not fixed but a subset of nodes H of the graph is given as input together with an integer k, and the question is as follows: is there a placement of k robots at nodes in H such that the delivery is possible? We prove that this problem can be solved in polynomial time.

Keywords: Computational complexity · Energy sharing · Delivery
Mobile robots · Power-aware

1 Introduction

We consider algorithms for coordinated tasks performed by swarms of small inexpensive robots. There has been a lot of research interest on designing teams of simple robots that can perform a given task in collaborative fashion. The task we consider is the basic operation of moving an object or a package from its source to its target destination by one or more mobile robots. For example, the package could be a sample collected by a robotic sensor that needs to be delivered

Partially supported by National Science Centre (Poland) grant number 2015/17/B/ST6/01887 and the project ANR-ANCOR (anr-14-CE36-0002-01).

© Springer International Publishing AG 2017
A. Fernández Anta et al. (Eds.): ALGOSENSORS 2017, LNCS 10718, pp. 1–12, 2017.
https://doi.org/10.1007/978-3-319-72751-6_1

to a base station for analysis. One can imagine an automated postal delivery system where packages need to be delivered between sources and destinations using teams of robots or drones. We can model the sources and destinations as nodes of a graph and the *Delivery* problem consists of moving a single package from its source node to the target node. The main issue when using small robots is that they have a restricted supply of energy (e.g. a battery) and thus, a robot can move only a limited distance before running out of power. However, many small robots can cooperate to deliver a package from source to destination. Assuming that the robots start from different nodes of the graph, scheduling the moves of the robots for collaborative delivery is known to be a challenging task. Indeed, it was shown by Chalopin et al. [9] that even if the graph is a tree of n nodes, collaborative delivery from single source to a single destination using k robots having energy B each is NP-hard. Czyzowicz et al. [10] studied the problem when the robots can share their energy, i.e. a robot may give its unused energy to another robot. They showed that the problem can be easily solved in trees in polynomial time, but it remains NP-hard in general graphs. For all the above results, the parameters k and B have arbitrary (non-constant) values. Note that collaborative delivery for a constant value of k can be trivially solved by brute force manner in constant time. In this paper, we consider the problem when B is a small constant but k can have arbitrary values. This corresponds to the case of many small robots each operating with low power batteries such that each robot can move only for a constant number of steps in the graph. A robot with depleted energy can gather the unused energy of any other robot that it meets. However, no robot can have more than B units of energy at any time. Surprisingly, we show that even when $B = 2$, the smallest constant for which the problem is non-trivial, collaborative delivery is still NP-hard in general graphs. On the other hand, we provide an optimal polynomial-time algorithm for collaborative delivery with robots having $B = 2$ if we are allowed to choose the initial placement of robots among designated homebase nodes in the graph. Note that if we are allowed to place robots on any node of the graph, then there is a trivial optimal solution where all robots are placed on the shortest s-t path at intervals of distance 2. At the other extreme, if there is only one homebase node, then the solution is non-trivial.

Our Contributions. We completely solve the problem of collaborative delivery for $B = 2$ when energy sharing is allowed. We define two versions of the problem. In the first version of the problem called COLLABORATIVEDELIVERY *with Fixed Placement*, the initial placement of robots (i.e. the energy distribution is given as part of the problem). In the second version of the problem called COLLABORATIVEDELIVERY *with Chosen Placement*, a set of homebase nodes is given and the algorithm may choose the distribution of robots among the homebase nodes. We show that the first version of the problem is strongly NP-hard, while the second version of the problem admits a polynomial time solution and we present such a solution strategy. Proofs are omitted due to space constraints and will appear in the full version of the paper.

Related Work. Betke et al. [6] considered for the first time energy-constrained robots in the context of exploration of grid graphs by a robot who can return to its starting node for refueling. Awerbuch et al. [2] studied the same problem for general graphs. Duncan et al. [16] studied a similar model where the robot is tethered to its starting position with a rope of fixed length and they optimized the exploration time. Several other papers have considered robots with limited energy, or the goal of minimizing spent energy or maximum displacement of robots, e.g. in the context of exploration [12,14,17–20], formation [7,13], coverage [11,15], and broadcast/convergecast [1,10] problems.

In what concerns specifically collaborative delivery by energy-aware mobile robots, the problem has been considered in some recent works under various assumptions. In [8], the authors assume robots with limited energy that is consumed as they move. They prove that the problem of deciding whether delivery is feasible is NP-hard even if the robots are initially collocated, and they provide a 2-approximation algorithm for the optimization version of finding the minimum initial energy that can be given to all robots so that delivery becomes feasible, as well as exact, approximation, and resource-augmented algorithms for variants of the problem. In [9], the authors show that the problem is weakly NP-hard even on the line (with initially dispersed robots), and provide a quasi-, pseudo-polynomial algorithm under the assumption of integer numerical values in the problem instance. In [3], the authors consider the variant of the delivery probem in which the robots have to return to their respective starting positions and they prove that this problem is NP-hard for planar graphs but can be solved efficiently on trees and lines, in contrast to the non-returning version which is NP-hard on lines. They also give resource-augmented algorithms for returning delivery in general graphs and prove tight lower bounds on the resource augmentation for both the returning and the non-returning variant. In [4], the robots do not have a limited energy source, but instead they have different rates of energy consumption and the goal is to find a delivery schedule that minimizes the total energy spent. Moreover, there are several messages that need to be delivered from their respective source to their respective target. The authors study separately three subtasks that need to be solved in order to compute the optimal solution (collaboration of different robots on the same message, planning for a robot that works on multiple messages, and assignment of messages to robots) and they provide a polynomial-time (nonconstant) approximation algorithm for the problem. In this setting, [5] studies the design of truthful mechanisms in a game-theoretic model where the rate of energy consumption is information private to each robot.

The only previous work that considers *energy sharing* by mobile robots is [10], where this feature is introduced in the context of the delivery and convergecast problems. The authors show that both problems can be solved efficiently in trees, whereas they are NP-complete in general undirected and directed graphs. It is important to note that, in the model of [10], a robot may store an *unlimited* amount of energy as a result of receiving energy from other robots it encounters.

In other words, there is no battery capacity constraint for the robots, in contrast to the model that we study in the present paper.

2   Preliminaries

We now define precisely the *collaborative delivery* problem for a collection of energy sharing mobile robots. Given a simple undirected graph $G = (V, E)$, with two special nodes s (source) and t (target), and a collection of k mobile robots located initially in specific nodes of the graph, the objective is to decide whether there is schedule of robot moves that can deliver a package from s to t. Each robot has a constant energy budget and we denote its value by B. Traversing each edge consumes one unit of energy, thus, a fully charged robot can move a distance of B before running out of energy. When two robots are at the same node, one robot can transfer to the other robot, any integral part of its energy, with the only constraint that no robot can have more than B units of energy at any time.

In this paper, $B = 2$ for all robots. There is a unique package initially at node s that needs to be moved to node t. To simplify the discussion, we will assume that the system is synchronous (any synchronous strategy can also be implemented in an asynchronous system using appropriate waits). A robot r located at a node v at time j and having some positive energy, can perform any subset of the following actions:

- Pick up the package, if the package is present at v at time j.
- Transfer one unit of energy to another robot r' that is located at v at time j, if r' is not fully charged.
- Move to a neighboring node u, consuming one unit of energy and arriving at u at time $j + 1$.

A solution strategy is a sequence of steps as above, such that after the last step, the package is located at node t.

In the general version of the problem described below, the position of the robots is given by an adversary.

Problem 1 (COLLABORATIVEDELIVERY *with Fixed Placement*). CDX
Instance: $\langle G, s, t, k, h \rangle$, where $G = (V, E)$ is a simple undirected graph, $s, t \in V$ are, respectively, the source and target nodes, $h : \{1, \ldots, k\} \to V$ is the placement function that specifies the initial positions of the $k \geq 1$ robots.
Question: Does there exist a solution strategy for moving the package from s to t, when each robot start with $B = 2$ units of energy?

If the placement of robots among homebase nodes can be chosen by the algorithm, then we have the following version of the problem:

Problem 2 (COLLABORATIVEDELIVERY *with Chosen Placement*). CDC

Instance: $\langle G, s, t, k, H \rangle$, where $G = (V, E)$ is a simple undirected graph, $s, t \in V$ are, respectively, the source and target nodes, $k \geq 1$ is the number of mobile robots and $H \subset V$ is the set of homebase nodes.

Question: Does there exist a placement function $p : \{1, \ldots, k\} \to H$ and a corresponding solution strategy for moving the package from s to t, when the i-th robot starts at node $p(i)$ with $B = 2$ units of energy?

A path with nodes $v_1, \ldots, v_n$ and edges $\{v_i, v_{i+1}\}$, $i \in \{1, \ldots, n - 1\}$, is denoted by $(v_1, \ldots, v_n)$. The path graph with n vertices and $n - 1$ edges is denoted by P_n. The length of the shortest path between two nodes u, v of a graph G is denoted by $d_G(u, v)$, or simply $d(u, v)$ when there is no potential for confusion. If a robot is initially placed on a node v, then we write $a(v)$ to refer to this robot at any point of a strategy.

We obtain the following regarding the hardness of problem CDX:

Theorem 1. CDX *is* NP-*complete in the class of graphs with degree bounded by 5 and with each node being initially occupied by at most one robot.*

Theorem 2. CDX *is* NP-*complete in the class of graphs with diameter at most 42 with each node being initially occupied by at most one robot.*

3   An Efficient Algorithm for COLLABORATIVEDELIVERY with *Chosen Placement*

Let $I = \langle G, s, t, H \rangle$ be an instance of CDC, with $G = (V, E)$ and $H \subseteq V$. Recall that a solution to I is a strategy that enables a group of energy-exchanging robots starting from some or all of the nodes in H with battery capacity $B = 2$ to transfer the package from s to t. The cost of a solution is the total initial energy of the robots that are placed on nodes in H. If $u \in V$, we denote by h_u the node in H that is closest to u, breaking ties arbitrarily, and we denote by $z(u)$ the distance from u to h_u in G, i.e., $z(u) = d_G(h_u, u)$. Let $\hat{G}$ be a weighted complete digraph with vertex set V and arc set E', and let the weight of an arc $e = (u, v) \in E'$ be $w(e) = 2^{z(u) + d_G(u,v) - 1}$. If P is a directed path in $\hat{G}$, the *weight of* P is the sum of the weights of the arcs in P. This section is devoted to the proof of the following theorem:

Theorem 3. *An optimal solution to* $I = \langle G, s, t, H \rangle$ *has cost* $d_{\hat{G}}(s, t)$.

It suffices to show that there exists a directed s-t path in $\hat{G}$ with weight smaller than or equal to the cost of the optimal solution for I and, additionally, that every directed s-t path in $\hat{G}$ corresponds to some solution for I with cost equal to the weight of the path. The latter claim is given in the following lemma:

Lemma 1. *For every directed* s-t *path in* $\hat{G}$, *there exists a solution for* I *with cost equal to the weight of the path.*

In the rest of this section, we derive some structural properties of optimal solutions for I, which permit us to prove the former claim (cf. Lemma 5). In Sect. 3.1 we introduce our main tool in the analysis: an energy flow hypergraph that provides a way of presenting solutions to CDC. Then, Sects. 3.2 and 3.3 give a series of properties of this hypergraph, allowing us to finish the proof of Theorem 3 in Sect. 3.4. We assume that $s \neq t$, otherwise Theorem 3 holds trivially.

3.1   The Energy Flow Hypergraph

Given a solution $\mathcal{S}$ for I with cost $X > 0$, we represent $\mathcal{S}$ by a triple $\mathcal{S} = (\mathcal{V}, \mathcal{E}, \tilde{\mathcal{E}})$, where $(\mathcal{V}, \mathcal{E}) = \mathcal{H}$ is a directed hypergraph that represents the flow and eventual consumption of energy units (cf. Definition 1 below) and $\tilde{\mathcal{E}} \subseteq \mathcal{E}$ corresponds to the package moves under $\mathcal{S}$ (cf. Definition 2 below). The nodes of $\mathcal{H}$ are the energy arrival and extinction events of $\mathcal{S}$, as specified below.

We assume that the units of energy that are initially present at nodes in H receive distinct identities from 1 to X. We distinguish two types of *events* during the delivery under $\mathcal{S}$. An *arrival* event occurs whenever a robot with two units of energy i, j with $i < j$ moves from some node u of G to a neighbor v at time step t. We say that the unit of energy j is *wasted by* i during the event and that the unit of energy i *arrives* at v at time $t+1$. We denote this event as $(i, t+1, v)$. An *extinction* event occurs whenever a robot with one unit of energy i moves from some node u of G to a neighbor v at time step t. We say that the unit of energy i *wastes itself* during the event. We denote this event as $(\perp_i, t + 1, v)$.

We also consider as *arrival* events the appearance of the X units of energy at the homebases at time 0, and we denote them as $(i, 0, h_i)$, for $1 \leq i \leq X$, where $h_i \in H$ is the homebase where energy unit i was placed.

We are now ready to define the *energy flow hypergraph* $\mathcal{H} = (\mathcal{V}, \mathcal{E})$ in terms of the arrival and extinction events as follows:

Definition 1 (Energy flow hypergraph). *The vertex set $\mathcal{V}$ contains all of the arrival and extinction events, as specified above. The hyperarc set $\mathcal{E}$ contains $(\{(i, t_1, u), (j, t_2, u)\}, \{(i, t, v)\})$ if $t \geq 1$, the unit of energy i came from node u during the event (i, t, v), $j > i$ is the unit of energy consumed during the event (i, t, v), and in addition $t_1 < t$, $t_2 < t$, and i (resp. j) is not involved in any other events between times t_1 (resp. t_2) and t. Furthermore, $\mathcal{E}$ contains $(\{(i, t_1, u)\}, \{(\perp_i, t, v)\})$ if $t_1 < t$ and i is not involved in any other events between times t_1 and t.*

Definition 2 (Item moves). *The set $\tilde{\mathcal{E}}$ is defined as the subset of $\mathcal{E}$ that contains all of the hyperarcs that correspond to package moves. More precisely, a hyperarc $(\{(i, t_1, u), (j, t_2, v)\}, \{(i, t, v)\}) \in \tilde{\mathcal{E}}$ if the robot that arrived with energy unit i at v at time t was carrying the package. Similarly, a hyperarc $(\{(i, t_1, u)\}, \{(\perp_i, t, v)\}) \in \tilde{\mathcal{E}}$ if the robot that arrived with zero energy at v at time t (having wasted energy unit i) was carrying the package.*

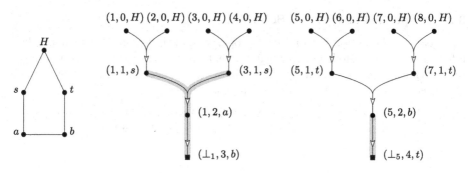

Fig. 1. An energy flow hypergraph constructed for a graph G shown on the left. The hypergraph consists of two components (the hyperarcs that correspond to package moves are highlighted): the first component dictates two robots to move from H to s and then one of those robots picks up the package and travels along path (s, a, b); the second component makes two robots to move from H to t and then one of them goes to b, picks up the package and returns to t.

We illustrate the energy flow hypergraph $\mathcal{H}$ corresponding to a simple solution for a CDC instance in Fig. 1. Note that the cost of $\mathcal{S}$ is given by the total number of energy units that "arrive" at nodes in H at time 0, which corresponds to the number of nodes of the form $(\cdot, 0, \cdot)$ in $\mathcal{H}$.

By construction, there is no cycle in $\mathcal{H}$. Moreover, the head of every hyperarc of $\mathcal{H}$ has size 1 and every node of $\mathcal{H}$ is contained in at most one hyperarc head and in at most one hyperarc tail. Therefore, $\mathcal{H}$ consists of a number of independent components $\mathcal{H}_1, \ldots, \mathcal{H}_\sigma$, each of which has a tree-like structure, as in Fig. 1.

Notation. If $e \in \mathcal{E}$, we denote by $\mathsf{head}(e)$ the unique node that is in the head of e. If $v \in \mathcal{V}$, we denote by $\mathfrak{g}(v)$ the node of G that is involved in the event v. If $e \in \mathcal{E}$, then we denote by $\mathfrak{g}_{\mathsf{tail}}(e)$ the node of G that is involved in the events in the tail of e (recall that, by definition of the hypergraph $\mathcal{H}$, all events in the tail of e must involve the same node of G), and by $\mathfrak{g}_{\mathsf{head}}(e)$ the node of G that is involved in the unique event in the head of e (i.e., $\mathfrak{g}_{\mathsf{head}}(e) = \mathfrak{g}(\mathsf{head}(e))$).

If $v \in \mathcal{V}$, let Δv denote the subgraph of $\mathcal{H}$ induced by the ancestors of v and v itself. Let $\mathsf{height}(v)$ denote the number of hyperarcs in the longest path that terminates at v. If $e \in \mathcal{E}$, we abuse the notation slightly and we denote by $\mathsf{height}(e)$ the height of $\mathsf{head}(e)$. If $v, v' \in \mathcal{V}$, we write $v \prec v'$ if v is an ancestor of v' and we write $v \sqsubset v'$ if v precedes v' temporally, i.e., $v = (i, t, x)$ and $v' = (i', t', x')$ with $t < t'$. Note that $v \prec v'$ implies $v \sqsubset v'$. As above, we extend the notation to arcs and we write $e \prec e'$ if $\mathsf{head}(e) \prec \mathsf{head}(e')$ and $e \sqsubset e'$ if $\mathsf{head}(e) \sqsubset \mathsf{head}(e')$.

If Δv contains x nodes of the form $(\cdot, 0, \cdot)$, then we say that Δv *incurs a cost of x*. This represents the energy units used by the solution in order to generate the event v. We also say that a component $\mathcal{H}_i$ incurs a cost equal to the cost incurred by its maximal node (under $\prec$). The cost of $\mathcal{S}$ is the sum of the costs incurred by the components of $\mathcal{H}$.

3.2   Properties of Optimal Solutions

The goal of this section is to prove a property of optimal solutions that can be informally stated as follows: every component of the hypergraph corresponding to the solution contains exactly one chain of item moves and the last hyperarc of this chain is an extinction event of the component (Fig. 2).

Proposition 1. *For every solution* $S = (V, \mathcal{E}, \tilde{\mathcal{E}})$ *in which there exist arcs* $f, g \in \tilde{\mathcal{E}}$ *with* $f \prec g$, *there exists a solution* $S' = (V, \mathcal{E}, \tilde{\mathcal{E}}')$ *with* $\tilde{\mathcal{E}}' = \tilde{\mathcal{E}} \setminus \{e : f \sqsubset e \sqsubset g\} \cup \{e : f \prec e \prec g\}$.

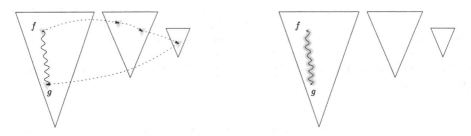

Fig. 2. Illustration of Proposition 1. Triangles represent components of the solution hypergraph. The solution S is shown on the left (in which some arcs of the path from f to g are not package moves—those package moves can be possibly in different components; the dotted arrows represent the time succession between package moves that take the package from f to g) and the corresponding S' is shown on the right.

By repeated application of Proposition 1, we obtain the following:

Corollary 1. *For every solution* $S = (V, \mathcal{E}, \tilde{\mathcal{E}})$, *there exists a solution* $S' = (V, \mathcal{E}, \tilde{\mathcal{E}}')$ *and a partition of* $\tilde{\mathcal{E}}'$ *into sets* $\tilde{\mathcal{E}}_1', \ldots, \tilde{\mathcal{E}}_\tau'$ *such that, for every* i, *the hyperarcs of* $\tilde{\mathcal{E}}_i'$ *form a chain in* $\mathcal{H}$ *and, for every* $e \in \tilde{\mathcal{E}}_i'$ *and* $e' \in \tilde{\mathcal{E}}_j'$ *with* $i < j$, *we have* $e \sqsubset e'$, $e \not\prec e'$, *and* $e' \not\prec e$.

Note that Proposition 1 and Corollary 1 apply to *any* solution (not necessarily an optimal one). Furthermore, in both statements, the obtained solution S' has the same energy flow hypergraph as S, and therefore it has the same cost as S.

Lemma 2. *For every optimal solution* $S = (V, \mathcal{E}, \tilde{\mathcal{E}})$ *and for every component* $\mathcal{H}_i$ *of* $\mathcal{H} = (V, \mathcal{E})$ *with maximum (under* $\prec$) *hyperarc* r_i, *we have* $r_i \in \tilde{\mathcal{E}}$ *and* head(r_i) *is an extinction event.*

By applying Corollary 1 to an arbitrary optimal solution, we obtain the following corollary in view of Lemma 2:

Corollary 2. *There exists an optimal solution* $S = (V, \mathcal{E}, \tilde{\mathcal{E}})$ *such that every component* $\mathcal{H}_i$ *of* $\mathcal{H} = (V, \mathcal{E})$ *with maximum (under* $\prec$) *hyperarc* r_i *contains exactly one chain of package moves whose last hyperarc is* r_i *and, in addition,* head(r_i) *is an extinction event.*

3.3　Canonical Nodes

Given a solution $\mathcal{S} = (\mathcal{V}, \mathcal{E}, \tilde{\mathcal{E}})$ for I, let $v \in \mathcal{V}$ such that Δv does not contain any hyperarc in $\tilde{\mathcal{E}}$. Intuitively, if v is not an extinction event, then the sole function of Δv in the solution is to bring one unit of energy to $\mathfrak{g}(v)$. It thus makes sense that, if $\mathcal{S}$ is optimal, then the energy units that participate in the events of Δv travel along shortest paths from their respective homebases to $\mathfrak{g}(v)$. If v satisfies these conditions, then we say that v is *canonical*. The following definition captures this notion:

Definition 3 (Canonical nodes). *Given a solution* $\mathcal{S} = (\mathcal{V}, \mathcal{E}, \tilde{\mathcal{E}})$, *a node* $v \in \mathcal{V}$ *is called* canonical *if either* $\mathsf{height}(v) = 0$, *or* $\mathsf{height}(v) = h + 1$ *for some* $h \geq 0$ *(in this case,* $v = \mathsf{head}(e)$ *for some* $e \in \mathcal{E}$) *and all of the following hold: (i) for every node* u *in the tail of* e, u *is canonical and* $\mathsf{height}(u) = h$, *(ii)* $e \notin \tilde{\mathcal{E}}$, *and (iii)* $z(\mathfrak{g}(v)) = 1 + z(\mathfrak{g}_{\mathsf{tail}}(e))$.

The two propositions below follow easily by induction on the height of v. Recall that, by definition of $\mathcal{H}$, if $\mathsf{height}(v) = 0$ then $\mathfrak{g}(v) \in H$.

Proposition 2. *If* v *is canonical, then* $z(\mathfrak{g}(v)) = \mathsf{height}(v)$.

Proposition 3. *If* v *is canonical and it is not an extinction event, then the cost incurred by* Δv *is* $2^{\mathsf{height}(v)}$.

We can now prove that, in every optimal solution, every node v whose Δv does not contain any package move is canonical.

Lemma 3. *For every optimal solution* $\mathcal{S} = (\mathcal{V}, \mathcal{E}, \tilde{\mathcal{E}})$ *and for every* $v \in \mathcal{V}$, v *is canonical or* Δv *contains a hyperarc in* $\tilde{\mathcal{E}}$.

3.4　Completing the Proof of Theorem 3

In the following, let $\mathcal{S}^{\star} = (\mathcal{V}, \mathcal{E}, \tilde{\mathcal{E}})$ be an optimal solution as guaranteed by Corollary 2, with the maximum number σ of components of $\mathcal{H} = (\mathcal{V}, \mathcal{E})$. Let $(\mathcal{H}_i)_{i=1,\ldots,\sigma}$ be an enumeration of the components of $\mathcal{H}$ in temporal order of their extinction events. For $i \in \{1, \ldots, \sigma\}$, component $\mathcal{H}_i$ is responsible for moving the package along a path $P_i = (u_{i,0}, u_{i,1}, \ldots, u_{i,\rho_i})$ in G, where $u_{1,0} = s$, $u_{\sigma,\rho_\sigma} = t$, and $u_{i,\rho_i} = u_{i+1,0}$ (for $i < \sigma$).

Lemma 4. *For every* i, j *in the ranges* $1 \leq i \leq \sigma$ *and* $0 \leq j < P_i - 1$, $z(u_{i,j+1}) = 1 + z(u_{i,j})$.

Corollary 3. *The cost incurred by component* $\mathcal{H}_i$ *is* $2^{z(u_{i,0}) + \rho_i - 1}$.

Let Q be the directed s-t path in $\hat{G}$ that consists of the arcs $(u_{1,0}, u_{1,\rho_1})$, $(u_{2,0}, u_{2,\rho_2}), \ldots, (u_{\sigma,0}, u_{\sigma,\rho_\sigma})$. By definition of $\hat{G}$, the i-th arc of Q has weight $2^{z(u_{i,0}) + d_G(u_{i,0}, u_{i,\rho_i}) - 1}$. However, P_i is a path in G from $u_{i,0}$ to u_{i,ρ_i} and its length is ρ_i. Therefore, $d_G(u_{i,0}, u_{i,\rho_i}) \leq \rho_i$. In view of Corollary 3, we conclude that the weight of the i-th arc of Q is at most equal to the cost of $\mathcal{H}_i$ and thus the total weight of the arcs of Q is at most equal to the cost of $\mathcal{S}^{\star}$. We have proved the following lemma, which concludes the proof of Theorem 3:

Lemma 5. *There exists a directed s-t path in $\hat{G}$ with weight at most equal to the cost of an optimal solution to I.*

4   Concluding Remarks

Our work reveals an interesting differentiation in the complexity of the collaborative delivery problem by robots with battery capacity $B = 2$, depending on whether the energy allocation to the homebases (starting nodes of robots) is given as part of the input on the one hand, or the allocation can be chosen as part of the solution on the other hand.

As we showed, the problem with fixed allocation of energy units to the homebases is NP-complete. However, we proved in Sect. 3 that the delivery problem in which one is given the total available energy and is asked if it is possible to distribute this energy to robots at the homebases in order to achieve delivery is solvable in polynomial time. In fact, what we proved is that the underlying optimization problem, i.e., finding the minimum amount of energy that can be distributed to the homebases so that delivery is feasible, is solvable in polynomial time by reduction to a shortest path computation in a complete directed graph.

A natural question is how to handle greater battery capacities $B \geq 3$. While we expect that our NP-completeness reduction generalized to $B \geq 3$, the situation is less clear when it comes to the question of computing the energy allocation to the homebases as part of the solution. A straightforward adaptation of our algorithm from Sect. 3 would be to reduce the problem to computing the shortest s-t path in a directed graph $\hat{G}$ similar to the one we construct in Sect. 3, except that the weight of an arc (u, v) would be equal to the minimum amount of energy required by robots with capacity B to traverse the path $h_u \rightsquigarrow u \rightsquigarrow v$, where h_u is the nearest homebase to u. Unfortunately, this algorithm is no longer guaranteed to produce an optimal solution for $B \geq 3$ (see Fig. 3).

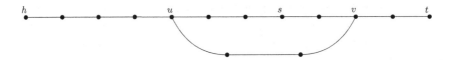

Fig. 3. An example of a graph in which the straightforward adaptation of our algorithm from Sect. 3 to $B = 3$ does not give an optimal solution. Here, $H = \{h\}$ and the shortest s-t path in $\hat{G}$ is $s \rightarrow v \rightarrow t$, with each arc having a weight of 41 for a total cost of 82. However, the optimal solution has a cost of 81: 27 fully charged robots start from h and they reach u with 16 remaining energy units in total. At u, the robots split into two groups with 8 units of energy each. The first group goes to s and then to v from the top branch, picking up the package from s on the way. The second group goes to v from the bottom branch, picks up the package, and continues until t.

The reason is that several nice properties of the optimal solutions for $B = 2$ no longer hold for $B \geq 3$. In particular, since the hyperarcs in the energy flow hypergraph can now have up to two nodes in their heads, each component can now have more than one bottommost nodes and it can contain more than one chains of package moves. This is exactly the case in the example of Fig. 3, where the energy flow hypergraph of the optimal solution has only one component, which contains two chains of package moves. Furthermore, it is no longer the case that all of the nodes in the tail of a given hyperarc have the same height. This can be seen even in cases where the optimal solution consists of only one component with only one chain of package moves (see Fig. 4).

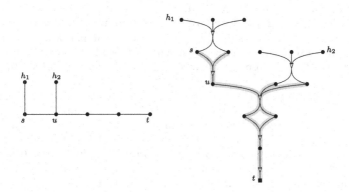

Fig. 4. An example of a graph (*left*, with $H = \{h_1, h_2\}$) in which the energy flow hypergraph of the optimal solution (*right*) contains a hyperarc with nodes of different heights in its tail.

References

1. Anaya, J., Chalopin, J., Czyzowicz, J., Labourel, A., Pelc, A., Vaxès, Y.: Converge-cast and broadcast by power-aware mobile agents. Algorithmica **74**(1), 117–155 (2016)
2. Awerbuch, B., Betke, M., Rivest, R.L., Singh, M.: Piecemeal graph exploration by a mobile robot. Inf. Comput. **152**(2), 155–172 (1999)
3. Bärtschi, A., Chalopin, J., Das, S., Disser, Y., Geissmann, B., Graf, D., Labourel, A., Mihalák, M.: Collaborative delivery with energy-constrained mobile robots. In: Suomela, J. (ed.) SIROCCO 2016. LNCS, vol. 9988, pp. 258–274. Springer, Cham (2016). https://doi.org/10.1007/978-3-319-48314-6_17
4. Bärtschi, A., Chalopin, J., Das, S., Disser, Y., Graf, D., Hackfeld, J., Penna, P.: Energy-efficient delivery by heterogeneous mobile agents. In: Vollmer, H., Vallée, B. (eds.) 34th Symposium on Theoretical Aspects of Computer Science, STACS 2017. LIPIcs, 8–11 March 2017, Hannover, Germany, vol. 66, pp. 10:1–10:14. Schloss Dagstuhl - Leibniz-Zentrum fuer Informatik (2017)
5. Bärtschi, A., Graf, D., Penna, P.: Truthful mechanisms for delivery with mobile agents. CoRR abs/1702.07665 (2017)
6. Betke, M., Rivest, R.L., Singh, M.: Piecemeal learning of an unknown environment. Mach. Learn. **18**(2), 231–254 (1995)

7. Biló, D., Disser, Y., Gualá, L., Mihal'ák, M., Proietti, G., Widmayer, P.: Polygon-constrained motion planning problems. In: Flocchini, P., Gao, J., Kranakis, E., Meyer auf der Heide, F. (eds.) ALGOSENSORS 2013. LNCS, vol. 8243, pp. 67–82. Springer, Heidelberg (2014). https://doi.org/10.1007/978-3-642-45346-5_6

8. Chalopin, J., Das, S., Mihal'ák, M., Penna, P., Widmayer, P.: Data delivery by energy-constrained mobile agents. In: Flocchini, P., Gao, J., Kranakis, E., Meyer auf der Heide, F. (eds.) ALGOSENSORS 2013. LNCS, vol. 8243, pp. 111–122. Springer, Heidelberg (2014). https://doi.org/10.1007/978-3-642-45346-5_9

9. Chalopin, J., Jacob, R., Mihalák, M., Widmayer, P.: Data delivery by energy-constrained mobile agents on a line. In: Esparza, J., Fraigniaud, P., Husfeldt, T., Koutsoupias, E. (eds.) ICALP 2014 Part II. LNCS, vol. 8573, pp. 423–434. Springer, Heidelberg (2014). https://doi.org/10.1007/978-3-662-43951-7_36

10. Czyzowicz, J., Diks, K., Moussi, J., Rytter, W.: Communication problems for mobile agents exchanging energy. In: Suomela, J. (ed.) SIROCCO 2016. LNCS, vol. 9988, pp. 275–288. Springer, Cham (2016). https://doi.org/10.1007/978-3-319-48314-6_18

11. Czyzowicz, J., Kranakis, E., Krizanc, D., Lambadaris, I., Narayanan, L., Opatrny, J., Stacho, L., Urrutia, J., Yazdani, M.: On minimizing the maximum sensor movement for barrier coverage of a line segment. In: Ruiz, P.M., Garcia-Luna-Aceves, J.J. (eds.) ADHOC-NOW 2009. LNCS, vol. 5793, pp. 194–212. Springer, Heidelberg (2009). https://doi.org/10.1007/978-3-642-04383-3_15

12. Das, S., Dereniowski, D., Karousatou, C.: Collaborative exploration by energy-constrained mobile robots. In: Scheideler, C. (ed.) Structural Information and Communication Complexity. LNCS, vol. 9439, pp. 357–369. Springer, Cham (2015). https://doi.org/10.1007/978-3-319-25258-2_25

13. Demaine, E.D., Hajiaghayi, M., Mahini, H., Sayedi-Roshkhar, A.S., Oveisgharan, S., Zadimoghaddam, M.: Minimizing movement. ACM Trans. Algorithms 5(3), 1–30 (2009)

14. Dereniowski, D., Disser, Y., Kosowski, A., Pająk, D., Uznański, P.: Fast collaborative graph exploration. Inf. Comput. 243, 37–49 (2015)

15. Dobrev, S., Durocher, S., Hesari, M.E., Georgiou, K., Kranakis, E., Krizanc, D., Narayanan, L., Opatrny, J., Shende, S.M., Urrutia, J.: Complexity of barrier coverage with relocatable sensors in the plane. Theoret. Comput. Sci. 579, 64–73 (2015)

16. Duncan, C.A., Kobourov, S.G., Kumar, V.S.A.: Optimal constrained graph exploration. In: 12th ACM Symposium on Discrete Algorithms, SODA 2001, pp. 807–814 (2001)

17. Dynia, M., Korzeniowski, M., Schindelhauer, C.: Power-aware collective tree exploration. In: Grass, W., Sick, B., Waldschmidt, K. (eds.) ARCS 2006. LNCS, vol. 3894, pp. 341–351. Springer, Heidelberg (2006). https://doi.org/10.1007/11682127_24

18. Dynia, M., Lopuszański, J., Schindelhauer, C.: Why robots need maps. In: Prencipe, G., Zaks, S. (eds.) SIROCCO 2007. LNCS, vol. 4474, pp. 41–50. Springer, Heidelberg (2007). https://doi.org/10.1007/978-3-540-72951-8_5

19. Fraigniaud, P., Gąsieniec, L., Kowalski, D.R., Pelc, A.: Collective tree exploration. Networks 48(3), 166–177 (2006)

20. Ortolf, C., Schindelhauer, C.: Online multi-robot exploration of grid graphs with rectangular obstacles. In: 24th ACM Symposium on Parallelism in Algorithms and Architectures, SPAA 2012, pp. 27–36 (2012)

Data Collection in Population Protocols with Non-uniformly Random Scheduler

Joffroy Beauquier[1], Janna Burman[1(✉)], Shay Kutten[2], Thomas Nowak[1], and Chuan Xu[1(✉)]

[1] LRI, Université Paris Sud, CNRS, Université Paris Saclay, Orsay, France
{joffroy.beauquier,janna.burman,thomas.nowak,chuan.xu}@lri.fr
[2] Technion - Israel Institute of Technology, Haifa, Israel
kutten@ie.technion.ac.il

Abstract. Contrary to many previous studies on population protocols using the uniformly random scheduler, we consider a more general non-uniform case. Here, pair-wise interactions between *agents* (moving and communicating devices) are assumed to be drawn *non-uniformly* at random. While such a scheduler is known to be relevant for modeling many practical networks, it is also known to make the formal analysis more difficult.

This study concerns *data collection*, a fundamental problem in mobile sensor networks (one of the target networks of population protocols). In this problem, pieces of information given to the agents (e.g., sensed values) should be delivered eventually to a predefined sink node without loss or duplication. Following an idea of the known deterministic protocol TTF solving this problem, we propose an adapted version of it and perform a complete formal analysis of execution times in expectation and with high probability (w.h.p.).

We further investigate the non-uniform model and address the important issue of energy consumption. The goal is to improve TTF in terms of energy complexity, while still keeping good time complexities (in expectation and w.h.p.). Namely, we propose a new parametrized protocol for data collection, called *lazy* TTF, and present a study showing that a good choice of the protocol parameters can improve energy performances (compared to TTF), at a slight expense of time performance.

1 Introduction

Population protocols have been introduced in [1] as a model for passively mobile sensor networks (cf. the journal version [2]). In this model, tiny indistinguishable agents with bounded memory move unpredictably and interact in pairs. That is, when two agents are sufficiently close to each other, they can communicate (i.e., interact). During an interaction, they exchange and update their respective states according to a transition function (the protocol). Such successive interactions contribute to the realization of some global task.

The fact that agent moves are unpredictable is usually modeled by assuming the uniformly random scheduler [2–5]. That is, the interactions between any

© Springer International Publishing AG 2017
A. Fernández Anta et al. (Eds.): ALGOSENSORS 2017, LNCS 10718, pp. 13–25, 2017.
https://doi.org/10.1007/978-3-319-72751-6_2

two agents are drawn uniformly at random. However, for some practical sensor networks, this assumption may be unrealistic. Consider, for instance, agents moving at different speeds. In this case, an agent interacts more frequently with a faster agent than with a slower one. In other networks, certain agents may be frequently prevented from communicating with some others, because they move in different limited areas, or disfunction from time to time, etc. In all these examples, the interactions are clearly not uniformly random. There are thus strong arguments for enhancing the basic model.

This paper initiates the study of non-uniform schedulers in the context of population protocols. Considering the scheduler as the generator of sequences of pairwise interactions, non-uniform means that the next interacting pair (i, j) is chosen with a non-uniform probability $P_{i,j}$, depending on i and j.

As a supplementary justification for studying a non-uniform scheduler, notice that many experimental and analytical studies of different (finite boundary) mobile sensor networks show and exploit (respectively) the assumption that the *inter-contact* time of two agents (the time period between two successive interactions of the same two mobile agents) is distributed exponentially (cf. [6–9]). Similarly, under a non-uniformly random scheduler, it appears that the inter-contact time $T_{i,j}$, of any two agents i and j, follows a geometric distribution $(P[T_{i,j} = t] = (1 - P_{i,j})^{t-1} P_{i,j})$, which is the discrete analogue of the exponential case (observed in practical mobile networks).

The counterpart of considering a non-uniform scheduler is a more complex analysis. Though, it remains feasible in certain cases, as it is shown in this paper. To illustrate this point, consider a fundamental task for mobile sensors, *data collection* (or data gathering). In this task, each agent has initially an input value (for instance, a sensed value). Each value must be gathered exactly once (as a multi-set) by a special agent which we call the base station. In the context of population protocols (assuming non-random schedulers), several data collection protocols have been proposed and their complexity in time has been studied [10]. Notice that the analysis there was only for the worst case. We are not aware of any previous results concerning the average complexity of these protocols. The current paper presents protocols that basically use the simple ideas of the TTF (Transfer To the Faster) protocol of [10]. The new protocols are adapted to a non-uniform scheduler and improve energy consumption, as explained further.

First, consider the original version of TTF. It uses a deterministic parameter called *cover time*, which is an upper bound on the time, counted in the number of global interactions, for an agent to interact with all the others. The data transfer between the agents in TTF depends on the comparison of cover times of two interacting agents. Here we follow this idea. However, as the scheduler is probabilistic, we adapt the corresponding definition of the cover time to be the *expected* (instead of the maximum) number of interactions for an agent to interact with every other agent (see Sect. 2).

The complexity analysis starts with the proofs of two lower bounds on the expected convergence time of any protocol solving data collection (Sect. 3). Then, an analysis of execution times in expectation and with high probability (w.h.p.), for the new version of TTF, is given (Sect. 4). The complexity in expectation indicates how the protocol is good in average, while the complexity w.h.p. tells how it is good almost all the time. We obtain explicit bounds, thus justifying the relevance of the enhanced model in protocol analysis and its operability.

We further investigate the non-uniform model by addressing also energy complexity, which is known to be a crucial issue for sensor networks. The goal is to improve energy consumption of TTF, while keeping good time complexity. For that, we propose a new parametrized protocol, called *lazy* TTF (Sect. 5). As opposed to TTF, it does not execute necessarily the transition of TTF resulting from an interaction. Instead, during an interaction (i, j), TTF is executed with probability p_i (depending on agent i, playing the role of *initiator* in the interaction). Analysis and the corresponding numerical study show that a good choice of the parameters p_i results in lower energy consumption. To find such parameters, we formulate and solve a polynomial-time optimization program. The resulting optimized lazy TTF is compared to TTF in respect with time and energy complexity (Sect. 6). For this analysis, we adopt the energy scheme proposed for population protocols in [11].

Due to the lack of space, most of the proofs and the survey on additional related work have been moved to the report [12].

2   Model and Definitions

Population Protocols. The system is represented by an *interaction graph* $G = (\mathbf{A}, \mathbf{E})$, a table $\mathbf{T}$ of *transition rules* and a *scheduler* $S(P)$. All are defined below.

A set $\mathbf{A}$ consists of n anonymous agents and is also called a population. An agent $i \in \mathbf{A}$ represents a finite state sensing and communicating mobile device, which can be seen as a finite state machine. The size of the population n is unknown to the agents. Among the agents, there is a distinguishable one called the *base station* (BST), which can be as powerful as needed, in contrast with the resource-limited agents. The non-BST agents are also called *mobile*. Each agent has a state that is taken from a finite set of states which is the same for all mobile agents, but possibly different for the base station.

A directed edge $(i, j) \in \mathbf{E}$ intuitively represents a possible interaction between two agents. That is, if such an edge exists, then the scheduler (see below) is allowed to schedule an event, called interaction, between i, called then the *initiator*, and j, called the responder for that event. In this work, we consider only complete interaction graphs. What happens in the interaction event is now described.

When two agents i, in state p, and j, in state q, interact (meet), they execute a *transition* $(p, q) \rightarrow (p', q')$. As a result, i changes its state from p to p' and j from q to q'. The table $\mathbf{T}$ of all the *transition rules* defines the population protocol. A protocol (respectively, its transition rules) are called *deterministic*, if for every pair of states (p, q), there is exactly one (p', q') such that $(p, q) \rightarrow (p', q')$. Otherwise, they are *non-deterministic*. Note that, as interactions are supposed to be asymmetric (with one agent acting as the initiator and the other as the responder), the transition rules for (p, q) and (q, p) may be different.

A *configuration* of the system is defined by the vector of agents' states. If, in a given configuration C, a configuration C' can be obtained by executing one transition of the protocol (between two interacting agents), it is denoted by $C \rightarrow C'$. An *execution* of a protocol is a sequence of configurations $C_0, C_1, C_2, \ldots$ such that C_0 is the *initial configuration* and for each $i \geq 0$, $C_i \rightarrow C_{i+1}$. We consider the number of interactions in an execution as the time reference, i.e., each interaction adds one time unit to the global time. This is similar to the *step complexity*, a common measure in population protocols (cf. [2,13]) and in distributed computing in general [14].

The sequence of the corresponding interactions in an execution is provided by an external entity called scheduler.

Non-uniformly Random Scheduler. Such a scheduler, denoted by $S(P)$, is defined by a matrix of probabilities $P \in \mathbb{R}^{n \times n}$. During an execution, $S(P)$ chooses the next pair of agents (i, j) to interact (taking i as initiator and j as responder) with the probability $P_{i,j}$. Notice that, in the case of the matrix with entries $P_{i,j} = 1/n(n-1)$ for $i \neq j$, and $P_{i,i} = 0$, the scheduler chooses each pair of agents uniformly at random for each next interaction (i.e., the scheduler is uniformly random).

The matrix P satisfies $\sum_{i=1}^{n} \sum_{j=1}^{n} P_{i,j} = 1$ and $\forall i \in \{1, \ldots, n\}, P_{i,i} = 0$, since interactions are pairwise. Moreover, for any edge (i, j) in the interaction graph G, $P_{i,j} > 0$. As the considered here G is complete, *every* pair of agents is chosen infinitely often with probability 1.

For a given P, one can compute the *expected* (finite) time for a given agent i to meet all the others. We call it *cover time of agent i* and denote it by cv_i. By resolving the coupon collector's problem with a non-uniform distribution [15], we obtain the cover time of each agent: $\mathrm{cv}_i = \int_0^\infty (1 - \prod_{j \neq i}(1 - e^{-(P_{i,j} + P_{j,i})t}))dt$. Similarly to [10], for two agents i and j, if $\mathrm{cv}_i < \mathrm{cv}_j$, we say that i is *faster* than j, and j is *slower* than i. If $\mathrm{cv}_i = \mathrm{cv}_j$, i and j are said to be in the same *category* of cover times. We denote by m the number of different categories of cover times. We emphasize that agents are not assumed to know their cvs (to conform with the finite state population protocol model). Instead, we do assume that two interacting agents can compare their respective cvs. For instance, this can be implemented by comparing categories instead of *cv*s, in applications where the overall number of categories is likely to be uniformly bounded.

Data Collection. Each agent, except the base station, owns initially a constant input value. Eventually, every input value has to be delivered to the base station, and exactly once (as a multi-set). When this happens, we say that a *terminal configuration* or simply *termination* has been reached. A protocol is said *to solve* data collection if termination is reached in every execution of the protocol.

In the sequel, when describing or analyzing a protocol, the term "transfer an input value (or token) from agent i to j" means copy it to j's memory and erase it from the memory of i. In particular, this prevents loss or duplication of input values. Moreover, in this preliminary study, we make the assumption that every agent has enough memory to store n values. This assumption is common in the literature [4, 16].

Time Complexity Measures. The *convergence time* of a data collection protocol $\mathcal{P}$ can be evaluated in two ways: first, in terms of expected time until termination, denoted by $T_{\mathrm{E}}(\mathcal{P})$, and second, in terms of time until termination w.h.p.[1], denoted by $T_{whp}(\mathcal{P})$.

Remark 1. The notion of parallel time, which is common when considering the uniformly random scheduler (cf. [3, 17]), is not used in this paper. When using this measure of time, it is assumed that each agent participates in an expected number $\Theta(1)$ of interactions per time unit. With the uniformly random scheduler, this time measure is asymptotically equal to the number of interactions divided by n. However, with non-uniformly random scheduler, this is no more true.

3   Lower Bounds on the Expected Convergence Time

We now give two nontrivial lower bounds on the expected convergence time of data collection protocols. The first one (Theorem 1) only depends on the number of agents. The second one (Theorem 2) depends on the specific values of the probability matrix P used by the scheduler. The bounds are incomparable in general. To obtain the bounds, we observe that, for performing data collection, each agent has to interact at least once (otherwise, its value simply won't be delivered), and we compute the expected time ensuring that. The proof of Theorem 1 uses an analogy with a generalization of the classical coupon collector's problem, which we introduce next.

Let k be a positive integer. Given a probability distribution $(p_1, \ldots, p_k)$ on $[k] = \{1, \ldots, k\}$, the corresponding k-coupon collector's problem is defined by its *coupon sequence* $(X_1, X_2, \ldots)$ of independent and identically distributed (i.i.d.) random variables with $\mathbb{P}(X_t = i) = p_i$ for all $i \in [k]$ and all $t \geq 0$. The k-coupon collector's problem's *expected time* is the expectation of the earliest time T such that $\{X_1, \ldots, X_T\} = [k]$, i.e., all coupons were collected at least once.

More generally, given a set $\mathcal{A}$ of subsets of $[k]$ such that $\bigcup_{A \in \mathcal{A}} A = [k]$, and a probability distribution (p_A) on $\mathcal{A}$, the corresponding $\mathcal{A}$-group k-coupon collector's problem is defined by its *coupon group sequence* $(X_1, X_2, \ldots)$ of i.i.d. random variables with $\mathbb{P}(X_t = A) = p_A$ for all $A \in \mathcal{A}$ and all $t \geq 0$. Its *expected time* is the expectation of the earliest time T such that $\bigcup_{t=1}^{T} X_t = [k]$, i.e., all coupons were collected in at least one coupon group.

Given an integer $1 \leq g \leq k$, the g-group k-coupon collector's problem is the $\mathcal{A}$-group k-coupon collector's problem where $\mathcal{A} = \{A \subseteq [k] \mid |A| = g\}$. This generalization of the classical coupon collector's problem has been studied, among others, by Stadje [18], Adler and Ross [19], and Ferrante and Saltalamacchia [20].

The following lemma characterizes the probability distributions that lead to a minimal expected time for the group coupon collector's problem. To the best

---

[1] An event $\varXi$ is said to occur w.h.p., if $\mathbb{P}(\varXi) \geq 1 - \frac{1}{n^c}$, where $c \geq 1$.

of our knowledge, this is a new result which generalizes the characterization in the classical coupon collector's problem [15,21], for which it is known that the uniform distribution leads to the minimal expected time.

Lemma 1. *The expected time of any $\mathcal{A}$-group k-coupon collector's problem is greater than or equal to the $\mathcal{B}$-group k-coupon collectors problem with uniform probabilities where $\mathcal{B} \subseteq \mathcal{A}$ is of minimal cardinality such that $\bigcup \mathcal{B} = [k]$.*

In particular, the expected time of any g-group k-coupon collector's problem is $\Omega(k \log k)$ for every constant $g \geq 1$.

Theorem 1. *The expected convergence time of any protocol solving data collection with non-uniformly random scheduler is $\Omega(n \log n)$.*

Theorem 2. *The expected convergence time of any protocol solving data collection with random scheduler $S(P)$, is $\Omega(\max\limits_i \frac{1}{\sum_{j=1}^{n}(P_{i,j}+P_{j,i})})$.*

The next corollary considers a very simple protocol solving the data collection problem. In this protocol, agents transfer their values only when they interact with the base station. We consider it as a reference, to compare with other proposed protocols. The corollary follows from Theorem 2.

Corollary 1. *With random scheduler $S(P)$, the expected convergence time of the protocol solving data collection and where each agent transfers its value only to the base station is $\Omega(\max\limits_i 1/(P_{i,\mathrm{BST}} + P_{\mathrm{BST},i}))$.*

4   Protocol "Transfer to the Faster" (TTF)

Corollary 1 formalizes the straightforward observation that, if the only transfers performed by the agents are towards the base station, the convergence time depends on the slowest agent i. It can be very large, e.g. if $P_{i,\mathrm{BST}} + P_{\mathrm{BST},i} \ll 1/n^2$. Therefore, to obtain better time performances, we propose to study another data collection protocol based on the idea of the TTF protocol of [10]. In the sequel, the studied protocol is called TTF too, since its strategy is the same and there is no risk of ambiguity. The only difference is on the *definition* of the cover time parameter (Sect. 2) used by this strategy (as explained in the introduction).

The strategy of TTF is easy. When agent i meets a faster agent j, i transfers to j all the values it has in its memory (recall that transfer means to copy to the memory of the other and erase from its own). The intuition behind is that the faster agent j is more likely to meet the base station before i. Of course, whenever any agent i meets the base station, it transfers all the values it (still) has in its memory at that time to the base station. As a matter of fact, no transition depends on the actual value held by the agents. It depends only on the comparison between cover times, which are constants. Thus, the input values can be seen as tokens and the states of every agent can be represented by the number of tokens it currently holds. Recall, that in this study, it is assumed that each agent has enough memory for storing the tokens (i.e., an $O(n)$ memory), and each pair of agents interacts infinitely often (i.e., the interaction graph is complete).

The sequel concerns analytical results on the time performance of TTF. Firstly, we associate to each configuration a vector of non-negative integers representing the number of tokens held by each agent. Then, it is shown that the evolution of such vectors during executions can be expressed by a *stochastic linear system*. Next, $T_{\text{whp}}(\text{TTF})$ is expressed in terms of distances between the configuration vectors (Theorem 3) and, by applying stochastic matrix theory [22–24] an upper bound on $T_{\text{whp}}(\text{TTF})$ is obtained (Theorem 4). Finally, using this result, we obtain also an upper bound on the convergence time in expectation, $T_{\mathbb{E}}(\text{TTF})$ (Theorem 5).

Formally, we represent a configuration by a non-negative integer vector $x \in \mathbb{N}^n$ that satisfies $\sum_{i=1}^{n} x_i = n - 1$. By abusing the terminology, we sometimes call such a vector a configuration. We denote the configuration vectors' space by $\mathbb{V}$. By convention, the first element of x is the number of tokens held by the base station. Since, at the beginning of an execution, every mobile agent owns exactly one token and no token is held by the base station, the initial configuration is $x_{\text{init}} = \mathbf{1} - \mathbf{e}_1$, where $\mathbf{e}_i = (0, \ldots, 0, 1, 0, \ldots, 0)^T$ is the $n \times 1$ unit vector with the ith component equal to 1. The terminal configuration is $x_{\text{end}} = (n-1)\mathbf{e}_1$.

Let $x(t) \in \mathbb{V}$ be the discrete random integer vector that represents the configuration just after t^{th} interaction in executions of TTF. We can see that $\mathbb{P}(x(0) = x_{\text{init}}) = 1$, and since the base station never transfers tokens to others, $\mathbb{P}(x(t+1) = x_{\text{end}}) \geq \mathbb{P}(x(t) = x_{\text{end}})$. Moreover, since at any moment there is a positive probability for delivering any of the tokens to the base station, $\lim_{t \to \infty} \mathbb{P}(x(t) = x_{\text{end}}) = 1$. Furthermore, the time complexities of TTF can be formalized using $x(t)$ by $T_{\mathbb{E}}(\text{TTF}) = \sum_{t=1}^{\infty} t \cdot (\mathbb{P}(x(t) = x_{\text{end}} \wedge x(t-1) \neq x_{\text{end}}))$ and $T_{\text{whp}}(\text{TTF}) = \inf \left\{ t \mid \mathbb{P}(x(t) = x_{\text{end}}) \geq 1 - \frac{1}{n} \right\}$.

To evaluate these time complexities, we study the evolution of $x(t)$ during executions of TTF. Given time t, consider a transition rule applicable from a configuration represented by a vector v^t and resulting in a configuration with vector v^{t+1}. Suppose that at time t, the interaction (i, j) is chosen by the scheduler. If neither i nor j are the base station and i is faster than j ($\text{cv}_i < \text{cv}_j$), agent j transfers all its tokens to i. Thus, $v_i^{t+1} = v_i^t + v_j^t$ and $v_j^{t+1} = 0$. The relation between v^t and v^{t+1}, in this case, can be expressed by the linear equation $v^{t+1} = W(t+1)v^t$, where $W(t+1) = I + \mathbf{e}_i\mathbf{e}_j^T - \mathbf{e}_j\mathbf{e}_j^T \in \{0, 1\}^{n \times n}$. If $\text{cv}_i = \text{cv}_j$, no token is transferred and $v^{t+1} = v^t$. We still have $v^{t+1} = W(t+1)v^t$, but with $W(t+1) = I$. On the other hand, if j is the base station, $W(t+1) = I + \mathbf{e}_i\mathbf{e}_j^T - \mathbf{e}_j\mathbf{e}_j^T$, as agent i transfers all of its tokens to the base station.

As the pair of agents is chosen independently with respect to P, $W(t+1)$ can be seen as a *random* matrix such that with probability $P_{i,j} + P_{j,i}$:

$$W(t+1) = \begin{cases} I + \mathbf{e}_i\mathbf{e}_j^T - \mathbf{e}_j\mathbf{e}_j^T & \text{if } \text{cv}_i < \text{cv}_j \text{ or } i = 1 \text{ or } j = 1 \\ I & \text{if } \text{cv}_i = \text{cv}_j \end{cases} \tag{1}$$

By comparing the resulting probability distributions, we readily verify that the relation between $x(t)$ and $x(t+1)$, i.e., $x(t+1) = W(t+1)x(t)$, is a stochastic linear system with the matrices specified in (1).

Distance. Consider a function $d_\gamma(x) : \mathbb{V} \to \mathbb{R}$. It associates any x in $\mathbb{V}$ to a real number representing a "weighted" Euclidian norm distance between the configuration vector x and the vector representing a terminal configuration. That is, $d_\gamma(x) = ||(x - x_{\text{end}}) \circ \gamma||_2$, where $\gamma \in \mathbb{R}^n$ is a real vector, $\circ$ the entry-wise product, and $|| \cdot ||_2$ the Euclidean norm. The vector γ can be viewed as a weight vector. We choose γ in such a way that, if there is a transfer of tokens in interaction $t+1$, configuration v^t, then $d_\gamma(v^{t+1})$ is smaller than $d_\gamma(v^t)$. Intuitively this means that, when a transfer is performed, the resulting configuration is closer to termination.

Lemma 2. *Let i and j be two agents with $cv_i < cv_j$. Consider an interaction between i and j in a configuration represented by v^t and resulting in v^{t+1}. If $\gamma_j/\gamma_i \geq \sqrt{2n-3}$, then $d_\gamma(v^{t+1}) \leq d_\gamma(v^t)$.*

Theorem 3. *The convergence time with high probability of TTF, $T_{\text{whp}}(\text{TTF})$, is equal to $\inf \left\{ t \mid \mathbb{P} \left(\frac{d_\gamma(x(t))}{d_\gamma(x_{\text{init}})} < (2n)^{\frac{-(m-1)}{2}} \right) \geq 1 - 1/n \right\}$ if $\gamma_{\text{BST}} = 0$ and $\gamma_j/\gamma_i \geq \sqrt{2n}$ whenever $cv_i < cv_j$. Recall that $m \leq n$ denotes the number of cover time categories (Sect. 2).*

We are now ready to state and prove our main upper bound on the convergence time of TTF, $T_{\text{whp}}(\text{TTF})$ (Theorem 4). To prove it, we apply stochastic matrix theory to the stochastic linear system defined above for $x(t)$.

Without loss of generality, we assume that $cv_2 \leq cv_3 \leq \cdots \leq cv_n$. We choose $\gamma \in \mathbb{R}^n$ by setting $\gamma_1 = 0$, $\gamma_2 = 1$, and $\gamma_{i+1} = \gamma_i$, if $cv_{i+1} = cv_i$, and $\gamma_{i+1} = \gamma_i\sqrt{2n}$, if $cv_{i+1} > cv_i$. In particular, $\gamma_n = (2n)^{(m-1)/2}$.

Theorem 4. *With a non-uniformly random scheduler $S(P)$, the convergence time of TTF is at most $\frac{m \log 2n}{\log \lambda_2(\tilde{W})^{-1}}$ with high probability, where γ is defined above. $\Gamma_{i,j} = \gamma_i/\gamma_j$, $\tilde{W} = \sum_{i<j \wedge cv_i<cv_j} (P_{i,j} + P_{j,i})W_{ij}^{\Gamma^2} + \sum_{i<j \wedge cv_i=cv_j} (P_{i,j} + P_{j,i})I$, $W_{ij}^{\Gamma^2} = I + \Gamma_{i,j}(e_i e_j^T + e_j e_i^T) + (\Gamma_{i,j}^2 - 1)e_j e_j^T$, and $\lambda_2(A)$ denotes the modulus of the second largest eigenvalue of matrix A.*

Now, we study the performance of TTF with respect to the convergence time in expectation, i.e. $T_{\mathbb{E}}(\text{TTF})$.

Theorem 5. *The expected convergence time of the TTF protocol is $O\left(\frac{m \log n}{\log \lambda_2(\tilde{W})^{-1}}\right)$ where $\tilde{W}$ is the matrix defined in Theorem 4.*

5   Lazy TTF

The strategy of TTF may result in a long execution when an input value is transferred many times before being finally delivered to the base station. These transfers are certainly energy consuming. Then a natural issue is to transform TTF in order to save energy, while keeping the time complexity as low as possible. The idea is to prevent certain data transfers, for example, when it is more likely to meet soon a faster agent and thus possibly make fewer transfers in overall. We propose a simple protocol based on TTF, called *lazy* TTF. In contrast with TTF, lazy TTF does not necessarily execute the transition resulting from an interaction. It chooses randomly to execute it or not. Formally, during an interaction (i, j), with agent i acting as initiator, TTF is executed with probability p_i, where $p \in \mathbb{R}^n$ is a vector of probabilities.

Notice that the choice of executing TTF depends uniquely on the initiator i. In practical terms, an initiator represents an agent that, by sensing the environment, has detected another agent j. At this moment i takes the random decision (with probability p_i) whether a TTF transition should be executed and the interaction itself should take place, or not. In the latter case, not only the energy for the eventual data transfer is saved, but also the energy for establishing the interaction.

Observe that when p is the vector of all ones, lazy TTF behaves as TTF and its energy consumption is the same as for TTF. However, when p is the vector of all zeros, lazy TTF does not solve the problem of data collection as no value is ever transferred to the base station, but no energy is consumed for transferring of data or establishing interactions. Depending on p, time complexities of lazy TTF can be worse than of TTF, given the same scheduler. At the same time, longer executions of lazy TTF may be more energy efficient. Thus, there is a trade-off between time and energy performance depending on the values of p. We investigate the choice of p for obtaining good time/energy trade-off. Firstly, we give upper bounds on the time complexities of lazy TTF. Then, we introduce an optimization problem that takes p as a variable. Finally, numerical results in Sect. 6 demonstrate energy efficiency of lazy TTF, given the optimal p.

5.1   Convergence Time of Lazy TTF

To obtain an upper bound on the convergence time of lazy TTF, we show a particular equivalence of lazy TTF under scheduler $S(P)$ with TTF under scheduler $S(P \circ (p \cdot \mathbf{1}^T))$, where $\mathbf{1}$ is the vector of all ones and $\circ$ presents the entry-wise product. This equivalence is on the level of distribution of configurations of the two protocols. Precisely, as we show below, the random vector $x(t)$ for these two protocols is exactly the same, allowing to use Theorem 4 to obtain a time complexity upper bound for lazy TTF.

Let us express $x(t)$ in case of lazy TTF in a similar way as we did before for TTF in Sect. 4. First, $\mathbb{P}(x(0) = x_{\text{init}}) = 1$ is the same as for TTF. Then, $x(t+1) = W(t+1)x(t)$ and $W(t+1)$ can be seen as a *random* matrix such that, with probability $P_{i,j} \times p_i + P_{j,i} \times p_j$, $W(t+1)$ is as in Eq. 1. Notice that $x(t)$ in case of TTF under $S(P \circ (p \cdot \mathbf{1}^T))$ is expressed exactly in the same way (Sect. 4). Thus, by applying Theorem 4 for TTF under $S(P \circ (p \cdot \mathbf{1}^T))$, we obtain the upper bound on $T_{whp}(\text{lazy TTF}(p))$.

Theorem 6. *With a non-uniformly random scheduler $S(P)$, the convergence time with high probability of lazy TTF is at most $\frac{m \log 2n}{\log \lambda_2(\tilde{W})^{-1}}$,*

$$where \; \tilde{W} = \sum_{cv_i < cv_j} (P_{i,j}p_i + P_{j,i}p_j)W_{ij}^{\Gamma^2} + \sum_{cv_i < cv_j} (P_{i,j}(1 - p_i) + P_{j,i}(1 - p_j))I$$
$$+ \sum_{cv_i = cv_j} (P_{i,j} + P_{j_i})I, and \; W_{ij}^{\Gamma^2} = I + \Gamma_{i,j}(\mathbf{e}_i\mathbf{e}_j^T + \mathbf{e}_j\mathbf{e}_i^T) + (\Gamma_{i,j}^2 - 1)\mathbf{e}_j\mathbf{e}_j^T. \tag{2}$$

Then, the upper bound on $T_{\mathbb{E}}(\text{lazy TTF}(p))$ can be obtained in the same way as in Th. 5.

To summarize, note that, as executions of lazy TTF are equivalent to those of TTF under $S(P \circ (p \cdot \mathbf{1}^T))$ in the sense explained above, one can imagine that lazy TTF transforms the matrix of interaction probabilities "on the fly" (during executions). It can be also seen as if it transforms the interaction graph itself. Indeed, certain vectors p may make some pairs of agents to interact with extremely small probability (or not interact at all), thus effectively remove these pairs from the graph. This is illustrated in the report [12]. Next, we are looking for vectors p, optimizing an upper bound on the time performance of lazy TTF(p) to ensure a good time energy trade-off. Equivalently, we are looking for schedulers (matrices P) for which the original TTF is efficient in this sense.

Thus, the goal is to find a vector p minimizing the upper bound on $T_{whp}(\text{lazy TTF}(p))$ (Theorem 6). To that end, an optimization program OP_1, taking p as a variable, is proposed as follows:

$OP_1:\ \min\limits_{p \in \mathbb{R}^n} \lambda_2(\tilde{W})$ s.t Eq. 2, $0 \le p_i \le 1$.

By Theorem 6, minimizing the upper bound of $T_{whp}(\text{lazy TTF}(p))$ is equivalent to minimizing the second largest eigenvalue of $\tilde{W}$. Then, we reformulate OP_1 as a semi-definite program [25,26] OP_2 which is convex and can be solved in polynomial time.

$OP_2:\ \min\limits_{p \in \mathbb{R}^n, s} s$ s.t $sI - \tilde{W} \succeq 0$, Eq. 2, $0 \le p_i \le 1$.

Let $\hat{p}$ be the optimal solution of OP_2. We can see that if $\hat{p}$ is all ones vector, lazy TTF$(\hat{p})$ performs as TTF. Otherwise, lazy TTF$(\hat{p})$ outperforms TTF in terms of the upper bounds on time. This optimized upper bound ensures that lazy TTF$(\hat{p})$ converges in a reasonable time. In the next section, by the numerical results obtained for different small examples, we demonstrate the efficiency of lazy TTF$(\hat{p})$, in terms of energy consumption.

6   Numerical Results

6.1   The Relation Between $T_{\mathbf{whp}}$(TTF) and Its Upper Bound

The goal of this section is to justify the relevance of the method used here to obtain the optimal probability vector p for lazy TTF. To justify this, we show by simulations that the time upper bound value for TTF is well correlated with the exact value of its time complexity (calculated by Markov chains, for small systems). This implies the same correlation for lazy TTF, because the bounds in Theorems 4 and 6 are obviously well correlated too (one is obtained from the other; see Sect. 5). That is why the optimal probability vector p for the upper bound of lazy TTF is close to the optimal vector for the real (tight) convergence time.

From Theorem 4, we have an upper bound on time w.h.p. for TTF, denoted here by $T_{\mathrm{upp}}(\mathrm{TTF})$. In this section, we show the relation between $T_{\mathrm{upp}}(\mathrm{TTF})$ and $T_{\mathrm{whp}}(\mathrm{TTF})$. In our experiment, two systems of size 4 and 5 are considered and 100 schedulers are generated randomly for each system. Since the system is of small size, for each scheduler s, the exact value of $T^s_{\mathrm{whp}}(\mathrm{TTF})$ can be obtained by constructing the corresponding Markov Chain. The upper bound, $T^s_{\mathrm{upp}}(\mathrm{TTF})$, can be calculated by Theorem 4. Then, for every generated s, we plot $T^s_{\mathrm{whp}}(\mathrm{TTF})$ and $T^s_{\mathrm{upp}}(\mathrm{TTF})$ on the figure with x-axis for $T_{\mathrm{whp}}(\mathrm{TTF})$ and y-axis for $T_{\mathrm{upp}}(\mathrm{TTF})$.

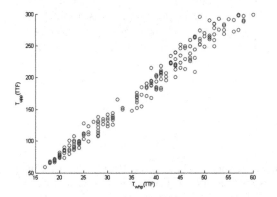

From Fig. 1, we can see that $T_{\mathrm{upp}}(\mathrm{TTF})$ has a nearly linear relation with $T_{\mathrm{whp}}(\mathrm{TTF})$. It means that $T_{\mathrm{upp}}(\mathrm{TTF})$ in Theorem 4 captures well the relation of the scheduler's behavior to the time performance of TTF in most of the cases. Moreover, it demonstrates that, for lazy TTF, minimizing $T_{whp}(\mathrm{lazy\,TTF}(p))$ in Sect. 5, is reasonable for improving the energy performance.

Fig. 1. Relation between $T_{\mathrm{whp}}(\mathrm{TTF})$ and $T_{\mathrm{upp}}(\mathrm{TTF})$.

6.2   Gaps on Time and Energy Between TTF and Lazy TTF($\hat{p}$)

For energy consumption analysis, we consider the energy model of [11] proposed for population protocols. In this model, an agent senses its vicinity by proximity sensor, consuming a negligible amount of energy [27]. Once the interaction is established, each participant consumes a fixed amount of energy $\mathcal{E}_{wkp}$ (mainly for switching on its radio, which is known to be very energy consuming; cf. [28]). Now, recall that, with lazy TTF, the choice of executing TTF depends on the probability p_i of the initiator i. If TTF should not be executed, the initiator does not proceed to establish the interaction neither (i.e., $\mathcal{E}_{wkp}$ is not spent), as explained in Sect. 5.

We study the expectation of the total energy consumption of a protocol $\mathcal{P}$, denoted $\mathcal{E}(\mathcal{P})$. According to the energy scheme explained above, $\mathcal{E}(\mathcal{P})$ is evaluated by the expected total energy spent for establishing all the interactions till convergence. It is proportional to the time expectation $T_{\mathbb{E}}(\mathcal{P})$. In particular, $\mathcal{E}(\text{TTF}) = 2T_{\mathbb{E}}(\text{TTF}) \cdot \mathcal{E}_{wkp}$ and $\mathcal{E}(\text{lazyTTF}(p)) = 2T_{\mathbb{E}}(\text{lazy TTF}(p)) \times \sum_i \sum_j (P_{i,j} p_i + P_{j,i} p_j) \times \mathcal{E}_{wkp}$.

For the systems of small size with a scheduler s, the exact values of $T_{\mathbb{E}}^s(\text{TTF})$ and $T_{\mathbb{E}}^s(\text{lazy TTF}(\hat{p}^s))$ can be calculated by constructing the corresponding Markov Chain. In the experiments, systems of size 4, 5, 6, 7 and 8 are considered and for each size n, 10000 different schedulers are generated randomly. Denote by $\mathcal{S}(n)$ the set of these schedulers. For each scheduler $s \in \mathcal{S}(n)$, $T_{\mathbb{E}}^s(\text{TTF})$, $\hat{p}^s$, $T_{\mathbb{E}}^s(\text{lazy TTF}(\hat{p}^s))$, $\mathcal{E}^s(\text{TTF})$ and $\mathcal{E}^s(\text{lazy TTF}(\hat{p}^s))$ are evaluated. Then, the gaps on time and on energy between lazy TTF$(\hat{p}^s)$ and TTF^s are denoted by $Gap(T_{\mathbb{E}}, n)$ and $Gap(\mathcal{E}, n)$, respectively, and are computed as follows.

$$Gap(T_{\mathbb{E}}, n) = \left(\sum_{s \in \mathcal{S}(n)} \frac{T_{\mathbb{E}}^s(\text{lazy TTF}(\hat{p}^s)) - T_{\mathbb{E}}^s(\text{TTF})}{T_{\mathbb{E}}^s(\text{TTF})} \right) / 10000 \text{ and}$$

$$Gap(\mathcal{E}, n) = \left(\sum_{s \in \mathcal{S}(n)} \frac{\mathcal{E}^s(\text{lazy TTF}(\hat{p}^s)) - \mathcal{E}^s(\text{TTF})}{\mathcal{E}^s(\text{TTF})} \right) / 10000.$$

Results appear in Table 1. In column 3, it can be seen that lazy TTF consumes less energy than TTF for all systems. Lazy TTF saves at least 15% of energy. The counterpart is (a slight) increase in the execution time, as shown in column 2.

Table 1. Gaps on time and energy.

Size n	$Gap(T_{\mathbb{E}}, n)$	$Gap(\mathcal{E}, n)$
4	11.60%	−15.32%
5	17.10%	−23.60%
6	22.04%	−30.79%
7	26.31%	−36.99%
8	27.41%	−39.07%

References

1. Angluin, D., Aspnes, J., Diamadi, Z., Fischer, M.J., Peralta, R.: Computation in networks of passively mobile finite-state sensors. In: Proceedings of 23rd Annual ACM Symposium on Principles of Distributed Computing, pp. 290–299 (2004)
2. Angluin, D., Aspnes, J., Diamadi, Z., Fischer, M.J., Peralta, R.: Computation in networks of passively mobile finite-state sensors. Distrib. Comput. **18**(4), 235–253 (2006)
3. Angluin, D., Aspnes, J., Eisenstat, D.: Fast computation by population protocols with a leader. In: Dolev, S. (ed.) DISC 2006. LNCS, vol. 4167, pp. 61–75. Springer, Heidelberg (2006). https://doi.org/10.1007/11864219_5
4. Alistarh, D., Gelashvili, R., Vojnović, M.: Fast and exact majority in population protocols. In: Proceedings of 2015 ACM Symposium on Principles of Distributed Computing, pp. 47–56. ACM (2015)
5. Aspnes, J., Beauquier, J., Burman, J., Sohier, D.: Time and space optimal counting in population protocols. In: OPODIS 2016, pp. 13:1–13:17 (2016)
6. Sharma, G., Mazumdar, R.R.: Scaling laws for capacity and delay in wireless ad hoc networks with random mobility. In: 2004 IEEE International Conference on Communications, vol. 7, pp. 3869–3873 (2004)
7. Cai, H., Eun, D.Y.: Crossing over the bounded domain: from exponential to power-law inter-meeting time in manet. In: Proceedings of 13th Annual ACM International Conference on Mobile Computing and Networking, pp. 159–170 (2007)
8. Zhu, H., Fu, L., Xue, G., Zhu, Y., Li, M., Ni, L.M.: Recognizing exponential inter-contact time in vanets. In: 2010 Proceedings of INFOCOM, pp. 1–5 (2010)

9. Gao, W., Cao, G.: User-centric data dissemination in disruption tolerant networks. In: Proceedings of IEEE INFOCOM 2011, pp. 3119–3127 (2011)
10. Beauquier, J., Burman, J., Clement, J., Kutten, S.: On utilizing speed in networks of mobile agents. In: Proceedings of 29th ACM SIGACT-SIGOPS Symposium on Principles of Distributed Computing, pp. 305–314 (2010)
11. Xu, C., Burman, J., Beauquier, J.: Power-aware population protocols. In: 2017 IEEE 37th International Conference on Distributed Computing Systems (ICDCS), pp. 2067–2074, June 2017
12. Beauquier, J., Burman, J., Kutten, S., Nowak, T., Xu, C.: Data collection in population protocols with non-uniformly random scheduler. Research report, July 2017. https://hal.archives-ouvertes.fr/hal-01567322
13. Alistarh, D., Aspnes, J., Eisenstat, D., Gelashvili, R., Rivest, R.L.: Time-space trade-offs in population protocols. In: Proceedings of 28th Annual ACM-SIAM Symposium on Discrete Algorithms, SODA 2017, pp. 2560–2579 (2017)
14. Tel, G.: Introduction to Distributed Algorithms, 2nd edn. Cambridge University Press, Cambridge (2000). https://doi.org/10.1017/CBO9781139168724
15. Flajolet, P., Gardy, D., Thimonier, L.: Birthday paradox, coupon collectors, caching algorithms and self-organizing search. Discret. Appl. Math. **39**(3), 207–229 (1992)
16. Guerraoui, R., Ruppert, E.: Names trump malice: tiny mobile agents can tolerate byzantine failures. In: Albers, S., Marchetti-Spaccamela, A., Matias, Y., Nikoletseas, S., Thomas, W. (eds.) ICALP 2009. LNCS, vol. 5556, pp. 484–495. Springer, Heidelberg (2009). https://doi.org/10.1007/978-3-642-02930-1_40
17. Angluin, D., Aspnes, J., Eisenstat, D.: A simple population protocol for fast robust approximate majority. Distrib. Comput. **21**, 87–102 (2008)
18. Stadje, W.: The collector's problem with group drawings. Adv. Appl. Probab. **22**(4), 866–882 (1990)
19. Adler, I., Ross, S.M.: The coupon subset collection problem. J. Appl. Probab. **38**, 737–746 (2001)
20. Ferrante, M., Saltalamacchia, M.: The coupon collector's problem. Mater. Matemàtics **2014**(2), 35 (2014)
21. Nakata, T.: Coupon collector's problem with unlike probabilities (2008, preprint)
22. Tsitsiklis, J.N., Bertsekas, D.P., Athans, M.: Distributed asynchronous deterministic and stochastic gradient optimization algorithms. In: American Control Conference 1984, pp. 484–489 (1984)
23. Jadbabaie, A., Lin, J., Morse, A.S.: Coordination of groups of mobile autonomous agents using nearest neighbor rules. IEEE Trans. Autom. Control **48**(6), 988–1001 (2003)
24. Ren, W., Beard, R.W., et al.: Consensus seeking in multiagent systems under dynamically changing interaction topologies. IEEE Trans. Autom. Control **50**(5), 655–661 (2005)
25. Vandenberghe, L., Boyd, S.: Semidefinite programming. SIAM Rev. **38**(1), 49–95 (1996)
26. Helmberg, C., Rendl, F., Vanderbei, R.J., Wolkowicz, H.: An interior-point method for semidefinite programming. SIAM J. Optim. **6**(2), 342–361 (1996)
27. Razzaque, M.A., Dobson, S.: Energy-efficient sensing in wireless sensor networks using compressed sensing. Sensors **14**(2), 2822–2859 (2014)
28. Rajendran, V., Obraczka, K., Garcia-Luna-Aceves, J.J.: Energy-efficient, collision-free medium access control for wireless sensor networks. Wirel. Netw. **12**(1), 63–78 (2006)

Parameterized Algorithms for Power-Efficient Connected Symmetric Wireless Sensor Networks

Matthias Bentert[1(✉)], René van Bevern[2,3],
André Nichterlein[1], and Rolf Niedermeier[1]

[1] Institut für Softwaretechnik und Theoretische Informatik,
TU Berlin, Berlin, Germany
{matthias.bentert,andre.nichterlein,rolf.niedermeier}@tu-berlin.de
[2] Novosibirsk State University, Novosibirsk, Russian Federation
rvb@nsu.ru
[3] Sobolev Institute of Mathematics, Siberian Branch of the Russian
Academy of Sciences, Novosibirsk, Russian Federation

Abstract. We study an NP-hard problem motivated by energy-efficiently maintaining the connectivity of a symmetric wireless sensor communication network. Given an edge-weighted n-vertex graph, find a connected spanning subgraph of minimum cost, where the cost is determined by letting each vertex pay the most expensive edge incident to it in the subgraph. We provide an algorithm that works in polynomial time if one can find a set of obligatory edges that yield a spanning subgraph with $O(\log n)$ connected components. We also provide a linear-time algorithm that reduces any input graph that consists of a tree together with g additional edges to an equivalent graph with $O(g)$ vertices. Based on this, we obtain a polynomial-time algorithm for $g \in O(\log n)$. On the negative side, we show that $o(\log n)$-approximating the difference d between the optimal solution cost and a natural lower bound is NP-hard and that there are presumably no exact algorithms running in $2^{o(n)}$ time or in $f(d) \cdot n^{O(1)}$ time for any computable function f.

Keywords: Monitoring areas and backbones
Parameterized complexity · Color coding · Data reduction
Parameterization above lower bounds · Approximation hardness
Spanning trees

1 Introduction

We consider a well-studied graph problem arising in the context of saving power in maintaining the connectivity of symmetric wireless sensor communication networks. Our problem, which falls into the category of survivable network design [22], is formally defined as follows (see Fig. 1 for an example).

Problem 1.1. Min-Power Symmetric Connectivity (MinPSC)

Input: A connected undirected graph $G = (V, E)$ with n vertices, m edges, and
edge weights (costs) $w: E \to \mathbb{N}$.

© Springer International Publishing AG 2017
A. Fernández Anta et al. (Eds.): ALGOSENSORS 2017, LNCS 10718, pp. 26–40, 2017.
https://doi.org/10.1007/978-3-319-72751-6_3

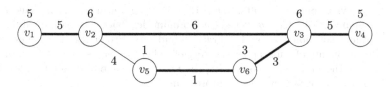

Fig. 1. A graph with positive edge weights and an optimal solution (bold edges). Each vertex pays the most expensive edge incident to it in the solution (the numbers next to the vertices). The cost of the solution is the sum of the costs paid by each vertex. Note that the optimal solution has cost 26 while a minimum spanning tree (using edge $\{v_2, v_5\}$ instead of $\{v_2, v_3\}$) has cost 27 (as a MINPSC solution).

Goal: Find a connected spanning subgraph $T = (V, F)$ of G that minimizes

$$\sum_{v \in V} \max_{\{u,v\} \in F} w(\{u, v\}).$$

We denote the minimum cost of a solution to an MINPSC instance $I = (G, w)$ by $\mathrm{Opt}(I)$. Throughout this work, *weights* always refer to edges and *costs* refer to vertices or subgraphs. For showing hardness results, we will also consider the *decision version* of MINPSC, which we call k-PSC. Herein the problem is to decide whether an input instance $I = (G, w)$ satisfies $\mathrm{Opt}(I) \leq k$.

Figure 1 reveals that computing a minimum-cost spanning tree may not yield an optimal solution for MINPSC (also see Erzin et al. [12] for further discussion concerning the relationship to minimum-cost spanning trees). In this work, we provide a refined computational complexity analysis by initiating parameterized complexity studies of MINPSC (and its decision version). In this way, we complement previous findings mostly concerning polynomial-time approximability [2,9,12], heuristics and integer linear programming [2,13,21], and computational complexity analysis for special cases [8,9,12,17].

Our Contributions. Our work is driven by asking when small input-specific parameter values allow for fast (exact) solutions in practically relevant special cases. Our two fundamental "use case scenarios" herein are monitoring areas and infrastructure backbones. Performing a parameterized complexity analysis, we obtain new encouraging exact algorithms together with new hardness results, all summarized in Table 1.

In Sect. 2, we provide an (exact) algorithm for MINPSC that works in polynomial time if one can find a set of *obligatory* edges that can be added to any optimal solution and yield a spanning subgraph with $O(\log n)$ connected components. In particular, this means that we show fixed-parameter tractability for MINPSC with respect to the parameter "number c of connected components in the spanning subgraph consisting of obligatory edges". Cases with small c occur, for example, in grid-like sensor arrangements, which arise when monitoring areas [28,29].

Table 1. Overview on our results, using the following terminology: n—number of vertices, m—number of edges, g—size of a minimum feedback edge set, d—difference between optimal solution cost and a lower bound (see Problem 4.1), c—number of connected components of subgraph consisting of obligatory edges (see Definition 2.3). MINPSC-AL is the problem of computing the minimum value of d (see Problem 4.1), d-PSC-AL is the corresponding decision problem.

	Problem	Result	Reference
Section 2	MINPSC	Solvable in $O(\ln(1/\varepsilon) \cdot (36e^2/\sqrt{2\pi})^c \cdot n^4/\sqrt{c})$ time with error probability at most ε	Theorem 2.5
	MINPSC	Solvable in $c^{O(c \log c)} \cdot n^{O(1)}$ time	Theorem 2.5
	MINPSC	Solvable in $O(3^n \cdot (n+m))$ time	Proposition 2.7
Section 3	MINPSC	Linear-time data reduction algorithm that guarantees at most $40g - 26$ vertices and $41g - 27$ edges	Theorem 3.1
Section 4	MINPSC-AL	NP-hard to approximate within a factor of $o(\log n)$	Theorem 4.2(i)
	d-PSC-AL	$W[2]$-hard when parameterized by d	Theorem 4.2(ii)
	k-PSC	Not solvable in $2^{o(n)}$ time unless ETH fails	Theorem 4.2(iii)

In Sect. 3, we provide a linear-time algorithm that reduces any input graph consisting of a tree with g additional edges to an equivalent graph with $O(g)$ vertices and edges (a partial kernel in terms of parameterized complexity, since the edge weights remain unbounded). Combined with the previous result, this yields fixed-parameter tractability with respect to the parameter g (also known as the feedback edge number of a graph), and, in particular, a polynomial-time algorithm for $g \in O(\log n)$. Such tree-like graphs occur when monitoring backbone infrastructure or pollution levels along waterways.

We provide some negative (that is, intractability) results in Sect. 4: We show that $o(\log n)$-approximating the difference d between the minimum solution cost and a natural lower bound is NP-hard. Moreover, we prove $W[2]$-hardness with respect to the parameter d, that is, there is presumably no algorithm running in $f(d) \cdot n^{O(1)}$ time for any computable function f. Finally, assuming the Exponential Time Hypothesis (ETH), we show that there is no $2^{o(n)}$-time algorithm for MINPSC.

Due to space constraints, all proofs are deferred to a full version[1].

2 Parameterizing by the Number of Connected Components Induced by Obligatory Edges

This section presents an algorithm that solves MINPSC efficiently if we can find *obligatory edges* that can be added to any optimal solution and yield a spanning subgraph with few connected components. This is the case, for example, when

---

[1] https://arxiv.org/abs/1706.03177.

sensors are arranged in a grid-like manner, which saves energy when monitoring areas [28,29]. To find obligatory edges, we use a lower bound $\ell(v)$ on the cost paid by each vertex v in the goal function of MINPSC (Problem 1.1).

Definition 2.1 (vertex lower bounds). Vertex lower bounds *are given by a function* $\ell \colon V \to \mathbb{N}$ *such that, for any solution* $T = (V, F)$ *of* MINPSC *and any vertex* $v \in V$, *it holds that*

$$\max_{\{u,v\} \in F} w(\{u,v\}) \geq \ell(v).$$

Example 2.2. A trivial vertex lower bound $\ell(v)$ is given by the weight of the lightest edge incident to v because v has to be connected to some vertex in any solution. Moreover, since any edge $\{u,v\}$ incident to a degree-one vertex u will be part of any solution, one can choose ℓ so that $\ell(u) = w(\{u,v\})$ and $\ell(v) \geq w(\{u,v\})$.

Clearly, coming up with good vertex lower bounds is a challenge on its own. Once we have vertex lower bounds, we can compute an *obligatory subgraph*, whose edges we can add to any solution without increasing its cost:

Definition 2.3 (obligatory subgraph). *The* obligatory subgraph G_ℓ *induced by vertex lower bounds* $\ell \colon V \to \mathbb{N}$ *for a graph* $G = (V, E)$ *consists of all vertices of* G *and all* obligatory edges $\{u,v\}$ *such that* $\min\{\ell(u), \ell(v)\} \geq w(\{u,v\})$.

The better the vertex lower bounds ℓ, the more obligatory edges they potentially induce, thus reducing the number c of connected components of G_ℓ. Yet already the simple vertex lower bounds in Example 2.2 may yield obligatory subgraphs with only a few connected components in some applications:

Example 2.4. Consider the vertex lower bounds ℓ from Example 2.2. If we arrange sensors in a grid, which is the most energy-efficient arrangement of sensors for monitoring areas [28,29], then G_ℓ has only one connected component. The number of connected components may increase due to sensor defects that disconnect the grid or due to varying sensor distances within the grid. The worst case is if the sensors have pairwise distinct distances. Then, G_ℓ has only one edge and $n - 1$ connected components.

The number c of connected components in G_ℓ can easily be exploited in an exact $O(n^{2c})$-time algorithm for MINPSC,[2] which runs in polynomial time for constant c, yet is inefficient already for small values of c. We will show, among other things, a randomized algorithm that runs in polynomial time for $c \in O(\log n)$:

---

[2] To connect the c components of G_ℓ, one has to add $c - 1$ edges. These have at most $2c - 2$ end points. One can try all n^{2c-2} possibilities for choosing these end points and check each resulting graph for connectivity in $O(n + m) \subseteq O(n^2)$ time.

Theorem 2.5. MINPSC *with vertex lower bounds ℓ is solvable*

(i) *in $O(\ln 1/\varepsilon \cdot (36e^2/\sqrt{2\pi})^c \cdot 1/\sqrt{c} \cdot n^4)$ time by a randomized algorithm with error probability at most ε for any given $\varepsilon \in (0,1)$, and*

(ii) *in $c^{O(c \log c)} \cdot n^{O(1)}$ time by a deterministic algorithm,*

where c is the number of connected components of the obligatory subgraph G_ℓ.

Remark 2.6. The deterministic algorithm in Theorem 2.5(ii) is primarily of theoretical interest, because it classifies MINPSC as *fixed-parameter tractable* parameterized by c. Practically, the randomized algorithm in Theorem 2.5(i) seems more promising.

The number of connected components of obligatory subgraphs has recently also been exploited in fixed-parameter algorithms for problems of servicing links in transportation networks [6,15,25,26], which led to practical results.

The rest of this section outlines the proof of Theorem 2.5. The proof also yields the following deterministic algorithm for MINPSC, which will be interesting in combination with the data reduction algorithm in Sect. 3. It is much faster than the trivial algorithm enumerating all of the possibly n^{n-2} spanning trees:

Proposition 2.7. MINPSC *can be solved in $O(3^n \cdot (m + n))$ time.*

Like some known approximation algorithms for MINPSC [2,17], our algorithms in Theorem 2.5 work by adding edges to G_ℓ in order to connect its c connected components. In contrast to these approximation algorithms, our algorithms will find an *optimal* set of edges to add. To this end, they work on a *padded* version $G_\ell^\bullet$ of the input graph G, in which each connected component of G_ℓ is turned into a clique. Then, it is sufficient to search for connected subgraphs of $G_\ell^\bullet$ that contain at least one vertex of each connected component of G_ℓ. We can always add the edges in G_ℓ to such subgraphs in order to obtain a connected spanning subgraph of G.

Definition 2.8 (padded graph, components). *Let $\ell \colon V \to \mathbb{N}$ be vertex lower bounds for a graph $G = (V, E)$. We denote the c connected components of the obligatory subgraph G_ℓ by $G_\ell^1, G_\ell^2, \dots, G_\ell^c$.*

The padded graph $G_\ell^\bullet = (V, E_\ell^\bullet)$ with edge weights $w_\ell^\bullet \colon E_\ell^\bullet \to \mathbb{N}$ is obtained from G with edge weights $w \colon E \to \mathbb{N}$ by adding zero-weight edges between each pair of non-adjacent vertices in G_ℓ^i for each $i \in \{1, \dots, c\}$.

To solve a MINPSC instance (G, w) with vertex lower bounds $\ell \colon V \to \mathbb{N}$, we have to add $c - 1$ edges to G_ℓ in order to connect its c connected components. These edges have at most $2c-2$ endpoints. Thus, we need to find a minimum-cost connected subgraph in $G_\ell^\bullet$ that

- contains at most $2c - 2$ vertices,
- contains at least one vertex of each connected component of G_ℓ,
- such that each of its vertices v pays at least the cost $\ell(v)$ that it would pay in any optimal solution to the MINPSC instance (G, w).

We will do this using the color coding technique introduced by Alon et al. [1]: randomly color the vertices of $G_\ell^\bullet$ using at most $2c - 2$ colors and then search for connected subgraphs of $G_\ell^\bullet$ that contain exactly one vertex of each color. Formally, we will solve the following auxiliary problem on $G_\ell^\bullet$.

Problem 2.9. MIN-POWER COLORFUL CONNECTED SUBGRAPH (MINPCCS)

Input: A connected undirected graph $G = (V, E)$, edge weights $w \colon E \to \mathbb{N}$, vertex colors $\mathrm{col} \colon V \to \mathbb{N}$, a function $\ell \colon V \to \mathbb{N}$ and a color subset $C \subseteq \mathbb{N}$.
Goal: Compute a connected subgraph $T = (W, F)$ of G such that col is a bijection between W and C and such that T minimizes

$$\sum_{v \in W} \max\Big\{\ell(v), \max_{\{u,v\} \in F} w(\{u, v\})\Big\}.$$

Note that, in the definition of MINPCCS, the function $\ell \colon V \to \mathbb{N}$ does not necessarily give vertex lower bounds, but makes sure that each vertex $v \in V$ pays at least $\ell(v)$ in any feasible solution to MINPCCS.

In contrast to the usual way of applying color coding, we cannot simply color the vertices of our input graph G *completely* randomly and then apply an algorithm for MINPCCS: One component of G_ℓ could contain all colors and, thus, a connected subgraph containing all colors does not necessarily connect the components of G_ℓ. Instead, we employ a trick that was previously applied mainly heuristically in algorithm engineering in order to increase the success probability of color coding algorithms [4,7,10]. Since we know that our sought subgraph contains at least one vertex of each connected component of G_ℓ, we color the connected components of G_ℓ using pairwise disjoint color sets. Herein, we first "guess" how many vertices c_i of each connected component G_ℓ^i of G_ℓ the sought subgraph will contain and use c_i colors to color each component G_ℓ^i. We thus arrive at the following algorithm for MINPSC:

Algorithm 2.10 (for MinPSC).

Input: A MINPSC instance $I = (G, w)$, vertex lower bounds $\ell \colon V \to \mathbb{N}$ for $G = (V, E)$, an upper bound $\varepsilon \in (0, 1)$ on the error probability.
Output: A solution for I that is optimal with probability at least $1 - \varepsilon$.

1. $c \leftarrow$ number of connected components of the obligatory subgraph G_ℓ.
2. **for each** $c_1, c_2, \ldots, c_c \in \mathbb{N}^+$ such that $\sum_{i=1}^{c} c_i \leq 2c - 2$ **do**
3.   choose pairwise disjoint $C_i \subseteq \{1, \ldots, 2c-2\}$ with $|C_i| = c_i$ for $i \in \{1, \ldots, c\}$.
4.   **repeat** $t := \ln \varepsilon / \ln(1 - \prod_{i=1}^{c} c_i! / c_i^{c_i})$ **times**
5.   **for** $i \in \{1, \ldots, c\}$, randomly color the vertices of component G_ℓ^i of G_ℓ using colors from C_i, let the resulting coloring be col$\colon V \to \mathbb{N}$.
6.   Solve MINPCCS instance $(G_\ell^\bullet, w_\ell^\bullet, \mathrm{col}, \ell, C)$ using dynamic programming.
7. let $T = (W, F)$ be the best MINPCCS solution found in any of the repetitions.
8. **return** $T' = (V, (F \cap E) \cup E_\ell)$.

The main ingredient of Algorithm 2.10 is the following $O(2^{|C|} \cdot (n+m)^2 + 3^{|C|} \cdot (m+n))$-time dynamic programming algorithm for MINPCCS used in line 6. It is inspired by an algorithm for finding signalling pathways in biological networks [24]. However, our case is complicated by the non-standard goal function in MINPCCS.

Algorithm 2.11 (for MinPCCS). Let $I := (G, w, \mathrm{col}, \ell, C)$ be an instance of MINPCCS, where $G = (V, E)$. Assume $w(e) = \infty$ for any $e \notin E$ and $\min \emptyset = \infty$. For any color set $C' \subseteq C$ and any pair of vertices $\{v, q\} \subseteq V$, let $D(v, q, C')$ be the cost of a feasible solution $T = (W, F)$ to the MINPCCS instance $(G, w, \mathrm{col}, \ell, C')$ that minimizes

$$\Phi(v, q, T) := \max\{\ell(v), w(\{v, q\})\} + \sum_{v' \in W \setminus \{v\}} \max\Big\{\ell(v'), \max_{\{u, v'\} \in F} w(\{u, v'\})\Big\}$$

under the constraints that $v \in W$ and

$$\max\{\ell(v), w(\{v, q\})\} \geq \max\Big\{\ell(v), \max_{\{u, v\} \in F} w(\{u, v\})\Big\}$$

(such a solution might not exist for some choices of v and q). Note that the only difference between $\Phi(v, q, T)$ and the goal function of MINPCCS (Problem 2.9) is that the vertex v pays exactly $\max\{\ell(v), w(\{v, q\})\}$.

The cost $\min_{\{v, q\} \subseteq V} D[v, q, C]$ of an optimal solution to I can then be computed as follows. Obviously, $D[v, q, C'] = \infty$ if $\mathrm{col}(v) \notin C'$. Moreover, $D[v, q, \{\mathrm{col}(v)\}] = \max\{\ell(v), w(\{v, q\})\}$. Finally, compute $D[v, q, C']$ with $|C'| \geq 2$ for subsets $C' \subseteq C$ of increasing cardinality, storing intermediate results, via

$$D[v, q, C'] = \min \left\{ \begin{array}{l} D[v, q, C_1] + D[v, q, C_2] - \max\{\ell(v), w(\{v, q\}) \\ \quad \text{for all } C_1 \subsetneq C' \text{ and } C_2 \subsetneq C' \\ \quad \text{such that } C_1 \cup C_2 = C' \text{ and } C_1 \cap C_2 = \{\mathrm{col}(v)\} \end{array} \right\}$$
$$\cup \left\{ \begin{array}{l} D[u, q', C' \setminus \{\mathrm{col}(v)\}] + \max\{\ell(v), w(\{v, q\}) \\ \quad \text{for all } u \in N(v) \text{ and } q' \in N(u) \\ \quad \text{such that } w(\{u, q'\}) \geq w(\{u, v\}) \\ \quad \text{and } w(\{v, q\}) \geq w(\{u, v\}) \end{array} \right\}.$$

3   Parameterizing by the Feedback Edge Number

This section studies the complexity of MINPSC parameterized by the feedback edge number—the minimum number of edges one has to delete in order to turn a graph into a tree. This parameter can be computed in linear time by computing a spanning tree and counting the remaining edges. The motivation for studying this parameter is twofold. From a theoretical point of view, the cost of any spanning tree of the input graph gives an upper bound on the cost of an optimal solution. The remaining edges are a *feedback edge set* whose size limits the freedom for improving this upper bound. In practice, graphs with small feedback edge number may appear when monitoring backbone infrastructure or waterways (for example, when deleting canals, the remaining, natural waterways usually form a forest [14]). Hence, it is natural to ask whether MINPSC is significantly easier on graphs with small feedback edge number than on general graphs. In this section, we answer this question in the affirmative by proving the following theorem.

Theorem 3.1. *In linear time, one can transform any instance* $I = (G, w)$ *of* MINPSC *with feedback edge number* g *into an instance* $I' = (G', w')$ *and compute a value* $a \in \mathbb{N}$ *such that* G' *has at most* $40g - 26$ *vertices,* $41g - 27$ *edges, and* $\mathrm{Opt}(I) = \mathrm{Opt}(I') + a$.

In terms of parameterized complexity theory, Theorem 3.1 gives a *partial kernel* [5] of linear size for k-PSC parameterized by the feedback edge number, leaving k and the edge weights unbounded.

By applying first Theorem 3.1 to shrink a MINPSC instance and then applying Proposition 2.7 to solve the shrunk instance, we can solve MINPSC in polynomial time on graphs with feedback edge number $g \in O(\log n)$:

Corollary 3.2. MINPSC *is solvable in* $2^{O(g)} + O(n + m)$ *time.*

We point out that, in practice, it seems more promising to solve the shrunk instance using Theorem 2.5 instead of Proposition 2.7, although it will not give a *provably* better bound than Corollary 3.2.

To prove Theorem 3.1, we first present data reduction rules for the following intermediate variant of MINPSC, which allows for annotating each vertex v with the minimum cost $\ell(v)$ it has to pay in any optimal solution. We will then show how to transform an instance of this variant to the original problem.

Problem 3.3. ANNOTATED MINPSC

Input: A connected undirected graph $G = (V, E)$, edge weights $w \colon E \to \mathbb{N}$, and vertex annotations $\ell \colon V \to \mathbb{N}$.
Goal: Find a connected spanning subgraph $T = (V, F)$ of G that minimizes

$$\sum_{v \in V} \max\Big\{ \ell(v), \max_{\{u,v\} \in F} w(\{u, v\}) \Big\}.$$

Note that ANNOTATED MINPSC can be seen as special case of MINPCCS (Problem 2.9) where each vertex is assigned a distinct color. Each instance of MINPSC (Problem 1.1) can be transformed into an equivalent instance of ANNOTATED MINPSC with $\ell(v) = 0$ for each vertex $v \in V$.

Our data reduction rules shrink the input graph and, at the same time, compute the value a as specified in Theorem 3.1. Initially, $a = 0$. The general approach is common to many results that upper-bound the size of the graph in terms of its feedback edge number [3,16,20,27]. To this end, it is sufficient to reduce the number of degree-one vertices and the lengths of paths of degree-two vertices. We will see that the second part—shrinking paths—is the challenging one.

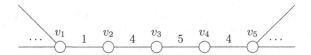

Fig. 2. A path where the most beneficial edge is not the heaviest edge. We assume that v_1 and v_5 are connected to the rest of the graph in such a way that v_1 has to pay at least 1 and v_5 has to pay at least 4. Omitting $\{v_3, v_4\}$ (the heaviest edge) results in the following optimal assignments: $v_2 : 4, v_3 : 4, v_4 : 4$ ($\sum = 12$) and omitting $\{v_2, v_3\}$ results in a better solution: $v_2 : 1, v_3 : 5, v_4 : 5$ ($\sum = 11$).

Like our dynamic programming algorithm for MINPCCS (see Algorithm 2.11), it is complicated by the nonlinear goal function of ANNOTATED MINPSC: even if we knew that an optimal solution does not contain all edges of a path, it is not obvious which edge of the path the optimal solution will skip (we will see that it is not always the heaviest edge). Our first data reduction rule for ANNOTATED MINPSC removes degree-one vertices.

Reduction Rule 3.4. Let v be a vertex with exactly one neighbor u. Then, set $\ell(u) := \max\{\ell(u), w(\{u, v\})\}$, delete v, and increase a by $\max\{w(\{u, v\}), \ell(v)\}$.

Henceforth, we assume that no degree-one vertices are left. Our second data reduction rule upper-bounds the length of paths of degree-two vertices. Let $P = (v_0, v_1, \ldots, v_h)$ be such a path with $h > 8$ and all inner vertices having degree two, that is, $\deg(v_i) = 2$ for $1 \leq i \leq h - 1$. Observe that at most one edge of the path is not in a solution—a connected spanning subgraph of G. Thus, there are two cases: either all edges of P or all but one edge of P are in the solution. We can encode this in a shorter path containing the edge that yields the highest benefit when omitted in a connected spanning subgraph. Remarkably, this is not necessarily the heaviest edge, as shown in Fig. 2. Besides such a most beneficial edge, we also need to keep the first and last edge in the path as the benefit of omitting them depends on the rest of the solution. We formalize this as follows.

Definition 3.5 (representative path, most beneficial edge). Let $P = (v_0, v_1, \ldots, v_h)$ with $h > 8$ be a path such that $\deg(v_0) > 2$, $\deg(v_h) > 2$, and $\deg(v_i) = 2$ for $i \in \{1, 2, \ldots, h - 1\}$. Let $\beta \colon E \to \mathbb{N}$ be a function defined by

$$\beta(\{v_j, v_{j+1}\}) := \max\Big\{0, w(\{v_i, v_{i+1}\}) - \max\big\{\ell(v_i), w(\{v_{i-1}, v_i\})\big\}\Big\} +$$
$$+ \max\Big\{0, w(\{v_i, v_{i+1}\}) - \max\big\{\ell(v_{i+1}), w(\{v_{i+1}, v_{i+2}\})\big\}\Big\}.$$

If $\beta(\{v_j, v_{j+1}\}) \geq \beta(\{v_k, v_{k+1}\})$, then we say that $\{v_j, v_{j+1}\}$ is more beneficial. For any most beneficial edge $\{v_i, v_{i+1}\}$ on P, that is, an edge maximizing β, we define the representative $\text{rep}(P) = (v_0, v_1, u_1, v_i, v_{i+1}, u_2, v_{h-1}, v_h)$ of P as a path with new vertices u_1 and u_2,

$$\ell(u_1) := \max \begin{cases} w(\{v_1, v_2\}), \\ w(\{v_{i-1}, v_i\}), \end{cases} \qquad \ell(u_2) := \max \begin{cases} w(\{v_{i+1}, v_{i+2}\}), \\ w(\{v_{h-2}, v_{h-1}\}), \end{cases}$$

and new incident edges of weights

$$w(\{v_1, u_1\}) := w(\{v_1, v_2\}), \qquad\qquad w(\{u_2, v_{h-1}\}) := w(\{v_{h-2}, v_{h-1}\}),$$
$$w(\{u_1, v_i\}) := w(\{v_{i-1}, v_i\}), \quad and \quad w(\{v_{i+1}, u_2\}) := w(\{v_{i+1}, v_{i+2}\}).$$

We will replace P by rep(P). At the same time, we have to adjust the value of a for Theorem 3.1. To this end, observe that each vertex v_j with $j \in \{2, \ldots h-2\} \setminus \{i, i+1\}$ pays $\max\{\ell(v_j), w(\{v_{j-1,v_j}\}), w(\{v_j, v_{j+1}\})\}$ before replacement and is deleted after replacement. Moreover, the new vertices u_1 and u_2 pay exactly $\ell(u_1)$ and $\ell(u_2)$, respectively, after replacement and are not part of the instance before replacement. Thus, we increase a by

$$\text{adj}(\text{rep}(P)) := -\ell(u_1) - \ell(u_2) + \sum_{\substack{2 \le j \le h-2 \\ j \notin \{i,i+1\}}} \max \begin{cases} \ell(v_j) \\ w(\{v_{j-1}, v_j\}) \\ w(\{v_j, v_{j+1}\}) \end{cases}$$

and the second data reduction rule works as follows.

Reduction Rule 3.6. Let $P = (v_0, v_1, \ldots, v_h)$ be a path with $h > 8$ such that $\deg(v_0) > 2$, $\deg(v_h) > 2$, and $\deg(v_i) = 2$ for $i \in \{1, 2, \ldots, h-1\}$. Replace P by rep(P) and increase a by adj(rep(P)).

So far, our data reduction rules turn an instance I of MINPSC into an instance I_A of Annotated MINPSC. Next, we show a result that any instance I_A of ANNOTATED MINPSC can be transformed back into an instance I' of MINPSC so that $\text{Opt}(I_A) + \sum_{v \in V} \ell(v) = \text{Opt}(I')$.

Lemma 3.7. *In linear time, one can transform any instance I_A of ANNO-TATED MINPSC into an instance I' of MINPSC so that $\text{Opt}(I_A) + \sum_{v \in V} \ell(v) = \text{Opt}(I')$. The instance I' contains at most $2n$ vertices and at most $n + m$ edges.*

We now give an upper bound on the size of graphs reduced by Reduction Rules 3.4 and 3.6. Combined with Lemma 3.7, this yields a proof of Theorem 3.1.

Proposition 3.8. *Let $I = (G, w, \ell)$ be an instance of ANNOTATED MINPSC and let g be the feedback edge number of G. If Reduction Rules 3.4 and 3.6 cannot be applied to I, then G contains at most $20g - 13$ vertices and $21g - 14$ edges.*

Theorem 3.1 can be proven by exhaustively applying Reduction Rules 3.4 and 3.6 and transforming the resulting instance into an instance of MINPSC using Lemma 3.7.

4   Parameterized Hardness and Inapproximability

In Sect. 2, we algorithmically exploited *vertex lower bounds*—lower bounds on the cost that each vertex has to pay in any optimal solution to MINPSC (see Definition 2.1). Vertex lower bounds immediately yield lower bounds on the total cost of optimal solutions. In this section, we show that the latter are much harder to exploit algorithmically.

For example, if the weights $w\colon E \to \mathbb{N}$ of the edges in a graph $G = (V, E)$ are at least one, then the vertex lower bounds given by $\ell(v) := \min_{\{u,v\} \in E} w(\{u, v\}) \geq 1$ for each vertex $v \in V$ (see Example 2.2) immediately yield a "large" lower bound of at least n on the cost of an optimal solution. This implies that even constant-factor approximation algorithms (e. g. the one by Althaus et al. [2]) can return solutions that are, in absolute terms, quite far away from the optimum. Furthermore, it follows that Proposition 2.7 already yields fixed-parameter tractability for k-PSC parameterized by the solution cost k.

A more desirable and stronger result would be a constant-factor approximation of the difference d between the optimal solution cost and a lower bound or a fixed-parameter tractability result with respect to the parameter d. However, we show that such algorithms (presumably) do not exist. Herein, we base some of our hardness results on the Exponential Time Hypothesis (ETH) as introduced by Impagliazzo and Paturi [18] and on the W-hierarchy from parameterized complexity theory [11]. The ETH is that 3-SAT cannot be solved in $2^{o(n+m)}$ time, where n and m are the number of variables and clauses in the input formula, respectively. Moreover, proving that a parameterized problem is W-hard with respect to some parameter k shows that there is (presumably) no $f(k) \cdot n^{O(1)}$-time algorithm, where n denotes the input size.

To state our hardness results, we use the following problem variant, which incorporates the lower bound.

Problem 4.1. MINPSC ABOVE LOWER BOUND (MINPSC-AL)

Input: A connected undirected graph $G = (V, E)$ and edge weights $w\colon E \to \mathbb{N}$.
Goal: Find a connected spanning subgraph $T = (V, F)$ of G that minimizes

$$\sum_{v \in V} \max_{\{u,v\} \in F} w(\{u, v\}) - \min_{\{u,v\} \in E} w(\{u, v\}). \tag{4.1}$$

For a MINPSC-AL instance $I = (G, w)$, we denote by $\mathrm{Opt}(I)$ the minimum value of (4.1) (we also refer to $\mathrm{Opt}(I)$ as the *margin* of I). For showing hardness results, we will also consider the *decision version* of the problem: By d-PSC-AL, we denote the problem of deciding whether an MINPSC-AL instance $I = (G, w)$ satisfies $\mathrm{Opt}(I) \leq d$.

Theorem 4.2.

(i) MINPSC-AL *is NP-hard to approximate within a factor of* $o(\log n)$.
(ii) d-PSC-AL *is W[2]-hard when parameterized by* d.
(iii) *Unless the Exponential Time Hypothesis fails,* k-PSC *and* d-PSC-AL *are not solvable in* $2^{o(n)}$ *time.*

Theorem 4.2(ii) shows that d-PSC-AL is presumably not solvable in $f(d) \cdot n^{O(1)}$ time, whereas it is easily solvable in $O(2^d \cdot n^{d+2})$ time, that is, in polynomial time for any fixed d.[3] Moreover, Theorem 4.2(iii) shows that our algorithm in Proposition 2.7 is asymptotically optimal.

We prove Theorem 4.2 using a reduction from MINIMUM SET COVER to MINPSC due to Erzin et al. [12].

Problem 4.3. MINIMUM SET COVER

Input: A universe $U = \{u_1, \ldots, u_n\}$ and a set family $\mathcal{F} = \{S_1, \ldots, S_m\}$ containing sets $S_i \subseteq U$.
Goal: Find a minimum cardinality *set cover* $\mathcal{F}' \subseteq \mathcal{F}$ (that is, $\bigcup_{S \in \mathcal{F}'} S = U$).

Again, in the corresponding decision version k-SET COVER, we are given a $k \in \mathbb{N}$ and want to decide whether there exists a set cover of size at most k. MINIMUM SET COVER is NP-hard to approximate within a factor of $o(\log n)$ [23]. Furthermore, k-SET COVER is W[2]-complete parameterized by the solution size k [11] and cannot be solved in $2^{o(n+m)}$ time unless the ETH fails [19].

We now present a slightly modified version of the reduction by Erzin et al. [12] from which Theorem 4.2 follows. Note that Erzin et al. [12] used edge weights zero and one in their reduction whereas we use positive integers as edge weights, namely one and two.

Transformation 4.4. Given an instance $I = (U, \mathcal{F})$ of MINIMUM SET COVER, construct an instance $I' = (G, w)$ of MINPSC-AL as follows. The graph G consists of a special vertex s, a vertex v_u for each element $u \in U$, and a vertex v_S for each set $S \in \mathcal{F}$. Denote the respective vertex sets by V_U and $V_{\mathcal{F}}$, that is, $V := \{s\} \uplus V_U \uplus V_{\mathcal{F}}$. There is an edge between each pair of vertices $v_u, v_S \in V$ if the set $S \in \mathcal{F}$ contains $u \in U$. Moreover, there is an edge $\{s, v_{S_i}\}$ for each $i \in \{1, \ldots, m\}$. The edge weights w are as follows. All edges incident to s get weight one. The remaining edges (between V_U and $V_{\mathcal{F}}$) get weight two.

5   Conclusion

We believe that both our randomized fixed-parameter algorithm exploiting vertex lower bounds and our data reduction rules (a partial kernelization for the parameter "feedback edge number") are worth implementing and testing. Empirical work is ongoing work; however, we believe that our algorithms are less suited for random test data (as typically used in published work so far) because our algorithms make explicit use of structure in the input which presumably occurs in some real-world monitoring instances.

---

[3] At most d vertices can pay more than their vertex lower bound. We can try all possibilities for choosing $i \leq d$ vertices, all $\binom{d}{i}$ possibilities to increase their total cost by at most d, and check whether the graph of the "paid" edges is connected. The algorithm runs in $\sum_{i=1}^{d} \binom{n}{i} \binom{d}{i} \cdot O(n + m) \subseteq O(2^d \cdot n^{d+2})$ time.

An important theoretical challenge is to find good vertex lower bounds for exploitation in Sect. 2. This goes hand in hand with identifying scenarios where (more) obligatory edges are given by the application (e.g., this may be the case in communication networks with designated hub nodes). Finally, we identified positive results for two natural network parameters; thus, the search for further useful parameterizations is a generic but nevertheless promising undertaking. Ideally, this should be driven by using data from real-world applications and analyzing their structural properties.

Acknowledgments. RvB was supported by the Russian Science Foundation, grant 16-11-10041, while working on Sect. 2. The results in Sects. 3 and 4 were obtained during a research stay of RvB at TU Berlin, jointly supported by TU Berlin, by the Russian Foundation for Basic Research under grant 16-31-60007 mol_a_dk, and by the Ministry of Science and Education of the Russian Federation under the 5-100 Excellence Programme.

References

1. Alon, N., Yuster, R., Zwick, U.: Color-coding. J. ACM **42**(4), 844–856 (1995)
2. Althaus, E., Călinescu, G., Mandoiu, I.I., Prasad, S.K., Tchervenski, N., Zelikovsky, A.: Power efficient range assignment for symmetric connectivity in static ad hoc wireless networks. Wirel. Netw. **12**(3), 287–299 (2006)
3. Bentert, M., Fluschnik, T., Nichterlein, A., Niedermeier, R.: Parameterized aspects of triangle enumeration. In: Klasing, R., Zeitoun, M. (eds.) FCT 2017. LNCS, vol. 10472, pp. 96–110. Springer, Heidelberg (2017). https://doi.org/10.1007/978-3-662-55751-8_9
4. Betzler, N., van Bevern, R., Fellows, M.R., Komusiewicz, C., Niedermeier, R.: Parameterized algorithmics for finding connected motifs in biological networks. IEEE/ACM Trans. Comput. Biol. **8**(5), 1296–1308 (2011)
5. Betzler, N., Guo, J., Komusiewicz, C., Niedermeier, R.: Average parameterization and partial kernelization for computing medians. J. Comput. Syst. Sci. **77**(4), 774–789 (2011)
6. van Bevern, R., Komusiewicz, C., Sorge, M.: A parameterized approximation algorithm for the mixed and windy capacitated arc routing problem: theory and experiments. Networks (2017, in press)
7. Bruckner, S., Hüffner, F., Karp, R.M., Shamir, R., Sharan, R.: Topology-free querying of protein interaction networks. J. Comput. Biol. **17**(3), 237–252 (2010)
8. Carmi, P., Katz, M.J.: Power assignment in radio networks with two power levels. Algorithmica **47**(2), 183–201 (2007)
9. Clementi, A.E., Penna, P., Silvestri, R.: On the power assignment problem in radio networks. Mob. Netw. Appl. **9**(2), 125–140 (2004)
10. Dost, B., Shlomi, T., Gupta, N., Ruppin, E., Bafna, V., Sharan, R.: Qnet: a tool for querying protein interaction networks. J. Comput. Biol. **15**(7), 913–925 (2008)

11. Downey, R.G., Fellows, M.R.: Fundamentals of Parameterized Complexity. Texts in Computer Science. Springer, Heidelberg (2013). https://doi.org/10. 1007/978-1-4471-5559-1
12. Erzin, A.I., Plotnikov, R.V., Shamardin, Y.V.: O nekotorykh polinomial'no razreshimykh sluchayakh i priblizhënnykh algoritmakh dlya zadachi postroyeniya optimal'nogo kommunikatsionnogo dereva. Diskretn. Anal. Issled. Oper. **20**(1), 12–27 (2013)
13. Erzin, A.I., Mladenovic, N., Plotnikov, R.V.: Variable neighborhood search variants for min-power symmetric connectivity problem. Comput. Oper. Res. **78**, 557–563 (2017)
14. Giacometti, A.: River networks. In: Complex Networks, Encyclopedia of Life Support Systems (EOLSS), pp. 155–180. EOLSS Publishers/UNESCO (2010)
15. Gutin, G., Wahlström, M., Yeo, A.: Rural postman parameterized by the number of components of required edges. J. Comput. Syst. Sci. **83**(1), 121–131 (2017)
16. Hartung, S., Komusiewicz, C., Nichterlein, A.: Parameterized algorithmics and computational experiments for finding 2-clubs. J. Graph Algorithms Appl. **19**(1), 155–190 (2015)
17. Hoffmann, S., Wanke, E.: Minimum power range assignment for symmetric connectivity in sensor networks with two power levels (2016). arXiv:1605.01752
18. Impagliazzo, R., Paturi, R.: On the complexity of k-SAT. J. Comput. Syst. Sci. **62**(2), 367–375 (2001)
19. Impagliazzo, R., Paturi, R., Zane, F.: Which problems have strongly exponential complexity? J. Comput. Syst. Sci. **63**(4), 512–530 (2001)
20. Mertzios, G.B., Nichterlein, A., Niedermeier, R.: Linear-time algorithm for maximum-cardinality matching on cocomparability graphs. In: MFCS 2017. LIPIcs, vol. 83, pp. 46:1–46:14, Schloss Dagstuhl – Leibniz-Zentrum fuer Informatik (2017)
21. Montemanni, R., Gambardella, L.: Exact algorithms for the minimum power symmetric connectivity problem in wireless networks. Comput. Oper. Res. **32**(11), 2891–2904 (2005)
22. Panigrahi, D.: Survivable network design problems in wireless networks. In: Proceedings of 22nd SODA, pp. 1014–1027. SIAM (2011)
23. Raz, R., Safra, S.: A sub-constant error-probability low-degree test, and a sub-constant error-probability PCP characterization of NP. In: Proceedings of 29th STOC, pp. 475–484. ACM (1997)
24. Scott, J., Ideker, T., Karp, R.M., Sharan, R.: Efficient algorithms for detecting signaling pathways in protein interaction networks. J. Comput. Biol. **13**(2), 133–144 (2006)
25. Sorge, M., van Bevern, R., Niedermeier, R., Weller, M.: From few components to an Eulerian graph by adding arcs. In: Kolman, P., Kratochvíl, J. (eds.) WG 2011. LNCS, vol. 6986, pp. 307–318. Springer, Heidelberg (2011). https://doi.org/10.1007/978-3-642-25870-1_28

26. Sorge, M., van Bevern, R., Niedermeier, R., Weller, M.: A new view on rural postman based on Eulerian extension and matching. J. Discrete Alg. **16**, 12–33 (2012)
27. Uhlmann, J., Weller, M.: Two-layer planarization parameterized by feedback edge set. Theoret. Comput. Sci. **494**, 99–111 (2013)
28. Zalyubovskiy, V.V., Erzin, A.I., Astrakov, S.N., Choo, H.: Energy-efficient area coverage by sensors with adjustable ranges. Sensors **9**(4), 2446–2460 (2009)
29. Zhang, H., Hou, J.C.: Maintaining sensing coverage and connectivity in large sensor networks. Ad Hoc Sens. Wirel. Netw. **1**(1–2), 89–124 (2005)

Fast Distributed Approximation for Max-Cut

Keren Censor-Hillel[(✉)], Rina Levy, and Hadas Shachnai

Computer Science Department, Technion, 3200003 Haifa, Israel
{ckeren,rinalevy,hadas}@cs.technion.ac.il

Abstract. Finding a maximum cut is a fundamental task in many computational settings, with a central application in wireless networks. Surprisingly, Max-Cut has been insufficiently studied in the classic distributed settings, where vertices communicate by synchronously sending messages to their neighbors according to the underlying graph, known as the $\mathcal{LOCAL}$ or $\mathcal{CONGEST}$ models. We amend this by obtaining almost optimal algorithms for Max-Cut on a wide class of graphs in these models. In particular, for any $\epsilon > 0$, we develop randomized approximation algorithms achieving a ratio of $(1 - \varepsilon)$ to the optimum for Max-Cut on bipartite graphs in the $\mathcal{CONGEST}$ model, and on *general* graphs in the $\mathcal{LOCAL}$ model.

We further present efficient *deterministic* algorithms, including a 1/3-approximation for Max-Dicut in our models, thus improving the best known (randomized) ratio of 1/4. Our algorithms make non-trivial use of the greedy approach of Buchbinder et al. (SIAM Journal Computing 44:1384–1402, 2015) for maximizing an unconstrained (non-monotone) submodular function, which may be of independent interest.

1 Introduction

Max-Cut is one of the fundamental problems in theoretical computer science. A *cut* in an undirected graph is a bipartition of the vertices, whose size is the number of edges crossing the bipartition. Finding cuts of maximum size in a given graph is among Karp's famous 21 NP-complete problems [25]. Since then, Max-Cut has received considerable attention, in approximation algorithms [17,19, 42,45], parallel computation [44], parameterized complexity [43], and streaming algorithms (see, e.g., [24]).

Max-Cut has a central application in *wireless mesh networks (WMNs)*. The capacity of WMNs that operate over a single frequency can be increased significantly by enhancing each router with multiple *transmit (Tx)* or *receive (Rx)* (*MTR*) capability. Thus, a node will not experience collision when two or more neighbors transmit to it. Yet, interference occurs if a node transmits and receives simultaneously. This is known as the *no mix-tx-rx* constraint. The set of links activated in each time slot, defining the capacity of an MTR WMN, is governed

---

K. Censor-Hillel—The research is supported in part by the Israel Science Foundation (grant 1696/14).

A. Fernández Anta et al. (Eds.): ALGOSENSORS 2017, LNCS 10718, pp. 41–56, 2017.
https://doi.org/10.1007/978-3-319-72751-6_4

by a link scheduler. As shown in [10], link scheduling is equivalent to finding Max-Cut in each time slot. A maximum cut contains the set of non-conflicting links that can be activated at the same time, i.e., they adhere to the no mix-tx-rx constraint. The induced bipartition of the vertices at each time slot defines a set of transmitters and a set of receivers in this slot. Link scheduling algorithms based on approximating Max-Cut, and other applications in wireless networks, can be found in [27,48–51].[1]

Surprisingly, Max-Cut has been insufficiently studied in the classic distributed settings, where vertices communicate by synchronously sending messages to their neighbors according to the underlying graph, known as the $\mathcal{LOCAL}$ or $\mathcal{CONGEST}$ models. Indeed, there are known distributed algorithms for Max-Cut using MapReduce techniques [5,35,36]. In this setting, the algorithms partition the ground set among m machines and obtain a solution using all the outputs. However, despite a seemingly similar title, our distributed setting is completely different.

In this paper we address Max-Cut in the classic distributed network models, where the graph represents a synchronous communication network. At the end of the computation, each vertex decides locally whether it joins the subset S or $\bar{S}$, and outputs 1 or 0, respectively, so as to obtain a cut of largest possible size.

It is well known that choosing a random cut, i.e., assigning each vertex to S or $\bar{S}$ with probability $1/2$, yields a $\frac{1}{2}$-approximation for Max-Cut, and a $\frac{1}{4}$-approximation for Max-Dicut, defined on directed graphs (see, e.g., [37,38]).[2] Thus, a local algorithm, where each vertex outputs 0 or 1 with probability $1/2$, yields the above approximation factors with no communication required. On the other hand, we note that a single vertex can find an optimal solution, once it has learned the underlying graph. However, this requires a number of communication rounds that depends *linearly* on global network parameters (depending on the exact model considered). This defines a tradeoff between time complexity and the approximation ratio obtained by distributed Max-Cut algorithms. The huge gap between the above results raises the following natural questions: How well can Max-Cut be approximated in the distributed setting, using a bounded number of communication rounds? Or, more precisely: How many communication rounds are required for obtaining an approximation ratio strictly larger than half, or even a *deterministic* $\frac{1}{2}$-approximation for Max-Cut?

To the best of our knowledge, these questions have been studied in our distributed network models only for a restricted graph class. Specifically, the paper [22] suggests a distributed algorithm for Max-Cut on d-regular triangle-free graphs, that requires a single communication round and provides a $(1/2 + 0.28125/\sqrt{d})$-approximation.

The key contribution of this paper is in developing two main techniques for approximating Max-Cut and Max-Dicut in distributed networks, with *any*

---

[1] Max-Cut naturally arises also in VLSI [9], statistical physics [4] and machine learning [47].

[2] In Max-Dicut we seek the maximum size edge-set crossing from S to $\bar{S}$.

communication graph. Below we detail the challenges we face, and our methods for overcoming them.

1.1   The Challenge

In the $\mathcal{LOCAL}$ model, where message sizes and the local computation power are unlimited, every standard graph problem can be solved in $O(n)$ communication rounds. For Max-Cut it also holds that finding an optimal solution requires $\Omega(n)$ communication rounds. This lower bound follows from Linial's seminal lower bound [31, Theorem 2.2] for finding a 2-coloring of an even-sized cycle. In an even cycle, the maximum cut contains all edges. Therefore, finding a Max-Cut is equivalent to finding a 2-coloring of the graph.

An approach that proved successful in many computational settings – in tackling hard problems – is to relax the optimality requirement and settle for approximate solutions. Indeed, in the distributed setting, many approximation algorithms have been devised to overcome the costs of finding exact solutions (see, e.g., [1–3,16,21,28–30,32,39], and the survey of Elkin [11]). Our work can be viewed as part of this general approach. However, we face crucial hurdles attempting to use the known sequential toolbox for approximating Max-Cut in the distributed setting.

As mentioned above, a $\frac{1}{2}$-approximation for Max-Cut can be obtained easily with no communication. While this holds in all of the above models, improving the ratio of $1/2$ is much more complicated. In the sequential setting, an approximation factor strictly larger than $1/2$ was obtained in the mid-1990's using semidefinite programming [17] (see Sect. 1.3). Almost two decades later, the technique was applied by [44] to obtain a parallel randomized algorithm for Max-Cut, achieving a ratio of $(1-\epsilon)0.878$ to the optimum, for any $\varepsilon > 0$. Adapting this algorithm to our distributed setting seems non-trivial, as it relies heavily on global computation. Trying to apply other techniques, such as local search, unfortunately leads to linear running time, because of the need to compare values of global solutions.

Another obstacle that lies ahead is the lack of *locality* in Max-Cut, due to strong dependence between the vertices. The existence of an edge in the cut depends on the assignment of both of its endpoints. This results in a chain of dependencies and raises the question whether cutting the chain can still guarantee a good approximation ratio.

1.2   Our Contribution

We develop two main techniques for approximating Max-Cut, as well as Max-Dicut. Our first technique relies on the crucial observation that the cut value is additive for edge-disjoint sets of vertices. Exploiting this property, we design *clustering-based* algorithms, in which we decompose the graph into small-diameter clusters, find an optimal solution within each cluster, and prove that the remaining edges still allow the final solution to meet the desired approximation ratio. An essential component in our algorithms is efficient graph decomposition to such small-diameter clusters connected by few edges (also known as a *padded partition*), inspired by a parallel algorithm of [34] (see also [12,13]).

For general graphs, this gives $(1 - \epsilon)$-approximation algorithms for Max-Cut and Max-Dicut, requiring $O(\frac{\log n}{\epsilon})$ communication rounds in the $\mathcal{LOCAL}$ model. For the special case of a bipartite graph, we take advantage of the graph structure to obtain an improved clustering-based algorithm, which does not require large messages. The algorithm achieves a $(1 - \epsilon)$-approximation for Max-Cut in $O(\frac{\log n}{\epsilon})$ rounds, in the more restricted $\mathcal{CONGEST}$ model.

For our second technique, we observe that the contribution of a specific vertex to the cut depends only on the vertex itself and its immediate neighbors. We leverage this fact to make multiple decisions in parallel by independent sets of vertices. We find such sets using distributed coloring algorithms. Our *coloring-based* technique, which makes non-trivial use of the greedy approach of [7] for maximizing an unconstrained submodular function, yields deterministic $\frac{1}{2}$-approximation and $\frac{1}{3}$-approximation algorithms for Max-Cut and Max-Dicut, respectively, and a randomized $\frac{1}{2}$-approximation algorithm for Max-Dicut. Each of these algorithms requires $\tilde{O}(\Delta + \log^* n)$ communication rounds in the $\mathcal{CONGEST}$ model, where Δ is the maximal degree of the graph, and $\tilde{O}$ ignores polylogarithmic factors in Δ.

Finally, we present $\mathcal{LOCAL}$ algorithms which combine both of our techniques. Applying the coloring-based technique to low-degree vertices, and the clustering-based technique to high-degree vertices, allows as to design faster deterministic algorithms with approximation ratios of $\frac{1}{2}$ and $\frac{1}{3}$ for Max-Cut and Max-Dicut, respectively, requiring $\min\{\tilde{O}(\Delta + \log^* n), O(\sqrt{n})\}$ communication rounds (Table 1).[3]

Table 1. A summary of our results.

Algorithm Properties				Approximation Ratio		
Rounds	Deterministic	Model	Graph	Max-Cut	Max-Dicut	
no communication	✗	$\mathcal{CONGEST}$	any	1/2	1/4	folklore
$O(\log n/\epsilon)$	✗	$\mathcal{CONGEST}$	bipartite	$1 - \epsilon$	–	new
$O(\log n/\epsilon)$	✗	$\mathcal{LOCAL}$	any	$1 - \epsilon$	$1 - \epsilon$	new
$\tilde{O}(\Delta + \log^* n)$	✓	$\mathcal{CONGEST}$	any	1/2	1/3	new
$\tilde{O}(\Delta + \log^* n)$	✗	$\mathcal{CONGEST}$	any	1/2	1/2	new
$\min\{\tilde{O}(\Delta + \log^* n), O(\sqrt{n})\}$	✓	$\mathcal{LOCAL}$	any	1/2	1/3	new

1.3   Background and Related Work

The weighted version of Max-Cut is one of Karp's NP-complete problems [25]. The unweighted version that we study here is also known to be NP-complete [15].

While there are graph families, such as planar and bipartite graphs, in which a maximum cut can be found in polynomial time [18,19], in general graphs, even

---

[3] Due to space constraints, some of the results are omitted. A detailed version of this paper can be found in [8].

approximating the problem is NP-hard. In the sequential setting, one cannot obtain an approximation ratio better than $\frac{16}{17}$ for Max-Cut, or an approximation ratio better than $\frac{12}{13}$ for Max-Dicut, unless $P = NP$ [20, 46].

Choosing a random cut, i.e., assigning each vertex to S or $\bar{S}$ with probability $1/2$, yields a $\frac{1}{2}$-approximation for Max-Cut, and $\frac{1}{4}$-approximation for Max-Dicut. In the sequential setting there are also deterministic algorithms yielding the above approximation ratios [40, 42]. For 20 years there was no progress in improving the $1/2$ approximation ratio for Max-Cut, until (in 1995) Goemans and Williamson [17] achieved the currently best known ratio, using semidefinite programming. They present a 0.878-approximation algorithm, which is optimal assuming the Unique Games Conjecture holds [26]. In the same paper, Goemans and Williamson also give a 0.796-approximation algorithm for Max-Dicut. This ratio was improved later by Matuura et al. [33], to 0.863. Using spectral techniques, a 0.53-approximation algorithm for Max-Cut was given by Trevisan [45]. In [23] Kale and Seshadhri present a combinatorial approximation algorithm for Max-Cut using random walks, which gives a $(0.5 + \delta)$-approximation, where δ is some positive constant that appears also in the running time of the algorithm. In particular, for $\tilde{O}(n^{1.6}), \tilde{O}(n^2)$ and $\tilde{O}(n^3)$ times, the algorithm achieves approximation factors of $0.5051, 0.5155$ and 0.5727, respectively.

Max-Cut and Max-Dicut can also be viewed as special cases of submodular maximization, which has been widely studied. It is known that choosing a solution set S uniformly at random yields a $\frac{1}{4}$-approximation, and a $\frac{1}{2}$-approximation for a general and for symmetric submodular function, respectively [14]. These results imply the known random approximation ratios for Max-Cut and Max-Dicut. Buchbinder et al. [7] present determinstic $\frac{1}{2}$-approximation algorithms for both symmetric and asymmetric submodular functions. These algorithms assume that the submodular function is accessible through a black box returning $f(S)$ for any given set S (known as the *value oracle* model).

In recent years, there is an ongoing effort to develop distributed algorithms for submodular maximization problems, using MapReduce techniques [5, 35, 36]. Often, the inputs consist of large data sets, for which a sequential algorithm may be inefficient. The main idea behind these algorithms is to partition the ground set among m machines, and have each machine solve the problem optimally independently of others. After all machines have completed their computations, they share their solutions. A final solution is obtained by solving the problem once again over a union of the partial solutions. The algorithms achieve performance guarantees close to the sequential algorithms while decreasing the running time, where the running time is the number of communication rounds among the machines. As mentioned above, these algorithms do not apply to our classic distributed settings.

2 Preliminaries

The Max-Cut problem is defined as follows. Given an undirected graph $G = (V, E)$, one needs to divide the vertices into two subsets, $S \subset V$ and $\bar{S} = V \setminus S$, such that the size of the cut, i.e., the number of edges between S and the complementary subset $\bar{S}$, is as large as possible. In the Max-Dicut problem, the given graph $G = (V, E)$ is directed, and the cut is defined only as the edges which are directed from S to $\bar{S}$. As in the Max-Cut problem, the goal is to obtain the largest cut.

Max-Cut and Max-Dicut can be described as the problem of maximizing the submodular function $f(S) = |E(S, \bar{S})|$, where for Max-Dicut $f(S)$ counts only the edges directed from S to $\bar{S}$. Given a finite set X, let 2^X denote the power set of X. A function $f : 2^X \to \mathbb{R}$ is *submodular* if it satisfies the following equivalent conditions:

(i) For any $S, T \subseteq X$: $f(S \cup T) + f(S \cap T) \leq f(S) + f(T)$.
(ii) For any $A \subseteq B \subseteq X$ and $x \in X \setminus B$: $f(B \cup \{x\}) - f(B) \leq f(A \cup \{x\}) - f(A)$.

For Max-Cut and Max-Dicut, the submodular function also satisfies: for any pair of disjoint sets $S, T \subseteq X$ such that $E_{S \times T} = \{(u, v) | u \in S, v \in T\} = \emptyset$, $f(S) + f(T) = f(S \cup T)$. Note that for Max-Cut, the function is also symmetric, i.e., $f(S) = f(\bar{S})$.

Model: We consider a distributed system, modeled by a graph $G = (V, E)$, in which the vertices represent the computational entities, and the edges represent the communication channels between them. We assume that each vertex v has a unique identifier $id(v)$ of size $O(\log n)$, where $n = |V|$.

The communication between the entities is synchronous, i.e., the time is divided into rounds. In each round, the vertices send messages simultaneously to all of their neighbors and make a local computation based on the information gained so far. This is the classic $\mathcal{LOCAL}$ model [41], which focuses on analyzing how locality affects the distributed computation. Therefore, message sizes and local computation power are unlimited, and the complexity is measured by the number of communication rounds needed to obtain a solution. It is also important to study what can be done in the more restricted $\mathcal{CONGEST}$ model [41], in which message sizes are $O(\log n)$.

We assume that each vertex has preliminary information including the size of the network $n = |V|$, its neighbors, and the maximal degree of the graph Δ.[4]

Each vertex runs a local algorithm to solve the Max-Cut problem. Along the algorithm, each vertex decides locally whether it joins S or $\bar{S}$, and outputs 1 or 0, respectively. We define the *solution* of the algorithm as the set of all outputs. Note that individual vertices do not hold the entire solution, but only their local information. The solution *value* is defined as the size of the cut induced by the solution. We show that this value approximates the size of the maximum cut.

3   Clustering-Based Algorithms

In this section we present clustering-based algorithms for Max-Cut and Max-Dicut. Our technique relies on the observation that Max-Cut is a collection of

---

[4] This assumption is needed only for the $(\Delta + 1)$-coloring algorithm [6] used in Sect. 4; it can be omitted (see [6]), increasing the running time by a constant factor.

edges having their endpoints in different sets; therefore, it can be viewed as the union of cuts in the disjoint parts of the graph.

Given a graph $G = (V, E)$, we first eliminate a small fraction of edges to obtain small-diameter connected components. Then, the problem is solved optimally within each connected component. For general graphs, this is done by gathering the topology of the component at a single vertex. For the special case of a bipartite graph, we can use the graph structure to propagate less information. Since the final solution, consisting of the local decisions of all vertices, is at least as good as the sum of the optimal solutions in the components, and since the fraction of eliminated edges is small, we prove that the technique yields a $(1 - \epsilon)$-approximation.

3.1   A Randomized Distributed Graph Decomposition

We start by presenting the randomized distributed graph decomposition algorithm. The algorithm is inspired by a parallel graph decomposition by Miller et al. [34] that we adapt to the distributed.[5] The PRAM algorithm of [34] generates a *strong padded partition* of a given graph, namely, a partition into connected components with strong diameter $O(\frac{\log n}{\beta})$, for some $\beta \leq 1/2$, such that the fraction of edges that cross between different clusters of the partition is at most β. As we show below, the distributed version guarantees the same properties with high probability and requires only $O(\frac{\log n}{\beta})$ communication rounds in the $\mathcal{CONGEST}$ model.

The distributed version of the graph decomposition algorithm proceeds as follows: Let δ_v be a random value chosen by vertex v from an exponential distribution with parameter β. Define the *shifted distance* from vertex v to vertex u as $dist_\delta(u, v) = dist(u, v) - \delta_u$. Along the algorithm, each vertex v finds a vertex u within its $\frac{k \log n}{\beta}$-neighborhood that minimizes $dist_\delta(u, v)$, where k is a constant. We define this vertex as v's *center*. This step implies the difference between the parallel and the distributed decomposition. Indeed, in the parallel algorithm, each vertex chooses its center from the *entire* ground set V. We show that our modified process still generates a decomposition with the desired properties. Furthermore, w.h.p. the distributed algorithm outputs a decomposition identical to the one created by the parallel algorithm. A pseudocode of the algorithm is given in Algorithm 1.

We prove that the fraction of edges between different components is small. In order to do so, we bound the probability of an edge to be between components, i.e., the probability that the endpoints of the edge choose different centers. We consider two cases for an edge $e = (u, v)$. In the first case, we assume that both u and v choose the center that minimizes their shifted distance, $dist_\delta$, over all the vertices in the graph. In other words, if the algorithm allowed each vertex to learn the entire graph, they would choose the same center as they did in our

---

[5] Our algorithm can be viewed as one phase of the distributed algorithm presented by Elkin et al. in [12] with some necessary changes.

Algorithm 1. Distributed Decomposition, *code for vertex v*

1: $0 < \beta < 1, k > 2$.
2: choose δ_v at random from $Exp(\beta)$
3: $center = id(v)$
4: $dist_{\delta_{min}} = -\delta_v$
5: **for** $\frac{k \log n}{\beta}$ iterations **do**
6: 　　send $(dist_{\delta_{min}}, center)$
7: 　　**for** every $(dist'_{\delta_{min}}, center')$ received from $u \in N(v)$ **do**
8: 　　　　**if** $\left(dist'_{\delta_{min}} + 1 < dist_{\delta_{min}}\right)$ OR $\left((dist'_{\delta_{min}} + 1 = dist_{\delta_{min}})\right.$ AND $\left.(center' < center)\right)$ **then**
9: 　　　　　　$center \leftarrow center'$
10: 　　　　　　$dist_{\delta_{min}} \leftarrow dist'_{\delta_{min}} + 1$
11: 　　　　**end if**
12: 　　**end for**
13: **end for**
14: output $center$

---

algorithm. In the second case, we assume that at least one of u and v chooses differently if given a larger neighborhood.

Define the *ideal* center of a vertex v as $argmin_{w \in V} dist_\delta(w, v)$. In the next lemma, we upper bound the probability that a vertex does not choose its ideal center.

Lemma 3.1. *Let v' be the ideal center of vertex v, then the probability that $dist(v', v) > \frac{k \log n}{\beta}$, i.e., vertex v does not join its ideal center, is at most $\frac{1}{n^k}$.*

Proof. Since v' is the ideal center of vertex v, we have that $dist_\delta(v', v) \leq dist_\delta(v, v)$. Therefore, $dist(v', v) - \delta_{v'} \leq dist(v, v) - \delta_v = -\delta_v \leq 0$, which implies that $dist(v', v) \leq \delta_{v'}$. That is, the distance between each vertex v to its ideal center v' is upper bounded by $\delta_{v'}$, and hence $\Pr\left[dist(v', v) > \frac{k \log n}{\beta}\right] \leq \Pr\left[\delta_{v'} > \frac{k \log n}{\beta}\right]$. Using the cumulative exponential distribution, we have that $\Pr\left[\delta_{v'} > \frac{k \log n}{\beta}\right] = \exp\left(-\frac{k \cdot \beta \log n}{\beta}\right) = \exp\left(-k \log n\right) \leq \frac{1}{n^k}$. □

Corollary 3.2. *The Distributed Decomposition algorithm generates a decomposition identical to the decomposition generated by the parallel decomposition algorithm with probability at least $1 - \frac{1}{n^{k-1}}$*

Define an *exterior* edge as an edge connecting different vertex components, and let F denote the set of exterior edges. Let $A_{u,v}$ denote the event that both u and v choose their ideal centers.

Lemma 3.3. *The probability that an edge $e = (u, v)$ is an exterior edge, given that u and v choose their ideal centers, is at most β.*

The lemma follows directly from [34], where indeed the algorithm assigns to each vertex its ideal center. We can now bound the probability of any edge to be an exterior edge.

Lemma 3.4. *The probability that an edge $e = (u, v)$ is in F is at most $\beta + \frac{2}{n^k}$.*

We can now prove the performance guarantees of the Distributed Decomposition algorithm. Recall that the *weak diameter* of a set $S = \{u_1, u_2, ...u_l\}$ is defined as $\max_{(u_i, u_j) \in S} dist(u_i, u_j)$.

Theorem 3.5. *The Distributed Decomposition algorithm requires $O(\frac{\log n}{\beta})$ communication rounds in the $\mathcal{CONGEST}$ model, and partitions the graph into components, such that in expectation there are $O(\beta m)$ exterior edges. Each of the components is of weak diameter $O(\frac{\log n}{\beta})$, and with high probability also of strong diameter $O(\frac{\log n}{\beta})$.*

Proof. Clearly, as each vertex chooses a center from its $\frac{k \log n}{\beta}$-neighborhood, the distance between two vertices that choose the same center, i.e., belong to the same component, over the graph G, is at most $O(\frac{\log n}{\beta})$. Therefore, the weak diameter of every component is at most $O(\frac{\log n}{\beta})$. By Corollary 3.2, with probability at least $1 - \frac{1}{n^{k-1}}$, the algorithm outputs a partition identical to the one output by the parallel algorithm, and therefore with the same properties, which implies that the strong diameter of every component is at most $O(\frac{\log n}{\beta})$ as well.

Using the linearity of expectation and Lemma 3.4, we have that $\mathbb{E}[|F|] \leq \sum_{e \in E} \left(\beta + \frac{2}{n^k}\right) = \beta m + \frac{2m}{n^k}$. Since $m \leq n^2$, for any $k > 2$, $\mathbb{E}[|F|] \leq O(\beta m)$. Finally, as can be seen from the code, the algorithm requires $O(\frac{\log n}{\beta})$ communication rounds. □

3.2 A Randomized $(1 - \epsilon)$-Approximation Algorithm for Max-Cut on a Bipartite Graph

Clearly, in a bipartite graph, the maximum cut contains all of the edges. Such a cut can be found by selecting arbitrarily a root vertex, and then simply putting all the vertices of odd depth in one set and all the vertices of even depth in the complementary set. However, this would require a large computational time in our model, that depends on the diameter of the graph. We overcome this by using the above decomposition, and finding an optimal solution within each connected component. In each component C, we find an optimal solution in $O(D_c)$ communication rounds, where D_c is the diameter of C. First, the vertices in each component search for the vertex with the lowest id.[6] Then, the vertex with the lowest id joins S or $\bar{S}$ with equal probability and sends its decision to its neighbors. When a vertex receives a message from one of its neighbors, it joins the opposite set, outputs its decision, and sends it to its neighbors. Since finding the optimal solution within each component does not require learning the entire component topology, the algorithm is applicable in the more restricted

---

[6] This can be done by running a BFS in parallel from all vertices. Each vertex propagates the information from the root with lowest id it knows so far, and joins its tree. Thus, at the end of the process, we have a BFS tree rooted at the vertex with the lowest id.

$\mathcal{CONGEST}$ model. The algorithm yields a $(1 - \epsilon)$-approximation for the Max-Cut problem on a bipartite graph in $O(\frac{\log n}{\epsilon})$ communication rounds with high probability.

Theorem 3.6. *Bipartite Max-Cut is a randomized $(1 - \epsilon)$-approximation for Max-Cut, requiring $O(\frac{\log n}{\epsilon})$ communication rounds in the $\mathcal{CONGEST}$ model w.h.p.*

Algorithm 2. Bipartite Max-Cut

1: G=(V,E)
2: apply Distributed Decomposition to G, with $\beta = \epsilon, k > 2$
3: **for** each component C obtained by the decomposition **do**
4:     build a BFS tree from the vertex v with the lowest id
5:     assign v to S or $\bar{S}$ with equal probability, assign the rest of the vertices to alternating sides
6: **end for**

3.3 A Randomized $(1 - \epsilon)$-Approximation Algorithm for General Graphs

We present below a $(1 - \epsilon)$-approximation algorithm for Max-Cut in general graphs, using $O(\frac{\log n}{\epsilon})$ communication rounds. As before, the algorithm consists of a decomposition phase and a solution phase. While the decomposition phase works in the $\mathcal{CONGEST}$ model, the algorithm suits for the $\mathcal{LOCAL}$ model, since for general graphs, the generated components are not necessarily sparse, and learning the components topology is expensive in the $\mathcal{CONGEST}$ model.

Algorithm 3. Decomposition-Based Max-Cut

1: G=(V,E)
2: apply Distributed Decomposition on G, with $\beta = \epsilon/2, k > 2$
3: **for** each component C obtained by the decomposition **do**
4:     gather the component topology at the vertex $v \in C$ with the lowest id.
5:     let v find an optimal solution and determine the value output by the component's vertices.
6: **end for**

Theorem 3.7. *Decomposition-Based Max-Cut is a randomized $(1 - \epsilon)$-approximation for Max-Cut, requiring $O(\frac{\log n}{\epsilon})$ communication rounds in the $\mathcal{LOCAL}$ model.*

Proof. Let $OPT(G)$ be the set of edges that belong to some maximum cut in G, and let $ALG(G)$ be the set of edges in the cut obtained by Decomposition-Based Max-Cut. Let S_u be the component induced by the vertices which choose u as their center, and denote by S the set of components that algorithm Distributed Decomposition constructs. Then $\mathbb{E}\left[|ALG(G)|\right] \geq \mathbb{E}\left[\sum_{S_u \in S}|OPT(S_u)|\right] \geq |OPT(G)| - \beta m \geq |OPT(G)| - 2\beta|OPT(G)| = (1-\epsilon)|OPT(G)|$. The last inequality follows from the fact that for every graph G it holds that $|OPT(G)| \geq \frac{m}{2}$.

The graph decomposition requires $O(\frac{\log n}{\epsilon})$ communication rounds, and outputs components with weak diameter at most $O(\frac{\log n}{\epsilon})$. Therefore, finding the optimal solution within each component takes $O(\frac{\log n}{\epsilon})$ as well. The time bound follows.   □

By taking $\beta = \epsilon/4$, one can now obtain a $(1 - \epsilon)$-approximation algorithm for Max-Dicut. The difference comes from the fact that for Max-Dicut it holds that $|OPT(G)| \geq \frac{m}{4}$ for every graph G. The rest of the analysis is similar to the analysis for Max-Cut. Hence, we have

Theorem 3.8. *Decomposition-Based Max-Dicut is a randomized $(1 - \epsilon)$-approximation for Max-Dicut, requiring $O(\frac{\log n}{\epsilon})$ communication rounds in the $\mathcal{LOCAL}$ model.*

4 Coloring-Based Algorithms

Many of the sequential approximation algorithms for Max-Cut perform n iterations. Each vertex, in its turn, makes a greedy decision so as to maximize the solution value. We present distributed greedy algorithms that achieve the approximation ratios of the sequential algorithms much faster. We first prove that the greedy decisions of vertices can be done locally, depending only on their immediate neighbors. Then we show how to parallelize the decision process, such that in each iteration an independent set of vertices completes. The independent sets are generated using $(\Delta+1)$-coloring; then, for $(\Delta+1)$ iterations, all vertices of the relevant color make their parallel independent decisions. All algorithms run in the $\mathcal{CONGEST}$ model (see [8]).

5 A Deterministic $\mathcal{LOCAL}$ Algorithm

Our coloring-based algorithms may become inefficient for high degree graphs, due to the strong dependence on Δ. Consider a clique in this model. The above algorithms require a linear number of communication rounds, while learning the entire graph and finding an optimal solution requires only $O(1)$ communication rounds in the $\mathcal{LOCAL}$ model. Indeed, there is a tradeoff between the graph diameter and the average degree of its vertices. Based on this tradeoff, we propose a faster, two-step, deterministic algorithm for Max-Cut that requires $min\{\tilde{O}(\Delta + \log^* n), O(\sqrt{n})\}$ communication rounds in the $\mathcal{LOCAL}$ model. The pseudocode is given in Algorithm 4.

We call a vertex v a *low-degree* vertex, if $deg(v) < \sqrt{n}$, and a *high-degree* vertex, if $deg(v) \geq \sqrt{n}$. Define G_{low}, and G_{high} as the graphs induced by the low-degree vertices and the high-degree vertices, respectively. The idea is to solve the problem separately for G_{low} and for G_{high}.

In the first step, the algorithm deletes every high-degree vertex, if there are any, and its adjacent edges, creating G_{low}. The deletion means that the low-degree vertices ignore the edges that connect them to high-degree vertices and do not communicate over them. Then, the algorithm approximates the Max-Cut on G_{low}, using one of the coloring-based algorithms described in Sect. 4.

In the second step, the problem is solved optimally within each connected component in G_{high}. However, the high-degree vertices are allowed to communicate over edges which are not in G_{high}. As we prove next, the distance in the original graph G between any two vertices which are connected in G_{high} is upper bounded by $O(\sqrt{n})$. Hence, the number of rounds needed for this part of the algorithm is $O(\sqrt{n})$.

Algorithm 4. Fast Distributed Greedy Max-Cut

1: run Distributed Greedy Max-Cut on G_{low}
2: **for** each connected component in G_{high} **do**
3:     learn the component topology in G, including all its adjacent edges
4:     let the vertex with the lowest id find an optimal solution, and determine the output for each vertex in its component
5: **end for**
6: output the vertices decisions

Lemma 5.1. *Assume u, v are connected in G_{high}, then the distance between u and v in the original graph G is at most $3\sqrt{n}$.*

Theorem 5.2. *Fast Distributed Greedy Max-Cut yields a $\frac{1}{2}$-approximation to Max-Cut, using $min\{\tilde{O}(\Delta + \log^* n), O(\sqrt{n})\}$ communication rounds in the $\mathcal{LOCAL}$ model.*

Proof. We first prove the approximation ratio. Since Distributed Greedy Max-Cut is applied on G_{low}, at least half of the edges of G_{low} are in the cut. Given the decisions of vertices in G_{low}, the algorithm finds an optimal solution for all vertices in G_{high}. Note that running Distributed Greedy Max-Cut on the high-degree vertices of G, would give at least half of the remaining edges. This is due to the fact that the algorithm makes sequential greedy decisions. Therefore, an optimal solution for the high-degree vertices guarantees at least half of the edges in $G \setminus G_{low}$, implying the approximation ratio.

Applying Distributed Greedy Max-Cut on G_{low} requires $\tilde{O}(\Delta_{low} + \log^* n)$ communication rounds, where $\Delta_{low} = min\{\Delta, \sqrt{n}\}$. Using Lemma 5.1 we have that each high degree vertex can communicate with every high-degree vertex connected to it in G_{high}, using at most $O(\sqrt{n})$ communication rounds. Hence, Steps $2. - 4.$ of the algorithm take $O(\sqrt{n})$ communication rounds. We note that when $\Delta < \sqrt{n}$, the algorithm terminates after the first step. Thus, the algorithm requires $min\{\tilde{O}(\Delta + \log^* n), O(\sqrt{n})\}$ communication rounds. □

Using the above technique, we obtain a fast, deterministic algorithm for the Max-Dicut problem, by replacing the call to Distributed Greedy Max-Cut in Step 1 with a call to Distributed Greedy Max-Dicut. Using the same arguments as in the analysis for the Max-Cut algorithm, we have:

Theorem 5.3. *Fast Distributed Greedy Max-Dicut yields a $\frac{1}{3}$-approximation to Max-Dicut, using $min\{\tilde{O}(\Delta + \log^* n), O(\sqrt{n})\}$ communication rounds in the $\mathcal{LOCAL}$ model.*

Acknowledgements. We thank Roy Schwartz and Shay Kutten for stimulating discussions and for helpful comments on the paper.

References

1. Åstrand, M., Floréen, P., Polishchuk, V., Rybicki, J., Suomela, J., Uitto, J.: A local 2-approximation algorithm for the vertex cover problem. In: Keidar, I. (ed.) DISC 2009. LNCS, vol. 5805, pp. 191–205. Springer, Heidelberg (2009). https://doi.org/10.1007/978-3-642-04355-0_21
2. Åstrand, M., Suomela, J.: Fast distributed approximation algorithms for vertex cover and set cover in anonymous networks. In: Proceedings of the Twenty-Second Annual ACM Symposium on Parallelism in Algorithms and Architectures, pp. 294–302. ACM (2010)
3. Bar-Yehuda, R., Censor-Hillel, K., Schwartzman, G.: A distributed $(2+\epsilon)$-approximation for vertex cover in O($\log\Delta/\epsilon$ log log Δ) rounds. In: Proceedings of the 2016 ACM Symposium on Principles of Distributed Computing, PODC 2016, Chicago, IL, USA, 25–28 July 2016, pp. 3–8 (2016)
4. Barahona, F., Grötschel, M., Jünger, M., Reinelt, G.: An application of combinatorial optimization to statistical physics and circuit layout design. Oper. Res. **36**(3), 493–513 (1988)
5. da Ponte Barbosa, R., Ene, A., Nguyen, H.L., Ward, J.: A new framework for distributed submodular maximization. arXiv preprint http://arxiv.org/abs/1507.03719 (2015)
6. Barenboim, L.: Deterministic $(\delta+ 1)$-coloring in sublinear (in δ) time in static, dynamic and faulty networks. In: Proceedings of the 2015 ACM Symposium on Principles of Distributed Computing, pp. 345–354. ACM (2015)
7. Buchbinder, N., Feldman, M., Naor, J., Schwartz, R.: A tight linear time (1/2)-approximation for unconstrained submodular maximization. SIAM J. Comput. **44**(5), 1384–1402 (2015)
8. Censor-Hillel, K., Levy, R., Shachnai, H.: Fast distributed approximation for max-cut. arXiv preprint http://arxiv.org/abs/1707.08496 (2017)
9. Chang, K., Du, D.C.: Efficient algorithms for layer assignment problem. IEEE Trans. Comput.-Aided Des. Integr. Circuits Syst. **6**(1), 67–78 (1987)
10. Chin, K.W., Soh, S., Meng, C.: Novel scheduling algorithms for concurrent transmit/receive wireless mesh networks. Comput. Netw. **56**(4), 1200–1214 (2012)

11. Elkin, M.: Distributed approximation: a survey. ACM SIGACT News **35**(4), 40–57 (2004)

12. Elkin, M., Neiman, O.: Distributed strong diameter network decomposition. In: Proceedings of the 2016 ACM Symposium on Principles of Distributed Computing, pp. 211–216. ACM (2016)

13. Elkin, M., Neiman, O.: Efficient algorithms for constructing very sparse spanners and emulators. In: Proceedings of the Twenty-Eighth Annual ACM-SIAM Symposium on Discrete Algorithms, SODA 2017, Barcelona, Spain, Hotel Porta Fira, 16–19 January, pp. 652–669 (2017)

14. Feige, U., Mirrokni, V.S., Vondrák, J.: Maximizing non-monotone submodular functions. SIAM J. Comput. **40**(4), 1133–1153 (2011)

15. Garey, M.R., Johnson, D.S., Stockmeyer, L.: Some simplified NP-complete graph problems. Theoret. Comput. Sci. **1**(3), 237–267 (1976)

16. Ghaffari, M., Kuhn, F.: Distributed minimum cut approximation. In: Afek, Y. (ed.) DISC 2013. LNCS, vol. 8205, pp. 1–15. Springer, Heidelberg (2013). https://doi.org/10.1007/978-3-642-41527-2_1

17. Goemans, M.X., Williamson, D.P.: Improved approximation algorithms for maximum cut and satisfiability problems using semidefinite programming. J. ACM **42**(6), 1115–1145 (1995)

18. Grötschel, M., Pulleyblank, W.R.: Weakly bipartite graphs and the max-cut problem. Oper. Res. Lett. **1**(1), 23–27 (1981)

19. Hadlock, F.: Finding a maximum cut of a planar graph in polynomial time. SIAM J. Comput. **4**(3), 221–225 (1975)

20. Håstad, J.: Some optimal inapproximability results. J. ACM (JACM) **48**(4), 798–859 (2001)

21. Henzinger, M., Krinninger, S., Nanongkai, D.: A deterministic almost-tight distributed algorithm for approximating single-source shortest paths. In: Proceedings of the 48th Annual ACM SIGACT Symposium on Theory of Computing, pp. 489–498. ACM (2016)

22. Hirvonen, J., Rybicki, J., Schmid, S., Suomela, J.: Large cuts with local algorithms on triangle-free graphs. arXiv preprint arXiv:1402.2543 (2014)

23. Kale, S., Seshadhri, C.: Combinatorial approximation algorithms for maxcut using random walks. arXiv preprint arXiv:1008.3938 (2010)

24. Kapralov, M., Khanna, S., Sudan, M.: Streaming lower bounds for approximating max-cut. In: Proceedings of the Twenty-Sixth Annual ACM-SIAM Symposium on Discrete Algorithms, pp. 1263–1282. SIAM (2015)

25. Karp, R.M.: Reducibility among combinatorial problems. In: Miller, R.E., Thatcher, J.W., Bohlinger, J.D. (eds.) Complexity of Computer Computations, pp. 85–103. Springer, Heidelberg (1972). https://doi.org/10.1007/978-1-4684-2001-2_9

26. Khot, S., Kindler, G., Mossel, E., O'Donnell, R.: Optimal inapproximability results for max-cut and other 2-variable CSPs? SIAM J. Comput. **37**(1), 319–357 (2007)

27. Komurlu, C., Bilgic, M.: Active inference and dynamic Gaussian Bayesian networks for battery optimization in wireless sensor networks. In: AI for Smart Grids and Smart Buildings, Papers from the 2016 AAAI Workshop, Phoenix, Arizona, USA (2016)

28. Kuhn, F., Moscibroda, T.: Distributed approximation of capacitated dominating sets. Theory Comput. Syst. **47**(4), 811–836 (2010)

29. Kuhn, F., Moscibroda, T., Wattenhofer, R.: Local computation: lower and upper bounds. J. ACM (JACM) **63**(2), 17 (2016)

30. Lenzen, C., Pignolet, Y.A., Wattenhofer, R.: Distributed minimum dominating set approximations in restricted families of graphs. Distrib. Comput. **26**(2), 119–137 (2013)

31. Linial, N.: Locality in distributed graph algorithms. SIAM J. Comput. **21**(1), 193–201 (1992)

32. Lotker, Z., Patt-Shamir, B., Pettie, S.: Improved distributed approximate matching. In: Proceedings of the Twentieth Annual Symposium on Parallelism in Algorithms and Architectures, pp. 129–136. ACM (2008)

33. Matuura, S., Matsui, T.: 0.863-approximation algorithm for MAX DICUT. In: Goemans, M., Jansen, K., Rolim, J.D.P., Trevisan, L. (eds.) APPROX/RANDOM -2001. LNCS, vol. 2129, pp. 138–146. Springer, Heidelberg (2001). https://doi.org/10.1007/3-540-44666-4_17

34. Miller, G.L., Peng, R., Xu, S.C.: Parallel graph decompositions using random shifts. In: Proceedings of the Twenty-Fifth Annual ACM Symposium on Parallelism in Algorithms and Architectures, pp. 196–203. ACM (2013)

35. Mirrokni, V., Zadimoghaddam, M.: Randomized composable core-sets for distributed submodular maximization. In: Proceedings of the Forty-Seventh Annual ACM on Symposium on Theory of Computing, pp. 153–162. ACM (2015)

36. Mirzasoleiman, B., Karbasi, A., Sarkar, R., Krause, A.: Distributed submodular maximization: identifying representative elements in massive data. In: Advances in Neural Information Processing Systems, pp. 2049–2057 (2013)

37. Mitzenmacher, M., Upfal, E.: Probability and Computing: Randomized Algorithms and Probabilistic Analysis. Cambridge University Press, Cambridge (2005)

38. Motwani, R., Raghavan, P.: Randomized Algorithms. Chapman & Hall/CRC, London (2010)

39. Nanongkai, D.: Distributed approximation algorithms for weighted shortest paths. In: Proceedings of the 46th Annual ACM Symposium on Theory of Computing, pp. 565–573. ACM (2014)

40. Papadimitriou, C., Yannakakis, M.: Optimization, approximation, and complexity classes. In: Proceedings of the Twentieth Annual ACM Symposium on Theory of Computing, pp. 229–234. ACM (1988)

41. Peleg, D.: Distributed Computing. SIAM Monographs on Discrete Mathematics and Applications, vol. 5 (2000)

42. Sahni, S., Gonzalez, T.: P-complete approximation problems. J. ACM (JACM) **23**(3), 555–565 (1976)

43. Saurabh, S., Zehavi, M.: $(k, n - k)$-MAX-CUT: an $\mathcal{O}^*(2^p)$-time algorithm and a polynomial kernel. In: Kranakis, E., Navarro, G., Chávez, E. (eds.) LATIN 2016. LNCS, vol. 9644, pp. 686–699. Springer, Heidelberg (2016). https://doi.org/10.1007/978-3-662-49529-2_51

44. Tangwongsan, K.: Efficient parallel approximation algorithms. Ph.D. thesis, School of Computer Science, Carnegie Mellon University (2011)

45. Trevisan, L.: Max cut and the smallest eigenvalue. SIAM J. Comput. **41**(6), 1769–1786 (2012)

46. Trevisan, L., Sorkin, G.B., Sudan, M., Williamson, D.P.: Gadgets, approximation, and linear programming. SIAM J. Comput. **29**(6), 2074–2097 (2000)

47. Wang, J., Jebara, T., Chang, S.F.: Semi-supervised learning using greedy max-cut. J. Mach. Learn. Res. **14**(Mar), 771–800 (2013)

48. Wang, L., Chin, K., Soh, S.: Joint routing and scheduling in multi-Tx/Rx wireless mesh networks with random demands. Comput. Netw. **98**, 44–56 (2016)

49. Wang, W., Liu, B., Yang, M., Luo, J., Shen, X.: Max-cut based overlapping channel assignment for 802.11 multi-radio wireless mesh networks. In: 2013 IEEE 17th International Conference on Computer Supported Cooperative Work in Design (CSCWD), pp. 662–667 (2013)
50. Xu, Y., Chin, K., Raad, R., Soh, S.: A novel distributed max-weight link scheduler for multi-transmit/receive wireless mesh networks. IEEE Trans. Veh. Technol. **65**(11), 9345–9357 (2016)
51. Xue, G., He, Q., Zhu, H., He, T., Liu, Y.: Sociality-aware access point selection in enterprise wireless LANs. IEEE Trans. Parallel Distrib. Syst. **24**(10), 2069–2078 (2013)

Barrier Coverage with Uniform Radii in 2D

Andrew Cherry[1], Joachim Gudmundsson[1(✉)], and Julián Mestre[1,2]

[1] School of Information Technologies, The University of Sydney, Sydney, Australia
joachim.gudmundsson@sydney.edu.au
[2] Facebook Inc., Menlo Park, USA

Abstract. Given a set B of disjoint line segments in the plane, the so-called *barriers*, and a set of n sensors with uniform range in the plane, the *barrier coverage* problem is to move the sensors so that they cover the segments in B, while minimizing the total movement of the sensors. In the 1D case when B contains a single barrier and all the sensors lie on B then the problem can be solved in $O(n \log n)$ time. In 2D very little is known about the complexity of the problem.

We consider the 2D setting and give a $\sqrt{2}$-approximation algorithm when B contains a single barrier, or a set of parallel barriers. We also give an approximation algorithm for arbitrarily oriented disjoint barriers.

1 Introduction

The problem of covering a set of barriers using sensors was originally motivated by intrusion detection and has been studied in numerous research papers [1,3, 5,7,9] and in two surveys [11,12]. A sensor can detect an intruder in a circular region of fixed range with center at the sensor, and the goal is to guard a region in the plane. Two cases have been considered in the literature. The *area coverage* case where the sensors have to monitor an entire region [6,8] and the *barrier coverage* case where only the perimeter of the region is monitored [1,3–5,7,9]. In this paper we will focus our attention on the barrier coverage problem.

As input we are given a set $B = \{b_1, \ldots, b_k\}$ of (closed) disjoint line segments, the so-called *barriers*, together with a set $S = \{1, \ldots, n\}$ of sensors and a location assignment $p : S \to \mathbb{R}^2$, where $p_i = (x_i, y_i)$ for each $i \in S$. Each sensor $i \in S$ is modelled as a unit disk centered at p_i. Following the model defined by Dobrev *et al.* [5], we say that a sensor located at q covers a barrier $b_j \in B$ if and only if $q \in b_j$. An assignment $\hat{p} : S \to \mathbb{R}^2$ is called a *covering assignment* if all barriers in B are completely covered; that is, for all $b_j \in B$, we have $b_j \subseteq \cup_{i:\hat{p}_i \in b_j} D(\hat{p}_i)$, where $D(q)$ is the unit disk centered at q. Figure 1(a) depicts an example of a covering assignment.

The *barrier coverage* problem is to find a covering assignment $\hat{p}$ of minimum cost, where the cost is equal to the sum of sensor movements, namely,

$$cost(\hat{p}) = \sum_{i=1}^{n} \text{dist}_2(p_i, \hat{p}_i),$$

where dist_2 is the usual Euclidean distance.

This work was supported by ARCs Discovery Projects funding scheme (DP150101134).

© Springer International Publishing AG 2017
A. Fernández Anta et al. (Eds.): ALGOSENSORS 2017, LNCS 10718, pp. 57–69, 2017.
https://doi.org/10.1007/978-3-319-72751-6_5

1.1 Background

For the barrier coverage problem three different optimisation criteria have been considered in the literature: minimise the sum of movements (min-sum), minimize the maximum movement (min-max) and, minimize the number of sensors that move (min-num). Most of the existing research has focused on the one-dimensional setting where B contains a single barrier and all the sensors lie on the line containing the barrier.

Min-Sum model in 1D. In this case the problem can be solved optimally in $O(n \log n)$ time when the sensors all have the same radius. If the sensors have different radius the problem is known to be NP-hard to approximate within a constant factor [4].

Gaspers et al. [9] strengthen this result and proved that no polynomial time $\rho^{1-\varepsilon}$-approximation algorithm exists unless $P = NP$, where ρ is the ratio between the largest radius and the smallest radius. Even if the number of intervals required to move in an optimal solution is small the problem turns out to be W[1]-hard. On the positive side they showed that a $((2 + \varepsilon)\rho + O(1/\varepsilon))$-approximation can be computed in $O(n^3/\varepsilon^2)$ time.

The Min-Max and Min-Num models in 1D. Czyzowicz et al. [4] considered the min-max version of the problem, where the aim is to minimize the maximum movement. If the sensors have unit radius they gave an $O(n^2)$ time algorithm. Chen et al. [3] improved the bound to $O(n \log n)$. In the same paper they presented an $O(n^2 \log n)$ time algorithm for the case when the sensors have different radius. For the min-num version Mehrandish et al. [8] showed that the problem can be solved in polynomial time using dynamic programming if the sensor radii are uniform, otherwise the problem is NP-hard.

The Min-Max and Min-Sum models in 2D. To the best of our knowledge there are only a few papers that consider the case when the sensors are originally placed anywhere in the plane.

Dobrev et al. [5] studied the min-sum and min-max versions with sensors of arbitrary ranges, initially located at arbitrary locations in the plane. They showed that the problem is NP-complete even for two barriers and that the min-max problem is NP-complete even for a single barrier (follows from results in [4]). In the same paper Dobrev et al. also gave algorithms for the restricted setting when the movement of the sensors have to be perpendicular to the barriers, that is, a sensor has to move to the closest point on a line containing a barrier. Also, the final positions of sensors have to be on the line containing the barrier. Using this model in the case of k parallel barriers they gave an $O(kn^{k+1})$ time algorithm. In the case when the input contains one vertical barrier and one horizontal barrier the problem becomes NP-hard even when restricted to perpendicular movement. They also showed an $O(n^{1.5})$ time algorithm for perpendicular movement when the problem is further restricted to the case when all sensors are located at integer positions and the sensors have unit radius.

Circular barriers in the plane were studied in [2,11]. Bhattacharya *et al.* [2] considered moving n sensors to the boundary of the circular barrier and gave an $O(n^{3.5} \log n)$ time for the min-max version. This bound was later improved to $O(n^{2.5} \log n)$ by Tan and Wu [10]. They also gave an $O(n^4)$ time algorithm that solves the min-sum version when the initial positions of the sensors are on the boundary of the circular barrier.

1.2   Our Results

In this paper we present approximation algorithms for the barrier coverage problem on the plane.

Theorem 1. *There is an $O(n^4)$ time $\sqrt{2}$-approximation for the setting where we have a single barrier on the plane.*

Theorem 2. *There is an $O(kn^{2k+2})$ time $\sqrt{2}$-approximation for the setting where we have k parallel barriers on the plane.*

Theorem 3. *There is an $O(kn^{2k+2})$ time $2\sqrt{2}(1 + 2\tan\alpha)$-approximation for the setting where we have k disjoint barriers on the plane, where α is the largest angle between a barrier and the x-axis.*

2   A Single Barrier

This section deals with the case where B consists of a single line segment b. For ease of presentation we assume without loss of generality that b is the horizontal segment between $(0,0)$ and some point $(L,0)$ (this can always be achieved by rotating and translating the input point set p). Furthermore, we assume that the sensors $S = \{1, \ldots, n\}$ are ordered from left to right (ties are broken arbitrarily).

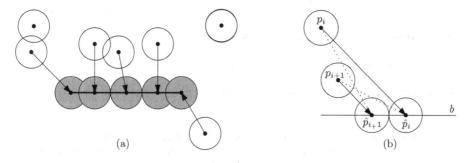

(a)   (b)

Fig. 1. (a) An instance of the covering assignment problem with seven sensors. The grey disks represent the new locations of sensors that move. The original location (p_i) and the new location ($\hat{p}_i$) are connected with a directed arrow. (b) An instance where the dotted lines represent an order preserving assignment, while the solid lines represent an optimal assignment.

We say that a covering assignment $\hat{p}$ is *order-preserving* if for every $\hat{p}_i = (\hat{x}_i, \hat{y}_i) \in B$ we have $\hat{x}_i < \hat{x}_j$ if and only if $i < j$.

If we knew that there is always an optimal solution that is order-preserving, perhaps we could exploit this fact to design a dynamic programming formulation for the problem. Unfortunately there are instances where an optimal covering assignment is not order-preserving. A simple example is shown in Fig. 1(b). The interesting question to ponder then is, how bad can an order-preserving covering assignment be in the worst case? The main result of this section is to show that there exists an order-preserving covering that is at most a factor $\sqrt{2}$ worse than an optimal solution and that we can compute such a solution efficiently.

To prove this we need to study the Manhattan distance cost of a covering assignment $\hat{p}$, namely,

$$cost_1(\hat{p}) = \sum_{i \in B} \text{dist}_1(\hat{p}_i, p_i).$$

The main reason to consider the dist_1-cost is that unlike the dist_2 case, there is always an order-preserving covering assignment that is optimal under dist_1.

Our algorithm and its correctness rests on the following key lemmas.

Lemma 1. *In the single-barrier case any $cost_1$-optimal order-preserving covering is $\sqrt{2}$-approximate under the dist_2-cost.*

Lemma 2. *For the single-barrier case there exists a set D of $O(n^2)$ discrete points on the barrier such that there exists a $cost_1$-optimal order-preserving covering assignment where sensors are only allowed to be placed in locations in D.*

Lemma 3. *For the single-barrier case there is an $O(n^4)$ time algorithm for finding a $cost_1$-optimal order-preserving covering assignment.*

We can combine these lemmas in a straightforward way to get a proof of Theorem 1, namely, an $O(n^4)$ time $\sqrt{2}$-approximation for the barrier coverage problem with one barrier. The rest of this section is devoted to proving these lemmas.

2.1 Order-Preserving Covering Assignments

We start with a simple observation on the structure of the $cost_1$-optimal solution.

Observation 1. *There is a $cost_1$-optimal covering assignment that is order-preserving.*

Proof. Assume we have a $cost_1$-optimal covering assignment $\hat{p}$ that is not order-preserving. Then there exists two sensors $i, j \in S$ such that $x_i \leq x_j$ and $\hat{x}_i > \hat{x}_j$. The contribution of i and j to $cost_1(\hat{p})$ is $|y_i| + |\hat{x}_i - x_i| + |y_j| + |\hat{x}_j - x_j|$.

Now consider a new covering assignment $\hat{p}'$ where sensors i and j are reversed, that is, $\hat{p}'_i = \hat{p}_j$ and $\hat{p}'_j = \hat{p}_i$. Note that the order of i and j is preserved under $\hat{p}'$. The contribution of i and j to $cost_1(\hat{p}')$ is $|y_i| + |\hat{x}_j - x_i| + |y_j| + |\hat{x}_i - x_j|$, which is equal to their contribution under $\hat{p}$ when both $\hat{x}_i$ and $\hat{x}_j$ lie to the right or to

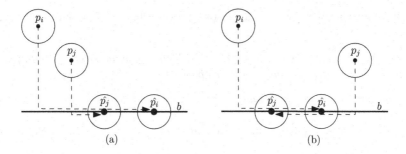

Fig. 2. Sensors in an $dist_1$-optimal covering.

the left of both x_i and x_j and less than otherwise. Figure 2 shows and example of both cases.

Therefore, $cost_1(\hat{p}') \leq cost_1(\hat{p})$. Repeating this process until complete order has been restored yields an order preserving covering that is optimal with respect to $cost_1$.   □

Using the above observation we can now give a proof of Lemma 1, namely that a $cost_1$-optimal order-preserving solution is $\sqrt{2}$-approximate under the $dist_2$ metric.

Proof (Proof of Lemma 1). Let opt^1 and opt^2 be optimal covering assignments minimizing the $dist_1$- and $dist_2$-cost, respectively. To differentiate the two we talk of $cost_1(\cdot)$ and $cost_2(\cdot)$. These two objectives are related as follows:

$$cost_2(opt^1) \leq cost_1(opt^1) \leq cost_1(opt^2) \leq \sqrt{2} \cdot cost_2(opt^2),$$

where the first inequality follows from the fact that $dist_1$ dominates $dist_2$, the second from the optimality of opt_1 under $cost_1$, and the third from the well-known relation between $dist_2$ and $dist_1$ in $\mathbb{R}^2$.   □

In the next two sections we will show how to compute a $dist_1$-optimal order-preserving covering assignment in time $O(n^4)$. First we need to impose more structure on such solutions.

2.2   Discrete Sensor Locations

Let $\hat{p}$ be an order preserving covering assignment. Consider two consecutive sensors $i, j \in S$ are said to be *neighbors* in a covering $\hat{p}$ if both cover the barrier (i.e., $\hat{p}_i, \hat{p}_j \in b$) and there is no other sensor covering the barrier between i and j. Furthermore, two neighbouring sensors are said to *overlap* if the distance between them is less than 2, otherwise they are *non-overlapping*. Notice that other than the leftmost and rightmost sensors, every sensor has exactly one neighbor to the left and one neighbor to the right.

Consider a sensor i covering the barrier in the solution $\hat{p}$. We say that the sensor's position $\hat{p}_i$ is an *anchor* if either $x_i = \hat{x}_i$ or p_i is the leftmost or rightmost point in b. Furthermore, we say a sensor i moves left if $x_i > \hat{x}_i$ and move right if $x_i < \hat{x}_i$.

Observation 2. *Let $\hat{p}$ be a* dist$_1$-*optimal order-preserving covering assignment. For any sensor i covering the barrier in $\hat{p}$, then*

1. *if i is not placed at $(0,0)$ and is right-moving, it will not overlap its right neighbour, and*
2. *if i is not placed at $(L,0)$ and is left-moving, it will not overlap its left neighbour.*

Proof. If $\hat{p}_i \neq (0,0)$ and p_i lies to the right of p_i then the cost of moving sensor i is $|y_i| + (\hat{x}_i - x_i)$. If it is overlapping with its right neighbour j in $\hat{p}$ then one could move i to the left by a distance $\min(\hat{x}_i - x_i, 2 - (\hat{x}_j - \hat{x}_i))$. This would not affect the coverage of the barrier but would reduce the cost of the movement of sensor i, contradicting the optimality of $\hat{p}$. Hence, a right-moving sensor, not at barrier position $(0,0)$, does not overlap its right neighbour in $\hat{p}$.

The case when a sensor moves left and $\hat{p}_i \neq (L,0)$ is symmetric.   □

This observation can be generalized to a chain of non-overlapping sensors.

Observation 3. *Let $\hat{p}$ be a* dist$_1$-*optimal order-preserving covering assignment. Then for any sequence of consecutive non-overlapping sensors that cover the barrier in $\hat{p}$,*

1. *if the sequence does not include a sensor at $(0,0)$ and has more right-moving than left-moving sensors, it will not overlap with the neighboring sensor to its right.*
2. *if the sequence does not include a sensor at $(L,0)$ and has more left-moving than right-moving sensors, it will not overlap with the neighboring sensor to its left.*

Proof. Given a sequence of consecutive non-overlapping sensors S' that cover the barrier, let $S'_L \subset S'$ be the sensors in S' that moved left and let $S'_R \subset S'$ be the sensors in S' that moved right. That is, $S'_L = \{i \in S' : x_i > \hat{x}_i\}$ and $S'_R = \{i \in S' : x_i < \hat{x}_i\}$.

If $|S'_L| < |S'_R|$ and there is an overlap between the right-most sensor in the sequence and its neighbour to the right then moving the entire sequence to the left a small distance such that the barrier is still covered will decrease the total cost which contradicts the assumption that $\hat{P}$ is a minimum cost covering assignment. Hence, if $|S'_L| < |S'_R|$ the sequence cannot overlap with its right neighbour.

The symmetric argument can be used for the case when $|S'_L| > |S'_R|$.   □

Note that the sensors in a sequence of neighboring non-overlapping sensors in $\hat{p}$ are separated by discrete distances. That is, if the leftmost sensor of the sequence is positioned at $\hat{x}_i$ then the following sensors' locations in the sequence are given by $(\hat{x}_i + 2k, 0)$, for $k = 1, 2, \ldots$. Below we will show key properties of an optimal order-preserving covering assignment $\hat{p}$ under dist$_1$.

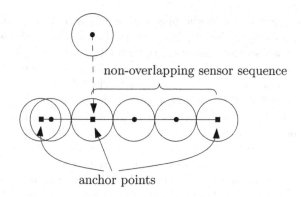

non-overlapping sensor sequence

anchor points

Fig. 3. A sequence of non-overlapping sensors in $\hat{P}$ together with three anchor points.

Observation 4. *There is a* dist$_1$*-optimal order-preserving covering assignment in which every interval between two consecutive anchors along b is covered by either*

Type 1: *a sequence of non-overlapping sensors that includes at least one of the anchors, or*

Type 2: *two sequences of non-overlapping sensors with the left sequence including the left anchor and the right sequence including the right anchor.*

Proof. Let $\hat{p}$ be a dist$_1$-optimal order-preserving covering assignment with the maximum number of anchor points. Let $\hat{p}_i$ and $\hat{p}_j$ be two consecutive anchors along b. If the number of sensors in $\hat{p}$ between $\hat{p}_i$ and $\hat{p}_j$ is zero then the sequence is trivially of type 1. If the number of sensors is at least one then we will have the five cases listed below. As above let S'_L be the sensors in the sequence that moved left and let S'_R be the sensors in the sequence that moved right.

Case 1: If the sensors are all right-moving then, according to Observation 2, the sequence will not overlap internally or with $\hat{p}_j$, hence it will be of Type 1.

Case 2: If the sensors are all left-moving then, according to Observation 2, the sequence will not overlap internally or with $\hat{p}_i$, hence it will be of Type 1.

Case 3: If the sequence contains no internal overlapping sensors and $|S'_R| \neq |S'_L|$ then, from Observation 3, we know that the sequence cannot overlap $\hat{p}_j$ ($|S'_R| > |S'_L|$) or the sequence cannot overlap $\hat{p}_i$ ($|S'_R| < |S'_L|$). Both situations result Type 1 sequences.

Case 4: If the sequence contains no internally overlapping sensors and $|S'_R| = |S'_L|$ then the sequence can be moved either left or right until one of the sensors in the sequence becomes an anchor which contradicts the assumption that $\hat{p}$ is an optimal solution with a maximum number of anchor points, or until the sequence does not overlap with either $\hat{p}_i$ or $\hat{p}_j$. This is a sequence of Type 1.

Case 5: If the sequence contains overlapping sensors then let $\hat{p}_i = p_{i_1}, \ldots, p_{i_k} = p_j$ be the sensor positions in the sequence ordered from left to right along b.

First consider the pairs of overlapping sensors in the sequence, and let $\hat{p}_{i_b}$ and $\hat{p}_{i_{b+1}}$ be the positions of the leftmost sensors that overlap in the sequence, see Fig. 4. If there is a second such overlap, then let $\hat{p}_{i_h}$ and $\hat{p}_{i_{h+1}}$ be the leftmost sensors that overlap in the sequence to the right of $\hat{p}_{i_{b+1}}$. Note that one can move the subsequence $\hat{p}_{i_{b+1}}, \ldots, \hat{p}_{i_h}$ either to the left or to the right without increasing the cost of the solution until (a) one sensor position becomes an anchor, or until (b) either $\hat{p}_{i_b}$ and $\hat{p}_{i_{b+1}}$ or $\hat{p}_{i_h}$ and $\hat{p}_{i_{h+1}}$ are non-overlapping. Since outcome (a) contradicts the assumption that the covering assignment has a maximum number of anchors only outcome (b) is possible. As a result the resulting sequence has one fewer overlap than the starting configuration. This argument can be used iteratively, with the result that one either gets case 3 or 4 (type 1), or a type 2 scenario where two sequences of non-overlapping sensors overlap in the intersection with the left sequence including the left anchor and the right sequence including the right anchor. ☐

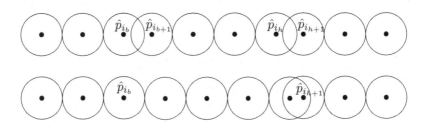

Fig. 4. (top) Illustrating the initial configuration for Case 5 in the proof of Observation 4. (bottom) The sequence after moving the sensors $\hat{p}_{i_{b+1}}, \ldots, \hat{p}_{i_h}$ to the right without increasing the cost of the covering assignment, resulting in one less overlap.

Let $A = \{x_i \,|\, 0 \le x_i \le L, i \in S\} \cup \{0, 1, L-1, L\}$ and let $D = \{a_i \pm 2 \cdot c \,|\, 0 \le a_i \pm 2 \cdot c \le L, a_i \in A, c \in \mathbb{Z}\}$. Note that the number of points in D is bounded by $O(n^2)$, since $L \le 2n$. Everything is in place to give the proof of Lemma 2, which states that there is a $cost_1$-optimal solution that places sensors only in locations in D.

Proof (Proof of Lemma 2). The set of potential sensor locations D is a finite set of locations on the barrier so the barrier can clearly be partitioned into intervals with a subset of elements of D as endpoints. By Observation 4, each interval will contain a (possibly empty) sequence of non-overlapping sensors that includes one or both of the anchors and so every sensor in the covering is located at $2 \cdot c$, $c \in \mathbb{Z}$, from some anchor sensor also in the covering. ☐

2.3 Dynamic Programming Formulation for a Single Barrier

In this section we develop a dynamic programming algorithm for finding a $cost_1$-optimal order-preserving covering assignment where each sensor lies on a point in D.

We follow the standard approach for a dynamic programming algorithm. Order the coordinates in D from left to right as $c_1, \ldots, c_m$, where $m = O(n^2)$. Let $M[i, j]$ denote the cost of an optimal order-preserving covering assignment of the part of b between 0 and c_j using only the sensors $1, \ldots, i$. For simplicity we will assume that a single point interval on a barrier does not have to be covered, hence, $M[0, 0] = 0$.

Assume that the algorithm has computed all $M[i', j']$-values for all subproblems smaller than $M[i, j]$. We can now calculate $M[i, j]$ as follows:

$$M[i,j] = \min_{c_r \in D \cap [c_{j-2}, c_{j+2})} \begin{cases} M[i-1, j] \\ M[i-1, r] + \text{dist}_1(p_i, (c_j - 1, 0)) & \text{if } x_i < c_j - 1 \\ M[i-1, r] + \text{dist}_1(p_i, (c_r + 1, 0)) & \text{if } x_i > c_r + 1 \\ M[i-1, r] + \text{dist}_1(p_i, (x_i, 0)) & \text{if } c_j - 1 \leq x_i \leq c_r + 1 \end{cases}$$

The dynamic programming checks all possible placements of sensor i in an attempt to find an optimal order-preserving covering assignment to the sub-barrier $[0, c_j]$. The first case is if the sensor i is not used in an optimal covering. For the second case the dynamic programming considers all possible solutions when p_i moves right, and hence will move a minimum distance to the right which is $\hat{x}_i = c_j - 1$. For the third case, the sensor i is assumed to move left a minimum distance, that is to position $\hat{x}_i = c_r + 1$. Finally, the last case considers the case when i becomes an anchor and $x_i = \hat{x}_i$.

We are ready to give the proof of Lemma 3, which states that we can compute a $cost_1$-optimal order-preserving covering assignment in $O(n^4)$ time.

Proof (Proof of Lemma 3). For the algorithm we use the above dynamic programming formulations. It is clear that the formulation computes a $cost_1$-optimal order preserving covering assignment with the additional constraint that sensors must be placed on locations in D. From Lemma 2 and Observation 1, we know that such a solution is also $cost_1$-optimal among all covering assignments, so the correctness of the algorithm follows.

For the running time, note that the algorithm considers $O(n^3)$ subproblems. Each subproblem can be solved in time proportional to the number of points of D within the interval $[c_{j-2}, c_{j+2})$, which is $O(n)$. This yields an overall running time of $O(n^4)$.   □

3   Parallel Barriers

In this section we relax the single barrier assumption to allow multiple parallel barriers $B = \{b_1, \ldots, b_k\}$. Without loss of generality we orient the plane as for the single-barrier case, choosing the left-most barrier endpoint to coincide with the origin, making this barrier and the x-axis collinear. Each barrier $b_t = ((x(t), y(t)), (x'(t), y'(t)))$ in B, $1 \leq t \leq k$, is now represented by a line segment parallel to the x-axis. All sensors assigned covering a given barrier will have the same y-value.

As in the single barrier case we assume that the sensors $S = \{1, \ldots, n\}$ are ordered from left to right (ties are broken arbitrarily). We now generalize our definition of order-preserving to multi-barrier. We say a covering assignment $\hat{p}$ is order-preserving if for each barrier b_t the subset of sensors covering b_t are placed along b_t in the same order they appear in the input; that is, for any sensors $i, j \in S$ such that $\hat{p}_i, \hat{p}_j \in b_t$ then it must be the case that if $i < j$ then $\hat{x}_i \leq \hat{x}_j$.

The results from the single-barrier case can be generalized to the multiple-barrier case as follows.

Lemma 4. *In the parallel-barrier case any $cost_1$-optimal order-preserving covering is $\sqrt{2}$-approximate under the $dist_2$-cost.*

Lemma 5. *For the parallel-barrier case there exists a set D of $O(kn^2)$ discrete points on the barrier such that there exists a $cost_1$-optimal order-preserving covering assignment where sensors are only allowed to be placed in locations in D.*

Lemma 6. *For the parallel-barrier case there is an $O(kn^{2k+2})$ time algorithm for finding a $cost_1$-optimal order-preserving covering assignment.*

Lemma 4 follows directly from Lemma 1 and the new definition of order-preserving to the parallel-barrier case. Lemma 5 follows by applying Lemma 2 to each of the k barriers in the instance. Finally, in the next subsection we show how to adapt the proof of the Lemma 3 to show Lemma 6.

Given these three lemmas, it is straightforward to combine them to get a proof of Theorem 2, namely, an $O(kn^{2k+2})$ time $\sqrt{2}$-approximation for the barrier coverage problem with k parallel barriers.

3.1 Dynamic Programming Formulation for Parallel Barriers

We extend the dynamic programming formulation presented in Sect. 2.3 to accommodate problem instances with multiple parallel barriers. The dynamic program is very similar to the single barrier case, instead of considering all possible solutions for a sub-barrier of a single barrier the dynamic program has to consider all possible sub-barriers of all the barriers.

Let $A(t) = \{x_i \mid x(t) \leq x_i \leq x'(t), i \in S\} \cup \{x(t), x(t) + 1, x'(t) - 1, x'(t)\}$ and let $D(t) = \{a_i \pm 2 \cdot c \mid x(t) \leq a_i \pm 2 \cdot c \leq x'(t), a_i \in A(t), c \in \mathbb{Z}\}$.

Let $M[i, j_1, \ldots, j_k]$ be the cost of an optimal covering assignment of the part of each barrier $B_t \in B$ in the x-interval $[x(t), c_{t,j_t}]$, $1 \leq t \leq k$, using only the sensors $1, \ldots, i$. As for the single barrier case we will assume that a single point interval on a barrier does not have to be covered, hence, $M[0, x(1), \ldots, x(t)] = 0$.

Assume that the algorithm has computed all $M[i', j_1', \ldots, j_k']$ for all subproblems smaller than $M[i, j_1, \ldots, j_k]$.

$$M[i, j_1, \ldots, j_k] = \min_{c_r \in \cup_{t=1}^{k} D(t) \cap [c_{t,j_t} - 2, c_{t,j_t} + 2)}$$

$$
\begin{cases}
M[i-1, j_1, \ldots, j_k] & \\
M[i-1, c_r, \ldots, j_k] + \mathrm{dist}_1(p_i, (c_{1,j_1} - 1, y(t))) & \text{if } c_r \in A(1) \text{ and } x_i < c_{1,j_1} - 1 \\
\quad\vdots & \quad\vdots \\
M[i-1, j_1, \ldots, c_r] + \mathrm{dist}_1(p_i, (c_{k,j_k} - 1, y(k))) & \text{if } c_r \in A(k) \text{ and } x_i < c_{k,j_k} - 1 \\
M[i-1, c_r, \ldots, j_k] + \mathrm{dist}_1(p_i, (r+1, y(1))) & \text{if } c_r \in A(1) \text{ and } x_i > r+1 \\
\quad\vdots & \quad\vdots \\
M[i-1, j_1, \ldots, c_r] + \mathrm{dist}_1(p_i, (r+1, y(k))) & \text{if } c_r \in A(k) \text{ and } x_i > r+1 \\
M[i-1, c_r, \ldots, j_k] + \mathrm{dist}_1(p_i, (x_i, y(1))) & \text{if } c_r \in A(1) \text{ and } c_j - 1 \le x_i \le r+1 \\
\quad\vdots & \quad\vdots \\
M[i-1, j_1, \ldots, c_r] + \mathrm{dist}_1(p_i, (x_i, y(k))) & \text{if } c_r \in A(k) \text{ and } c_j - 1 \le x_i \le r+1
\end{cases}
$$

The dynamic programming checks all possible placements of sensor i in an attempt to find an optimal order-preserving covering assignment. The first case is if the sensor i is not used in an optimal covering. For the second set of cases the dynamic programming considers all possible solutions when p_i moves right, and hence will move a minimum distance to the right for each barrier. For the third set of cases, the sensor i is assumed to move left a minimum distance for each barrier. Finally, the last set of cases considers the case when i becomes an anchor on a barrier and $x_i = \hat{x}_i$.

We are ready to give the proof of Lemma 6, which states that we can compute a $cost_1$-optimal order-preserving covering assignment in $O(kn^{2k+2})$ time.

Proof (Proof of Lemma 6). For the algorithm we use the above dynamic programming formulation. The correctness of the algorithm follows immediately from Lemmas 4 and 5.

For the running time, note that the algorithm considers $O(n \cdot (n^2)^k)$ subproblems. Each subproblem can be solved in $O(kn)$, similar to the single barrier case. Summing up the running time of the algorithm is $O(kn^{2k+2})$. □

4 Non-parallel Barriers

Consider a set of non-parallel disjoint barriers B. Instead of using the $dist_1$ measure we will need to use a Manhattan distance measure for each of the barriers, which will be denoted the $dist_\perp$ measure. That is, we consider a form of rectilinear distance in which a sensor moves to a barrier using only movements perpendicular and parallel to that barrier. In the special case where a barrier is parallel to either axis the distance is the familiar Manhattan distance but note that, unlike the Manhattan distance, the $dist_\perp$ measure is invariant under rotation. As with the multiple parallel barrier case, we preserve the left-to-right order on each barrier but not necessarily across all barriers. Note that the definition of an order-preserving covering assignment is still valid for the non-parallel case.

Consider the minimum angle each barrier makes with the x-axis. Rotate the plane so that the largest of these angles is minimised, denote this angle by α. For example, for vertical and horizontal segments α would be $\pi/4$.

Lemma 7. *In the disjoint multiple barrier case any $cost_\perp$-optimal order-preserving covering is $2\sqrt{2}(1 + 2\tan\alpha)$-approximate under the $dist_2$-cost.*

Lemma 8. *For the disjoint multiple barrier case there exists a set D of $O(kn^2)$ discrete points along the barrier such that there exists a $cost_\perp$-optimal order-preserving covering assignment where sensors are only allowed to be placed in locations in D.*

Lemma 9. *For the disjoint multiple barrier case there is an $O(kn^{2k+2})$ time algorithm for finding a $cost_\perp$-optimal order-preserving covering assignment.*

Lemma 8 follows by applying Lemma 2 to each of the k barriers in the instance using the $dist_\perp$. The algorithm is analogous to that for computing an order-preserving covering assignment for multiple parallel barriers from Sect. 3, hence, Lemma 9 is identical to Lemma 6. It remains to prove Lemma 7 which will be done below using the following lemma (proof omitted):

Lemma 10. *Let B be a barrier with a minimum angle of α to the horizontal line. If two sensors in a $dist_\perp$-optimal covering assignment O of B are order-reversed then swapping their positions in the covering assignment increases their cost by at most a factor $(1 + 2\tan\alpha)$.*

Given a $dist_\perp$-optimal covering O, an order-preserving covering assignment can be created by exchanging sensors until a left-to-right order is obtained on each barrier.

Lemma 10 states that swap increases the sensors' cost by a factor of at most $(1 + 2\tan\alpha)$. In the worst case every sensor in O would need to be relocated. If the swaps were performed in sequential order – that is, move a sensor to its final location then move the sensor it dislocates and so on – until all are in order every sensor is handled at most twice and so we have that the total cost is at most $2(1 + 2\tan\alpha)$ times the cost of O. Finally, we apply Lemma 1 which completes the proof of Lemma 7, and hence also Theorem 3.

Remark 1. The results for disjoint multiple barriers can be modified to work for simple polygons such as squares. However, extra care has to be taken at the endpoints of the barriers, e.g., by modifying the dynamic programming.

References

1. Arora, A., Ramnath, R., Ertin, E., Sinha, P., Bapat, S., Naik, V., Kulathumani, V., Zhang, H., Cao, H., Sridharan, M., Kumar, S., Seddon, N., Anderson, C., Herman, T., Trivedi, N., Zhang, C., Nesterenko, M., Shah, R., Kulkarni, S.S., Aramugam, M., Wang, L., Gouda, M.G., Choi, Y.-R., Culler, D.E., Dutta, P., Sharp, C., Tolle, G., Grimmer, M., Ferriera, B., Parker, K.: ExScal: elements of an extreme scale wireless sensor network. In: 11th IEEE International Conference on Embedded and Real-Time Computing Systems and Applications (RTCSA), pp. 102–108 (2005)
2. Bhattacharya, B.K., Burmester, B., Hu, Y., Kranakis, E., Shi, Q., Wiese, A.: Optimal movement of mobile sensors for barrier coverage of a planar region. Theor. Comput. Sci. **410**(52), 5515–5528 (2009)

3. Chen, A., Kumar, S., Lai, T.: Local barrier coverage in wireless sensor networks. IEEE Trans. Mob. Comput. **9**(4), 491–504 (2010)
4. Czyzowicz, J., Kranakis, E., Krizanc, D., Lambadaris, I., Narayanan, L., Opatrny, J., Stacho, L., Urrutia, J., Yazdani, M.: On minimizing the sum of sensor movements for barrier coverage of a line segment. In: Nikolaidis, I., Wu, K. (eds.) ADHOC-NOW 2010. LNCS, vol. 6288, pp. 29–42. Springer, Heidelberg (2010). https://doi.org/10.1007/978-3-642-14785-2_3
5. Dobrev, S., Durocher, S., Eftekhari, M., Georgiou, K., Kranakis, E., Krizanc, D., Narayanan, L., Opatrny, J., Shende, S., Urrutia, J.: Complexity of barrier coverage with relocatable sensors in the plane. Theor. Comput. Sci. **579**, 64–73 (2015)
6. Huang, C.-F., Tseng, Y.-C.: The coverage problem in a wireless sensor network. Mob. Netw. Appl. **10**(4), 519–528 (2005)
7. Kumar, S., Lai, T.-H., Arora, A.: Barrier coverage with wireless sensors. In: Proceedings of the 11th Annual International Conference on Mobile Computing and Networking (MOBICOM), MobiCom 2005, pp. 284–298. ACM (2005)
8. Meguerdichian, S., Koushanfar, F., Potkonjak, M., Srivastava. M.B.: Coverage problems in wireless ad-hoc sensor networks. In: Proceedings of the 20th Annual Joint Conference of the IEEE Computer and Communications Societies INFOCOM, vol. 3, pp. 1380–1387 (2001)
9. Mestre, J., Gaspers, S., Gudmundsson, J., Rümmele, S.: Barrier coverage with non-uniform length to minimize aggregate movements (2017, submitted manuscript)
10. Tan, X., Wu, G.: New Algorithms for Barrier Coverage with Mobile Sensors. Springer, Heidelberg (2010). pp. 327–338
11. Tao, D., Wu, T.Y.: A survey on barrier coverage problem in directional sensor networks. IEEE Sens. J. **15**(2), 876–885 (2015)
12. Wu, F., Gui, Y., Wang, Z., Gao, X., Chen, G.: A survey on barrier coverage with sensors. Front. Comput. Sci. **10**(6), 968–984 (2016)

Rendezvous on a Line by Location-Aware Robots Despite the Presence of Byzantine Faults

Huda Chuangpishit[1,2], Jurek Czyzowicz[1],
Evangelos Kranakis[2(✉)], and Danny Krizanc[3]

[1] Départemant d'informatique, Université du Québec en Outaouais,
Gatineau, QC, Canada
[2] School of Computer Science, Carleton University, Ottawa, ON, Canada
kranakis@scs.carleton.ca
[3] Department of Mathematics and Computer Science, Wesleyan University,
Middletown, CT, USA

Abstract. A set of mobile robots is placed at points of an infinite line. The robots are equipped with GPS devices and they may communicate their positions on the line to a central authority. The collection contains an unknown subset of "spies", i.e., byzantine robots, which are indistinguishable from the non-faulty ones. The set of the non-faulty robots need to rendezvous in the shortest possible time in order to perform some task, while the byzantine robots may try to delay their rendezvous for as long as possible. The problem facing a central authority is to determine trajectories for all robots so as to minimize the time until the non-faulty robots have rendezvoused. The trajectories must be determined without knowledge of which robots are faulty. Our goal is to minimize the competitive ratio between the time required to achieve the first rendezvous of the non-faulty robots and the time required for such a rendezvous to occur under the assumption that the faulty robots are known at the start. We provide a bounded competitive ratio algorithm, where the central authority is informed only of the set of initial robot positions, without knowing which ones or how many of them are faulty. When an upper bound on the number of byzantine robots is known to the central authority, we provide algorithms with better competitive ratios. In some instances we are able to show these algorithms are optimal.

Keywords: Competitive ratio · Faulty · GPS · Line · Rendezvous Robot

1 Introduction

Rendezvous is useful for cooperative control in a distributed system, either when communication between distributed entities is restricted by range limitations or when it is required to speed up information exchanges in a distributed system. It is often presented as a consensus problem in which the agents have to agree

J. Czyzowicz and E. Kranakis—Research supported in part by NSERC Discovery grant.

A. Fernández Anta et al. (Eds.): ALGOSENSORS 2017, LNCS 10718, pp. 70–83, 2017.
https://doi.org/10.1007/978-3-319-72751-6_6

on the meeting point and time (see [28]) where by consensus we mean reaching an agreement regarding a certain quantity of interest that depends on the state of all the agents.

In this paper we consider the following version of the rendezvous problem. A population of mobile robots is distributed at points of an infinite line. The robots are equipped with GPS devices and are able to communicate their initial positions to a central authority. In order to perform some task, that the central authority shall assign to the robots, all of the non-faulty robots need to rendezvous (meet at the same point of the line). For this reason, the robots send to the central authority the coordinates of their positions on the line and the central authority assigns to each of them a route which eventually results in the rendezvous of all robots. A group of robots may attempt the task at any time in order to determine if all of the non-faulty robots have been brought together.

Unfortunately, an adversary has infected the population with "spies" - a collection of byzantine faulty robots, indistinguishable from the original ones, in order to delay the performance of the task for as long as possible. A byzantine robot may fail to report its position, report a wrong position or it may fail to follow its assigned route. Furthermore, a faulty robot may fail to help in performing the required task. As the central authority does not know the identity of the faulty robots it broadcasts travel instructions to all the robots.

We would like to define the strategy resulting in the smallest possible time of the rendezvous of all non-faulty robots. Our goal is to minimize the competitive ratio between the time required to achieve this first rendezvous of the non-faulty robots and the time required for such a rendezvous to occur under the assumption that the faulty robots are known at the start.

1.1 Our Model

A collection of n anonymous robots travel along a Cartesian line with maximum unit speed. Robots are equipped with GPS devices, so each of them is aware of the coordinate of its current position on the line. An unknown subset of f robots may turn out to be faulty. At some point in time, a task is identified that requires the coming together of all of the non-faulty robots at the same point on the line and this fact is broadcast to the robots by a central authority (CA). The robots stop what they are doing and report their positions to the CA. The CA computes trajectories for each of the robots and instructs them how to time their movement.

At this point the robots follow the trajectories provided. The movement of the robots continues until such time as all of the non-faulty robots meet for the first time and are able to perform the task, which ends the algorithm. We assume the time required to attempt the task is negligible in comparison to the time required for the robots to move between points. (As an example, imagine that the robots have chip cards, that are used to open a container carried by all robots. Using a secret-sharing scheme, the container is set to open only if $n - f$ or more of the keys are valid.) A failed attempt at the task may or may not identify those robots that are faulty (caused the attempt to fail). If identified as faulty, a robot need not continue on its trajectory. A successful attempt at the task means that all non-faulty robots are present and this is recognized by them and the central authority.

As stated, we assume that the robots report their correct locations at the beginning of the algorithm. We note that this need only be true of the non-faulty robots as in the worst case the robots could be anywhere and the algorithm must bring together all of them. It is possible that faulty robots may report initial locations that are incorrect and potentially adversely effect the lengths of the trajectories. Of course, this may result in their receiving trajectories that they cannot complete without being detected as faulty by the other robots. But as long as all non-faulty robots complete their trajectories the algorithm must ensure that they meet.

The message to the CA about a robot's position contains the robot's unique identity. We assume that the faulty robots cannot lie about their identity. Consequently, each faulty robot can send only one message about its position, otherwise it will be identified as faulty and ignored. Observe that, as the robot's identity, the CA could use the position communicated by the robots, and thus our approach could be extended to anonymous robots. This would require some extra conditions on the model (e.g., message uniqueness), so, for simplicity, we assume that our robots have unique identifiers.

We also assume that after the initial reporting of their positions, until the reporting of success with the task, there is no further communication between the robots themselves or the robots and the central authority. Again, this need only be true of the non-faulty robots. Any communication by robots during the execution of the trajectories is assumed to come from faulty robots and is ignored.

We note that the requirement of a central authority may be removed by allowing the robots to broadcast their initial positions to all other robots and each computing the same set of trajectories using the same algorithm.

A rendezvous algorithm specifies the trajectories of the robots as a function of time. We assume the robots have sufficient memory to carry out the instructions of the rendezvous algorithm. The competitive ratio of a given algorithm is the ratio of the time it takes the algorithm to enable rendezvous of all non-faulty robots divided by the time it takes the best off-line algorithm, with knowledge of which robots are faulty, to accomplish the same. Note: the time of the offline algorithm equals $D/2$, where D is the minimum diameter of the set of non-faulty robots. Indeed, these non-faulty robots could then meet at the mid point between the most distant ones in the set.

We assume that the task is such that $n - f$ non-faulty robots are necessary and sufficient to perform the task. Under this assumption, the task can be used to determine if all of the non-faulty robots are together. If a group of robots attempts the task and it succeeds, it contains all non-faulty robots. If it fails, then there exist more non-faulty robots outside the group.

Below we present algorithms which have no knowledge of f as well as others where an upper bound on f is provided. Depending upon that knowledge, different algorithms can achieve a better competitive ratio in different situations. We restrict our attention to the nontrivial case where at least two robots must rendezvous, i.e., $f \leq n - 2$.

1.2   Related Work

The mobile agent rendezvous problem has been studied extensively in many topologies (or domains) and under various assumptions on system synchronicity and capabilities of the agents [12,14,16,24] both as a dynamic symmetry breaking problem [30] as well as in operations research [2] in order to understand the limitations of search theory. A critical distinction in the models is whether the agents must all run the same algorithm, which is generally known as the *symmetric rendezvous problem* [3]. If agents can execute different algorithms, generally known as the *asymmetric rendezvous problem*, then the problem is typically much easier, though not always trivial.

Closely related to our research is the work of [9,10]. In [9] the authors study rendezvous of two anonymous agents, where each agent knows its own initial position in the environment, and the environment is a finite or infinite graph or a Euclidean space. They show that in the line and trees as well as in multi-dimensional Euclidean spaces and grids the agents can rendezvous in time $O(d)$, where d is the distance between the initial positions of the agents. In [10] the authors study efficient rendezvous of two mobile agents moving asynchronously in the Euclidean 2d-space. Each agent has limited visibility, permitting it to see its neighborhood at unit range from its current location. Moreover, it is assumed that each agent knows its own initial position in the plane given by its coordinates. The agents, however, are not aware of each other's position. Also worth mentioning is the work of [4] which studies the rendezvous problem of location-aware agents in the asynchronous case and whose proposed algorithm provides a route, leading to rendezvous.

The underlying domain which is traversed by the robots is a continuous curve (in our case an infinite line) and the robots may exploit a particular characteristic, e.g., different identifiers, speeds, or their initial location, to achieve rendezvous. For example, in several papers the robots make use of the fact that they have different speeds, as in the paper [18], as well as in the work on probabilistic rendezvous on a cycle [23]. Rendezvous on a cycle for multiple robots with different speeds is studied in [20], and rendezvous in arbitrary graphs for two robots with different speeds in [25].

There is also related work on gathering a collection of identical memoryless, mobile robots in one node of an anonymous ring whereby robots start from different nodes of the ring and operate in Look-Compute-Move cycles and have to end up in the same node [22], as well as oblivious mobile robots in the same location of the plane when the robots have limited visibility [19].

Fault tolerance has been extensively studied in distributed computing, though failures were usually related to static elements of the environment, like network nodes or links (e.g., see [26,27]), rather than to the mobile components. The unreliability of robots has been studied with respect to inaccurate robots' sensing or mobility devices (cf. [8,21,29]). Problems concerning faulty robots operating in a line environment have been studied in the context of searching in [13] and patrolling [11]. The questions of convergence or gathering involving faulty robots were investigated in [1,5,7,15,17]. To the best of our knowledge the rendezvous problem for location aware robots some of which may be faulty has never been considered by the research community in the past.

1.3 Our Results

Here is an outline of the results of the paper. In Sect. 2 we consider two general rendezvous algorithms for $n > 2$ robots with $f \leq n - 2$ faulty ones. Both algorithms assume no knowledge of the actual value of f and the second algorithm stops as soon as sufficiently many robots are available to perform the task. The competitive ratios of these algorithms are $f + 1$ and 12, respectively. We also prove a lower bound of 2 on the competitive ratio for arbitrary $n > 2$ and $1 \leq f \leq n - 2$. In Sect. 3 we provide algorithms for the case where the central authority possesses some knowledge concerning the number of faulty robots. For the case where the ratio of the number of faulty robots to the total number of robots is strictly less than $1/2$ we provide an optimal algorithm and when this number is strictly less than $2/3$ we give an algorithm that beats the general case algorithms above unless f is known to be less than 5. Next we provide optimal algorithms for the particular cases where $f \in \{1, 2\}$ in Sect. 4. The main result here is the case of $n = 4$ and $f = 2$ where we show the exact value of the competitive ratio is $1 + \phi$, where ϕ is the golden ratio. We end with a discussion of open problems. A full version of the paper can be found in [6].

2 General Results

In this section we present a rendezvous algorithm for n robots f of which are faulty with a competitive ratio of at most $\min\{f + 1, 12\}$. Neither of these algorithms require prior knowledge of f. We also show that the competitive ratio of any rendezvous algorithm is at least 2. We first observe that the assumptions of our model allow us to severely restrict the potential algorithms available to the CA. We can show the following lemma:

Lemma 1. *Consider a rendezvous algorithm A for n robots f of which are faulty with competitive ratio α. There exists a rendezvous algorithm B such that during the execution of B the movement of the robots follow these rules:*

(1) A robot does not change direction between meetings with other robots.
(2) The robots always move at full speed.

Moreover, the competitive ratio of B is less or equal to α.

We assume throughout the paper that the movement of the robots in any rendezvous algorithm follows rules (1) and (2) of Lemma 1.

2.1 Upper Bounds

The first rendezvous algorithm we present has a competitive ratio which is bounded above by the number of faulty robots plus one. It is interesting to note, that to obtain such competitive ratio no knowledge of the number of faulty robots is necessary. The idea of the algorithm can be summarized as follows. Consider the distances between consecutive robots on the line. The algorithm shrinks the shortest interval (between consecutive robots) in that the two robots at its endpoints meet at its midpoint while the rest of the robots "follow the shrinkage" depending on their location until all non-faulty robots meet (or sufficiently many of them in the case where the task does not require all non-faulty robots to be together to be performed). We prove the following theorem.

Theorem 1. *There is a rendezvous algorithm for $n > 2$ robots at most f of which are faulty whose competitive ratio is at most $f + 1$, where $f \leq n - 2$.*

We now describe a second general approach for rendezvous of n robots, which also works for any number f of faulty robots. Unlike the previous one, this algorithm has a competitive ratio independent of f, (it equals 12). The core of our approach is the Algorithm 1, presented in [9], which guarantees rendezvous of any two robots, at initial integer positions at distance d on the line, in time of at most $6d$. The idea of the algorithm is the following. Each robot gets an integer label corresponding to its initial position. The algorithm consists of a sequence of rounds, each round containing two stages. In the first round, odd-labelled robots move distance $1/2$ to the right in the first stage and then distance 1 to the left in the second stage. The even-labelled robots move distance $1/2$ to the left in the first stage and then distance 1 to the right in the second stage. Observe that each odd-label robot would meet its right neighbour at initial distance 1 in the first stage and its left neighbour at distance 1 in the second stage. At the end of the first round robots are in groups that from now on will travel together.

Algorithm 1. Rendezvous on the infinite line

1: Set $\ell = \frac{1}{2}$.
2: **for all** agents a **do**
3:     Set $label(a) =$ position of a on the line.
4: **for all** $i = 1, 2, 3, \ldots$ **do**
5:     **for all** agent a **do**
6:         **Stage 1.**
7:         **if** $odd(label(a))$ **then**
8:             move right distance ℓ
9:         **else**
10:             move left distance ℓ.
11:         **Stage 2.**
12:         **if** $odd(label(a))$ **then**
13:             move left distance 2ℓ.
14:         **else**
15:             move right distance 2ℓ.
16:         $\ell = 2\ell$
17:         $label(a) = \lfloor \frac{label(a)}{2} \rfloor$

All groups are then at even distances. In round two, the configuration of such groups on the line is scaled up by the factor of two and each group of robots meet neighbouring groups at distance 2 in the two corresponding stages. The process continues inductively and after round i, the groups are at integer positions being multiples of 2^i. It is possible to show that during round i, in its first stage meet all robots initially placed in any interval $[(2k-1)2^i, (2k-1)2^i)$, for some integer k, and in its second stage meet all the robots initially placed in any interval $[(2k)2^i, (2k+2)2^i)$, for some integer k. Let D be minimum diameter of the set of non-faulty robots required to rendezvous, and $i^* = \lceil \log_2 D \rceil$. It easy to see that all the non-faulty robots must meet in the first or the second stage of round $i^* + 1$. Moreover, the total distance travelled by each robot is linear in D. In [9] they show the following:

Theorem 2 ([9]). *For two agents a_1, a_2 starting at distance d (and at integer points) on the line, Algorithm 1 permits rendezvous within at most $6d$ time.*

The following lemma is an immediate consequence of Theorem 2.

Lemma 2. *Let a_1 and a_2 be two robots on the real line with integer starting positions at distance d. Then the rendezvous time of a_1 and a_2 in Algorithm 1 is at most $6d$.*

We now have all the required results to prove an upper bound of 12 on the competitive ratio of rendezvous of n robots f of which are faulty. Our approach is to approximate the initial positions of all robots by other ones which are at rational coordinates. Then the obtained configuration may be scaled up so that all initial robot positions are integers and Algorithm 1 may be applied. We show that for any $\epsilon > 0$ we can choose an approximation fine enough so that the competitive ratio does not exceed $12 + \epsilon$. We have the following theorem.

Theorem 3. *There exists a rendezvous algorithm for $n > 2$ robots, at most $f \le n - 2$ of which are faulty, which guarantees a competitive ratio less than $12 + \epsilon$, for any $\epsilon > 0$.*

As a corollary of Theorems 1 and 3 we can state the following.

Corollary 1. *There is a rendezvous algorithm for $n > 2$ robots at most $f \le n-2$ of which are faulty, with competitive ratio at most $\min\{12 + \epsilon, f + 1\}$, for any $\epsilon > 0$.*

2.2   Lower Bound

Next we show that any rendezvous algorithm for n robots, which include at least one which is faulty, must have a competitive ratio of at least 2.

Theorem 4. *For any $n > 2$ robots, any $1 \le f \le n - 2$ of which are faulty, the competitive ratio of any algorithm that achieves rendezvous of at least $n - f$ non-faulty robots is at least 2.*

Proof. (Theorem 4) Consider the following arrangement of the robots where $n-f$ robots are required to perform rendezvous: $\lceil \frac{f+1}{2} \rceil$ are located at position -1, $n - f - 1$ are located at the origin and $\lfloor \frac{f+1}{2} \rfloor$ are located at position 1. By Lemma 1, we can assume there is an optimal rendezvous algorithm in which all robots move at speed 1 for the first $1/2$ time unit. At that time, at least one of the robots, say r, starting at the origin must be at $-1/2$ or $1/2$. Wlog, assume it is at $-1/2$. Make all of robots starting at the origin non-faulty, one of the robots, say r', starting at 1 non-faulty, and the remaining f robots faulty. In order for $n - f$ non-faulty robots to meet, r and r' must meet which requires at least another $1/2$ time unit, i.e., the competitive ratio of the algorithm is at least 2. □

3   Bounded Number of Faults

In the previous section we proposed algorithms, whose competitive ratio did not depend on the knowledge of the number f of faulty robots. However, employing Corollary 1 to get the competitive ratio which is the best between the values 12 and $f + 1$ (cf. Theorems 1 and 3), we need to have knowledge of an upper bound on f. In this section we show, that having more precise knowledge on an upper bound on f allows us to obtain algorithms with more attractive competitive ratios. More exactly, we provide upper bounds for the competitive ratio of rendezvous algorithms where the number of faulty robots is known to be bounded by a fraction of the total number of robots.

The following theorem shows that if the majority of the robots are non-faulty then there is a rendezvous algorithm whose competitive ratio is at most 2. By Theorem 4, this is optimal.

Theorem 5. *Suppose that $n \geq 3$ and the number of faulty robots is $f \leq \frac{n-1}{2}$. Then there is a rendezvous algorithm with competitive ratio at most 2.*

As a consequence of this result and Theorem 1 we get the following corollary:

Corollary 2. *If the number of faulty robots is strictly less than the number of non-faulty robots then the competitive ratio for solving the rendezvous problem is exactly 2.*

In the sequel we consider the case $\frac{n-1}{2} < f < \frac{2}{3}(n - 1)$ and provide an algorithm that has a better guarantee than the general algorithm as long as our upper bound on f is greater than 4.

Theorem 6. *Suppose that $n \geq 3$ and there are at most f faulty robots. If $f \leq \frac{2}{3}(n - 1)$ then there is a rendezvous algorithm with competitive ratio at most 5.*

In the sequel, we present the proof of Theorem 6. First note that if $n \leq 8$. Then $f \leq \frac{2}{3}(8 - 1) = \frac{14}{3}$, and so $f \leq 4$. Therefore by Theorem 1, there is a rendezvous algorithm with competitive ratio 5. Thus, without loss of generality we can assume that $n \geq 9$.

Lemma 3. *Let $n \geq 9$ and $f < \frac{2}{3}(n-1)$ then there is a partition of the robots into three groups G_L, G_M, and G_R such that at least two of the groups G_L, G_M, and G_R contain a non-faulty robot.*

We are now ready to prove Theorem 6. We present a rendezvous algorithm for the case $f < \frac{2}{3}(n-1)$ whose competitive ratio is 5.

Proof. (Theorem 6) Let $f < \frac{2}{3}(n-1)$. As we discussed earlier we may assume that $n \geq 9$, as for the case $n \leq 8$ we obtain a competitive ratio of 5 by Theorem 1. Therefore we can use Lemma 3 to split the robots into three groups G_l, G_M, and G_R. Consider the following rendezvous algorithm:

Algorithm 2.

1: The robots broadcast their coordinates, and split into three groups as follows.
2:     G_L contains the $\lfloor \frac{n}{2} \rfloor - k - 1$ leftmost robots
3:     G_R contains the $\lceil \frac{n}{2} \rceil - k - 1$ rightmost robots.
4:     G_M contains the $2k + 2$ middle robots.
5: Let A_l and A_r be the leftmost and the rightmost robots of G_M, respectively. Moreover let m_l and m_r be the initial positions of A_l and A_r, respectively. For the robot A_l, sequence all the other robots based on their distances to A_l such that the robots with shorter distances appear earlier in the sequence, denote the sequence by S_l. Do the same for A_r, and let S_r denotes its corresponding sequence.
6: **while** the rendezvous has not occurred **do**
7:     The robots in G_L move at full speed to the right, and when they meet A_l stick to A_l.
8:     The robots in G_R move at full speed to the left, and when they meet A_r stick to A_r.
9:     The robots in interval $[m_l, \frac{m_l + m_r}{2})$ move towards A_l and when they meet A_l stick to it.
10:     The robots in interval $[\frac{m_l + m_r}{2}, m_r]$ move towards A_r, and when they meet A_r stick to it.
11:     The robot A_l moves to the robot next in the sequences S_l, until it meets A_r. Then it sticks to A_r.
12:     The robot A_r moves to the robot next in the sequences S_r, until it meets A_l. Then it sticks to A_l.
13:     When A_l and A_r meet they stick to each other. Then they sequence the robots based on their distances to the location of their meeting in such a way that the robots closer to the meeting point appear earlier in the sequence. Denote the sequence by S. The robots $A_l \cup A_r$ move to the next robot in the sequence S.

We now analyze the competitive ratio of the above algorithm. As seen in Fig. 1, define

- B_l: the rightmost robot in $[m_l, \frac{m_l + m_r}{2})$.
- B_r: the leftmost robot in $[\frac{m_l + m_r}{2}, m_r]$.
- C_l: the last robot in G_L that A_l meets before A_l moves to visit B_r.

- C_r: the last robot in G_R that A_r meets before A_r moves to visit B_l.
- d_1: the distance between A_l and B_l.
- d_2: the distance between A_r and B_r.
- d_3: the distance between A_l and C_l.
- d_4: the distance between A_r and C_r.
- x: the distance between A_l and A_r.

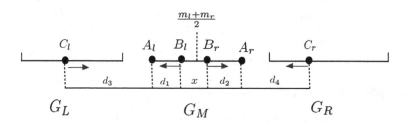

Fig. 1. Robot movement and the analysis of Algorithm 2.

The following inequalities follow immediately.

(1) A_l meets C_l before B_r: $d_3 \leq d_1 + x$.
(2) A_r meets C_r before B_l: $d_4 \leq d_2 + x$.
(3) Without loss of generality assume that $d_1 \leq d_2$.

Let M_l be the group of the robots which stick to A_l before a meeting with A_r, see Fig. 2. More precisely M_l contains C_l and all the robots to the right of C_l, and B_l and all the robots to the left of B_l. Similarly define M_r to be the group of the robots which stick to A_r before a meeting with A_l. Then M_r contains C_r and all the robots to its left, and B_r and all the robots to its right.

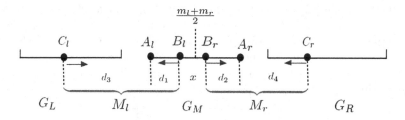

Fig. 2. Robot movement and the analysis of Algorithm 2.

Consider the following three cases:

Case 1. The rendezvous occurs among M_l or M_r: Without loss of generality assume that the rendezvous occurs among M_L. This implies that the non-faulty robots belong to G_L and the interval $[m_l, \frac{m_l+m_r}{2})$. Let a_l be the leftmost non-faulty robot of G_L and a_r be the rightmost non-faulty robot of $[m_l, \frac{m_l+m_r}{2})$. The rendezvous of Algorithm 2 occurs when A_l meets both a_r and a_l. Suppose that δ_1 and δ_2 are the distances between A_l, a_r and A_l, a_l respectively. Since A_l moves towards the closest robots then the rendezvous occurs at the time at most $\frac{3}{2}\max\{\delta_1, \delta_2\}$. Moreover $\max\{\delta_1, \delta_2\}$ is bounded above by the diameter of non-faulty robots. Therefore in this case the competitive ratio is at most 3.

Case 2. The rendezvous occurs at the time of the meeting of M_l and M_r: The meeting of M_l and M_r occurs when the robots C_l and C_r meet. The robot C_l moves to the right and the robot C_r moves to the left, and thus their meeting occurs at time

$$\frac{d_1 + d_2 + d_3 + d_4 + x}{2}.$$

By Inequalities (1), (2) and (3) we have

$$d_1 + d_2 + d_3 + d_4 + x \leq 2d_1 + 2d_2 + 3x$$
$$= 4d_1 + 2x_1 + 3x$$
$$= 4(d_1 + x) - x + 2x_1$$

By Lemma 3 we know that at least two of G_L, G_M, and G_R contain non-faulty robots. This implies that the diameter of the non-faulty robots, D, is at least $\min\{d_1+x+z, d_2+x+y\}$. By Inequality (3) we have that $D \geq d_1+x$. Therefore

$$CR \leq \frac{4(d_1 + x) - x + 2x_1}{d_1 + x}$$
$$= 4 + \frac{x_1 - x_2}{d_1 + x} \leq 5$$

Case 3. The rendezvous occurs after the meeting of M_l and M_r: This case occurs if there are non-faulty robots either to the left of C_l or to the right of C_r. First assume that there are non-faulty robots both to the left of C_l and to the right of C_r. Then the rendezvous occurs in at most two times the diameter of non-faulty robots. Therefore the competitive ratio is at most 2. Now suppose without loss of generality that there are only non-faulty robots to the right of C_r, and the rightmost non-faulty robot is R at distance δ from C_r. Then at the time of the meeting of M_l and M_r the distance between the robots in $M_l \cup M_r$ and R is δ. So it takes at most $\frac{3}{2}\delta$ for $M_l \cup M_r$ to meet R. Therefore the rendezvous occurs at time

$$\frac{d_1 + d_2 + d_3 + d_4 + x}{2} + \frac{3\delta}{2}$$

while the optimal time is at least $\frac{d_4+\delta}{2}$.

Note that A_r meets A_l before R, and thus $d_1 + d_2 + x \leq d_4 + \delta$. Moreover $d_3 \leq d_2 + x$. Therefore

$$\frac{d_1 + d_2 + d_3 + d_4 + x}{2} + \frac{3\delta}{2} \leq \frac{3(d_4 + \delta)}{2}$$

This implies that the competitive ratio in this case is at most 3.   □

4   Optimal Rendezvous Algorithms for at Most Two Faulty Robots

This section is dedicated to the study of optimal rendezvous algorithms when the number of faulty robots is small, i.e., for $f \in \{1, 2\}$.

The next theorem yields the competitive ratio for $f = 1$ fault and is an immediate consequence of Theorems 1 and 4.

Theorem 7. *For $n > 2$ robots with $f = 1$ faulty, the competitive ratio of the algorithm which shrinks the shortest interval is 2, and this is optimal.*

It remains to consider the competitive ratio for $f = 2$ faulty robots. By Corollary 2, the competitive ratio of the rendezvous problem for $n \geq 5$ robots with two faulty is exactly 2. Therefore the only unknown case concerning two faulty robots is when $n = 4$. In this section we present a rendezvous algorithm with optimal competitive ratio $1 + \phi$, where $\phi = \frac{1+\sqrt{5}}{2}$ is the golden ratio. We summarize the main result in the following two theorems. For the lower bound we prove:

Theorem 8. *Consider four robots exactly two of which are faulty. No rendezvous algorithm can have competitive ratio less than $1 + \phi$, where ϕ is the golden ratio.*

The proof of Theorem 8 is based on an exhaustive analysis and considers the competitive ratio of any potential algorithm solving the rendezvous problem for the four robots. For the upper bound we prove:

Theorem 9. *Consider four robots exactly two of which are faulty. There is a rendezvous algorithm for four robots two of which are faulty with competitive ratio at most $1 + \phi$, where ϕ is the golden ratio.*

The proof of Theorem 9 is a continuation of the proof of Theorem 8 leading to a specific algorithm whose competitive ratio is optimal for the rendezvous problem considered.

5   Conclusion

In this paper we considered the rendezvous problem for $n > 2$ robots on a line with $1 \leq f \leq n-2$ among them byzantine faulty. The robots were equipped with GPS devices and they could communicate their positions to a central authority. We designed several rendezvous algorithms and considered their competitive ratio depending on the knowledge the central authority has about the number of faulty robots. An interesting question remaining might be to improve the competitive of the algorithms presented. Another question concerns the model presented here which ignores any communication beyond the broadcasting of the initial positions of the robots. It might be of interest to consider algorithms in a "richer" communication model where the robots may broadcast information as they follow their trajectories. For example, one could consider a model where the faulty robots may crash and non-faulty robots may report not meeting them when expected.

References

1. Agmon, N., Peleg, D.: Fault-tolerant gathering algorithms for autonomous mobile robots. SIAM J. Comput. **36**(1), 56–82 (2006)
2. Alpern, S.: The rendezvous search problem. SIAM J. Control Optim. **33**(3), 673–683 (1995)
3. Alpern, S.: Rendezvous search: a personal perspective. Oper. Res. **50**(5), 772–795 (2002)
4. Bampas, E., Czyzowicz, J., Gąsieniec, L., Ilcinkas, D., Labourel, A.: Almost optimal asynchronous rendezvous in infinite multidimensional grids. In: Lynch, N.A., Shvartsman, A.A. (eds.) DISC 2010. LNCS, vol. 6343, pp. 297–311. Springer, Heidelberg (2010). https://doi.org/10.1007/978-3-642-15763-9_28
5. Bouzid, Z., Potop-Butucaru, M.G., Tixeuil, S.: Optimal byzantine-rezilient convergence in uni-dimensional robot network. Theor. Comput. Sci. **411**(34–36), 3154–3168 (2010)
6. Chuangpishit, H., Czyzowicz, J., Kranakis, E., Krizanc, D.: Rendezvous on a line by location-aware robots despite the presence of byzantine faults (2017). https://arxiv.org/pdf/1707.06776.pdf
7. Cohen, R., Peleg, D.: Convergence properties of the gravitational algorithm in asynchronous robot systems. SIAM J. Comput. **41**(1), 1516–1528 (2005)
8. Cohen, R., Peleg, D.: Convergence of autonomous mobile robots with inaccurate sensors and movements. SIAM J. Comput. **38**(1), 276–302 (2008)
9. Collins, A., Czyzowicz, J., Gąsieniec, L., Kosowski, A., Martin, R.: Synchronous rendezvous for location-aware agents. In: Peleg, D. (ed.) DISC 2011. LNCS, vol. 6950, pp. 447–459. Springer, Heidelberg (2011). https://doi.org/10.1007/978-3-642-24100-0_42
10. Collins, A., Czyzowicz, J., Gąsieniec, L., Labourel, A.: Tell me where I am so I can meet you sooner. In: Abramsky, S., Gavoille, C., Kirchner, C., Meyer auf der Heide, F., Spirakis, P.G. (eds.) ICALP 2010. LNCS, vol. 6199, pp. 502–514. Springer, Heidelberg (2010). https://doi.org/10.1007/978-3-642-14162-1_42
11. Czyzowicz, J., Gąsieniec, L., Kosowski, A., Kranakis, E., Krizanc, D., Taleb, N.: When patrolmen become corrupted: monitoring a graph using faulty mobile robots. In: Elbassioni, K., Makino, K. (eds.) ISAAC 2015. LNCS, vol. 9472, pp. 343–354. Springer, Heidelberg (2015). https://doi.org/10.1007/978-3-662-48971-0_30
12. Czyzowicz, J., Kosowski, A., Pelc, A.: Deterministic rendezvous of asynchronous bounded-memory agents in polygonal terrains. Theory Comput. Syst. **52**(2), 179–199 (2013)
13. Czyzowicz, J., Kranakis, E., Krizanc, D., Narayanan, L., Opatrny, J.: Search on a line with faulty robots. In: Proceedings of the 2016 ACM Symposium on Principles of Distributed Computing, PODC 2016, Chicago, IL, USA, 25–28 July 2016, pp. 405–414 (2016)
14. De Marco, G., Gargano, L., Kranakis, E., Krizanc, D., Pelc, A., Vaccaro, U.: Asynchronous deterministic rendezvous in graphs. Theor. Comput. Sci. **355**(3), 315–326 (2006)
15. Défago, X., Gradinariu, M., Messika, S., Raipin-Parvédy, P.: Fault-tolerant and self-stabilizing mobile robots gathering. In: Dolev, S. (ed.) DISC 2006. LNCS, vol. 4167, pp. 46–60. Springer, Heidelberg (2006). https://doi.org/10.1007/11864219_4
16. Dessmark, A., Fraigniaud, P., Kowalski, D., Pelc, A.: Deterministic rendezvous in graphs. Algorithmica **46**, 69–96 (2006)

17. Dieudonné, Y., Pelc, A., Peleg, D.: Gathering despite mischief. ACM Trans. Algorithms (TALG) **11**(1), 1 (2014)

18. Feinerman, O., Korman, A., Kutten, S., Rodeh, Y.: Fast rendezvous on a cycle by agents with different speeds. In: Chatterjee, M., Cao, J., Kothapalli, K., Rajsbaum, S. (eds.) ICDCN 2014. LNCS, vol. 8314, pp. 1–13. Springer, Heidelberg (2014). https://doi.org/10.1007/978-3-642-45249-9_1

19. Flocchini, P., Prencipe, G., Santoro, N., Widmayer, P.: Gathering of asynchronous robots with limited visibility. Theor. Comput. Sci. **337**(1), 147–168 (2005)

20. Huus, E., Kranakis, E.: Rendezvous of many agents with different speeds in a cycle. In: Papavassiliou, S., Ruehrup, S. (eds.) ADHOC-NOW 2015. LNCS, vol. 9143, pp. 195–209. Springer, Cham (2015). https://doi.org/10.1007/978-3-319-19662-6_14

21. Izumi, T., Souissi, S., Katayama, Y., Inuzuka, N., Défago, X., Wada, K., Yamashita, M.: The gathering problem for two oblivious robots with unreliable compasses. SIAM J. Comput. **41**(1), 26–46 (2012)

22. Klasing, R., Markou, E., Pelc, A.: Gathering asynchronous oblivious mobile robots in a ring. Theoret. Comput. Sci. **390**(1), 27–39 (2008)

23. Kranakis, E., Krizanc, D., MacQuarrie, F., Shende, S.: Randomized rendezvous algorithms for agents on a ring with different speeds. In: Proceedings of the 2015 International Conference on Distributed Computing and Networking, ICDCN 2015, Goa, India, 4–7 January 2015, pp. 9:1–9:10 (2015)

24. Kranakis, E., Krizanc, D., Markou, E.: The Mobile Agent Rendezvous Problem in the Ring: An Introduction. Synthesis Lectures on Distributed Computing Theory Series. Morgan & Claypool Publishers, San Rafael (2010)

25. Kranakis, E., Krizanc, D., Markou, E., Pagourtzis, A., Ramírez, F.: Different speeds suffice for rendezvous of two agents on arbitrary graphs. In: Steffen, B., Baier, C., van den Brand, M., Eder, J., Hinchey, M., Margaria, T. (eds.) SOFSEM 2017. LNCS, vol. 10139, pp. 79–90. Springer, Cham (2017). https://doi.org/10.1007/978-3-319-51963-0_7

26. Lamport, L., Shostak, R., Pease, M.: The byzantine generals problem. ACM Trans. Program. Lang. Syst. (TOPLAS) **4**(3), 382–401 (1982)

27. Lynch, N.A.: Distributed Algorithms. Morgan Kaufmann, Burlington (1996)

28. Olfati-Saber, R., Fax, J.A., Murray, R.M.: Consensus and cooperation in networked multi-agent systems. Proc. IEEE **95**(1), 215–233 (2007)

29. Souissi, S., Défago, X., Yamashita, M.: Gathering asynchronous mobile robots with inaccurate compasses. In: Shvartsman, M.M.A.A. (ed.) OPODIS 2006. LNCS, vol. 4305, pp. 333–349. Springer, Heidelberg (2006). https://doi.org/10.1007/11945529_24

30. Yu, X., Yung, M.: Agent rendezvous: a dynamic symmetry-breaking problem. In: Meyer, F., Monien, B. (eds.) ICALP 1996. LNCS, vol. 1099, pp. 610–621. Springer, Heidelberg (1996). https://doi.org/10.1007/3-540-61440-0_163

Querying with Uncertainty

Huda Chuangpishit[1], Kostantinos Georgiou[1], and Evangelos Kranakis[2(✉)]

[1] Department of Mathematics, Ryerson University, Toronto, ON, Canada
[2] School of Computer Science, Carleton University, Ottawa, ON, Canada
kranakis@scs.carleton.ca

Abstract. We introduce and study a new optimization problem on *querying with uncertainty*. k robots are required to locate a hidden item that is placed uniformly at random in one of n different locations, each associated with a probability p_i, $i = 1, \ldots, n$. If the item is placed in location i, a query trial by any of the robots reveals the item with probability p_i. Each robot j is assigned a subset A_j of the locations, and is allowed to perform a random walk among them, each time step querying the current location (being visited) for the item. We are interested in determining sets $\{A_j\}_{j=1,\ldots,k}$ so as to minimize the expected discovery time of the item. We measure the cost by the number of queries, while there is no cost for hopping from node to node.

Our first contribution is to prove a closed formula for the expected number of steps until the treasure is found when the robots execute unanimous queries. Then we focus on querying problems where the sets A_j are restricted to be either pairwise disjoint or identical. Our findings allow us to obtain optimal solutions, when sets A_j are exclusively pairwise disjoint, requiring time $n^{O(k)}$. In our second contribution, we devise an optimal polynomial time algorithm for querying with $k = 2$ robots even when the sets A_1, A_2 are allowed to overlap. All our algorithms are based on special concavity-type properties of the expected termination time when the robots execute unanimous queries, thus inducing special structural properties of optimal solutions for the general problem.

Keywords: Searching · Querying · Random walk · Partition
Assignment · Optimization

1 Introduction

Search in computer science aims to retrieve information which is stored in a given data structure. It involves an algorithm which describes the "trajectories" of each of the searchers (in the present paper they will be autonomous robots) within the data structure and a target representing the information and whose location–although unknown to the searcher(s)–can be recognized by the searchers

K. Georgiou and E. Kranakis—Research supported in part by NSERC Discovery grant.

© Springer International Publishing AG 2017
A. Fernández Anta et al. (Eds.): ALGOSENSORS 2017, LNCS 10718, pp. 84–97, 2017.
https://doi.org/10.1007/978-3-319-72751-6_7

when located (by a robot). There is vast literature on search algorithms (both deterministic and randomized) depending on the type of data structure or search domain, capabilities of the robots, and properties of the target of which it is worth mentioning the books [2, 4, 21].

In search, there are numerous situations when the searchers (be that humans or machines) have to make decisions whose outcomes may not be under the searcher's control. For example, when a drone needs to make routing decisions (in military strategy planning), or a guidance system must consult a forwarding database. These kinds of problems have been considered in artificial intelligence and typically involve some form of *combinatorial search* in which one is look-ing for a specific sub-structure of the given discrete structure which achieves a solution of the search problem.

In this paper we investigate the structure of search in an environment involv-ing uncertainties both in the actual location of the target as well as in recognizing the target. Assume there are n different "black" boxes and that a treasure is hid-den uniformly at random at one of them. Searchers may query the boxes to find the treasure that will be revealed with a certain probability (which depends on the box) but only if it is located there. We are interested in designing search algorithms for finding the treasure. As it is standard practice in this area, when uncertainty is involved a natural approach is to look for algorithms that have optimal behaviour, where this is measured either by the probability of success or the expected number of steps until the target is identified. In order to make the model more interesting, we also assume that each robot may only be assigned a specific subset of the boxes that it will be investigating (which will be part of the algorithmic decisions toward solving the underlying optimization problem). As we want to keep robots capabilities down to minimal, we also assume that robots may only perform random walks over the assigned subset of boxes.

1.1   Search Model

We describe in the sequel the essential aspects of our search model involving the domain being searched, the searchers, and the target.

There are n nodes (boxes) labeled $\{1, \ldots, n\} = [n]$ and a treasure is hid-den uniformly at random at one of them. These labels uniquely identify the nodes which can be distinguished as such by the robots. We are concerned with querying nodes using robots. There are k identical robots and each robot j is assigned a subset of the nodes A_j, called *query domain*, and this subset will usually be specified by the search algorithm. Note that the robots are allowed to share (i.e., search) the same query domain or part thereof. The agents query the nodes (as specified by their query domain) synchronously, independently and uniformly at random. For this, we assume that robots have identical clocks, and computation proceeds in rounds. Each round of querying consists of three steps: a robot (a) visits a node, (b) queries the node for the treasure, and (c) jumps to a new node. All that counts as one unit. Robots are also allowed to query the same node simultaneously at one unit time. Success is accomplished when one robot finds the treasure. Throughout this paper we use interchangeably the terms autonomous agents, robots, and searchers.

We distinguish three basic types of query strategies depending on the allowed intersection of the domain of nodes being queried by the robots.

- *Unanimous query:* A query strategy in which the query domain of all the agents consists of the entire set $[n]$ of nodes.
- *Partition query:* A query strategy in which every two query domains either coincide or are disjoint, and therefore the maximal collection of pairwise disjoint domains forms a partition of the set $[n]$. We further distinguish two more subtypes. When all domains are pairwise disjoint, we call the strategy *pure-partition query.* When all domains are of the same size, say n/t, and the same number of robots (k/t) perform in each domain, we call the strategy *t-uniform.* We refer to the collection of all t-uniform strategies as *uniform-partition.*
- *Overlap query:* A query strategy in which the union of the query domains of all the agents contains the entire set $[n]$ of nodes and query domains are allowed to have non empty intersection, (i.e. domain intersections and symmetric differences may be non-empty).

Note that the condition $\cup_{i=1}^{k} A_i = [n]$ only guarantees that the expected termination time is finite. We also adopt a model in which there is uncertainty not only in locating but also in identifying the target. Depending on the algorithm being considered, searches at a node may be executed by alternate robots. As such, robots may have to query a given node repeatedly so as to find (or identify) the target, note however that success is assured (at an unknown number of steps) only if the target is located at the node being queried. This gives rise to *the probability of success at node i:* Given that the treasure is at node i, the probability of success is denoted by p_i and it is equal to the probability that a robot successfully finds the treasure at the node i. The probabilities p_i are given as input to our problem. Moreover, the *expected search time* is the expected number of queries the robots perform until the treasure is found by any of the robots.

An interesting feature of our model is the uncertainty inherent in the fact that a query at a node, say i, may fail not only because the treasure is not located at node i but also because the probability that a query succeeds is only a certain probability $p_i > 0$ which depends on the node i. It follows that a robot cannot obtain any advantage in locating the treasure only by knowing the label of a node unless it can possibly have additional knowledge of the relative magnitude of the probability of success p_i at node i with respect to the probabilities p_j, for all other nodes $j \neq i$. Therefore a natural search approach is for the robots to execute random walks in which this type of knowledge may be exploited; for additional details see Sects. 3 and 4.

On input $p_1, \ldots, p_n$, any feasible solution to our search problem will be determined by the query domains $A_1, \ldots, A_k$ (also referred to as the *search strategy*) for the k robots, either in the unanimous, or the partition or the overlap variation of the problem. Finally, an *optimal solution* to our search problem is a search strategy $\mathcal{A}$ for querying n nodes with k robots such that the expected time until the treasure is found (for the first time) is minimized. Note that in our model the robots incur costs only for querying (with each query costing one unit); thus there is no cost when hopping from node to node.

1.2 Related Work

Search involves finding a target which is placed at an unknown location of a search domain while *searchers* (*autonomous mobile robots*) can move with certain maximum speed within the domain. One usually wants to minimize the search time required so that a robot finds the target. Searching has been studied extensively both in graph theoretic [15] as well as geometric settings [13].

Various search strategies have been studied involving static or moving targets, in discrete and continuous environments with or without knowledge of the location of one or several targets. A large number of such problems under various models has been considered and analyzed in the seminal book [2]. An important variant of the search problem is when searching for a target in an infinite line (known as linear search) [5]. Linear search is also known as a *single-lane cow-path* problem, as opposed to the *cow-path* problem where the target may be located in one of many possible paths. Optimal randomized algorithms for the cow-path problem can be found in [18,19]. Additional stochastic and game theoretic approaches to the search problem can be found in [3,7] as well as in the seminal papers of Beck [6] and Bellman [8] (and subsequent papers thereof) in which the authors attempted to minimize the competitive ratio in a stochastic setting.

A new line of research is emerging in which searchers may cooperate by exchanging messages and variants of linear search are being studied in a distributed setting. Two communication models that have been considered so far include wireless and face-to-face (F2F). In the former, the robots can communicate anywhere and anytime regardless of their distance, while in the latter they can communicate only if they are at the same location, at the same time. For example, [10] considered evacuation (this is linear group search, when the process is completed when the target is reached by the last robot visiting it) in the F2F communication model and proved that the competitive ratio of search is still bounded from below by 9. More interestingly, linear search for a team of cooperating robots where some fraction of the robots may exhibit either *crash faults* or where some robots may exhibit Byzantine faults have recently been studied in [12] and [11], respectively.

Most related to our current paper involves studies on *searching with uncertainty*. This type of search usually involves making decisions under an adversary (usually random) who is not under the searchers' control; this has been studied in various search domains and under various memory requirements for the robots. For example, in [20] the location of the target in the network is unknown, but information about its whereabouts can be obtained by querying the nodes (of the network), while [17] investigates memoryless search algorithms in a network with faulty advice. In [16] the authors study the problem of finding a destination node by a mobile agent in an unreliable network having the structure of an unweighted graph in which nodes of the network are able to give advice. Further, [1] studies nearest-neighbor queries in a probabilistic framework in which the location of each input node is specified as a probability distribution function. Finally, [14] studies memory lower bounds for randomized collaborative search and implications for biology (in the context of the Ants Nearby Treasure Search).

1.3 Outline and Results of the Paper

In Sect. 2 we are concerned with robots executing unanimous queries; (a) we prove an estimate on the number of queries to succeed with high probability (Theorem 1), and (b) prove a closed formula for the expected number of steps until the treasure is found (Theorem 2). This formula will prove important in our later analysis which involves partition queries in Sect. 3 whereby the search domains assigned to the robots form a partition of the entire domain. In Subsect. 3.1 we further restrict our attention on uniform-partition queries. The main results of this section are as follows; (c) a monotonicity result regarding uniform query strategies that indicates a divide-and-conquer phenomenon according to which, at a high level, it is more beneficial to partition the collection of nodes to as small search domains as possible (Theorem 3). The latter implies also a sublinear algorithm for determining the optimal uniform query strategy (Corollary 1), and an interesting structural property of optimal solutions to optimal partition query strategies (Corollary 2). Then in Subsect. 3.2 we study pure-partition query strategies, and we show that (d) optimal pure-partition strategies of n nodes with k robots can be obtained efficiently for every constant k (Theorem 4). As a corollary, we also show that (e) optimal partition strategies of even many nodes and 2 robots can be obtained in polynomial time (Theorem 6). In Sect. 4 we allow the robots to perform queries with overlapping domains, and we show (f) an optimal overlap query strategy of n nodes and 2 robots (Theorem 7). In Sect. 5 we briefly comment on the challenges toward generalizing our results. Finally, in Sect. 6 we provide the conclusion and discuss possible open problems for further research.

2   Unanimous Querying

In this section we are concerned with robots executing unanimous queries. First, we determine in Theorem 1 the number of such queries which are needed so that the treasure will be found by anyone of the k robots with high probability while in Theorem 2 we find an exact formula for the expected number of steps until the treasure is found.

Lemma 1. *Consider the problem of querying n nodes with k robots. Let r be a positive integer. If k robots perform unanimous querying for a time at least $(r - 1 + 3n \ln n)/k$ then every node will be queried at least r times, with high probability, i.e. at least $1 - \frac{1}{n}$.*

Proof (Lemma 1). Define $m := \lceil (r - 1 + 3n \ln n)/k \rceil$ and consider the first m steps of k synchronous robots, performing independent, random queries of the nodes. Let E_i be the event that node i has been queried less than r times. Define the event E that at the end of the m-th step of the k querying robots, some node i, where $1 \leq i \leq n$, has been queried less than r times. We estimate the probability of the event E. Using the union bound, we obtain that $\Pr[E] = \Pr\left[\bigcup_{i=1}^{n} E_i\right] \leq \sum_{i=1}^{n} \Pr[E_i]$. Consider the event E_i. Observe that k robots in m steps query the nodes at least mk times (not necessarily different nodes), where $r - 1 + 3n \ln n \leq mk \leq r - 1 + 3n \ln n + k$. Let $q := 1 - 1/n$. The probability that a given node i has been queried less than r times in mk queries of the robots is equal to $\sum_{j=0}^{r-1} q^{mk-j}(1-q)^j = q^{mk-r+1} \sum_{j=0}^{r-1} q^{r-1-j}(1-q)^j \leq q^{mk-r+1} \sum_{j=0}^{r-1} q^{r-1-j} = q^{mk-r+1}\frac{1-q^r}{1-q}$. It follows that

$$\Pr[E_i] \leq \left(1 - \frac{1}{n}\right)^{mk-r+1} \frac{1-q^r}{1-q} \leq n\left(1 - \frac{1}{n}\right)^{mk-r+1} \leq n\left(1 - \frac{1}{n}\right)^{3n \ln n}$$

since $mk - r + 1 \geq 3n \ln n$. Thus, $\Pr[E] \leq n^2 \left(1 - \frac{1}{n}\right)^{3n \ln n} \leq \frac{1}{n}$, for every $n \geq 1$. As a consequence, we conclude that k robots querying the n nodes independently, synchronously and randomly m times will query every node at least r times, with high probability ($\geq 1 - \frac{1}{n}$). $\qquad\square$

Using Lemma 1 we can prove the following theorem concerning the search time by k robots to find the treasure.

Theorem 1. *Consider the problem of querying n nodes with k robots. If k robots perform a unanimous query at least*

$$\frac{1}{k}\left(3n \ln n - 1 + \max_{i=1,\ldots,n}\left\{\frac{\ln n}{-\ln(1 - p_i)}\right\}\right) \tag{1}$$

times then the treasure will be found by one of the robots with high probability, i.e., at least $1 - \frac{1}{n}$.

Proof (Theorem 1). Let N_i be the random variable that counts the number of queries that the k robots make to node i before the treasure is found, where $1 \leq i \leq k$. As a consequence of Lemma 1 we obtain that if k robots query m times and $mk \geq r - 1 + \lceil 3n \ln n \rceil$ then $\Pr[\forall i (N_i \geq r)] \geq 1 - \frac{1}{n}$.

Let the treasure be placed at a node uniformly at random with probability $1/n$. If the treasure is located at node i, then the probability that a query succeeds is p_i. The probability that a robot finds the treasure in at most r queries is $\sum_{j=0}^{r-1} p_i(1 - p_i)^j = 1 - (1 - p_i)^r$. To ensure that the treasure is found at node i with high probability, we require that $1 - (1 - p_i)^r \geq 1 - 1/n$. This means that we must guarantee that $r \geq \frac{\ln n}{-\ln(1-p_i)}$. By selecting $r \geq \max_{i=1,\ldots,n}\left\{\frac{\ln n}{-\ln(1-p_i)}\right\}$ we can ensure that the treasure will be found with high probability regardless of where it is placed. $\qquad\square$

Next we compute the expected search time for the treasure to be found by a robot. The resulting formula will prove vital to our subsequent investigations.

Theorem 2. *Assume that k robots perform a unanimous query on n nodes. Then the expected number of steps until the treasure is found is given by the formula*

$$\frac{1}{n}\sum_{i=1}^{n}\frac{1}{1-(1-p_i/n)^k},\tag{2}$$

where for each node i, p_i is the probability that a query (by a robot) at node i will succeed to find the treasure.

Proof (Theorem 2). Let E_i denote the event that the treasure is located at node i. It is clear that $\Pr[E_i] = 1/n$. Let the random variable T_l count the number of steps until the l-th robot finds the treasure and define $T := \min_{l=1,\ldots,k} T_l$ to be the number of steps until the first robot (among the k) finds the treasure. Observe that $E[T] = \frac{1}{n}\sum_{i=1}^{n} E[T|E_i]$. Also, observe that $\Pr[T_l > s|E_i]$ is equal to the probability that the l-th robot does not find treasure after s attempts, given that the treasure is at node i. Therefore we have that

$$\Pr[T_l > s|E_i] = \Pr[l\text{-th robot does not find treasure after } s \text{ attempts}|E_i]$$

$$= \sum_{j=0}^{s} \Pr[l\text{-th robot has visited treasure } j \text{ times unsuccessfully}|E_i]$$

$$= \sum_{j=0}^{s}\binom{s}{j}\left(\frac{1-p_i}{n}\right)^j\left(1-\frac{1}{n}\right)^{s-j} = \left(1-\frac{1}{n}\right)^s\sum_{j=0}^{s}\binom{s}{j}\left(\frac{1-p_i}{n-1}\right)^j$$

$$= \left(1-\frac{1}{n}\right)^s\left(1+\frac{1-p_i}{n-1}\right)^s = (1-p_i/n)^s\,.$$

It is easy to see that in the summation above, the visitation to the treasure can be realized in $\binom{s}{j}$ different ways, and each of these j times happens and is unsuccessful with probability $(1-p_i)/n$, while for the remaining of the steps you do not visit the treasure, that happens with probability $(1-1/n)$.

$$E[T|E_i] = \sum_{s\geq 0}\Pr[T > s|E_i] = \sum_{s\geq 0}\Pr[\min\{T_l : 1 \leq l \leq k\} > s|E_i]$$

$$= \sum_{s\geq 0}(\Pr[T_1 > s|E_i])^k = \sum_{s\geq 0}(1-p_i/n)^{sk} = \frac{1}{1-(1-p_i/n)^k},$$

where the last equation follows from the fact that the random variables T_l $(1 \leq l \leq k)$ are independent and identical. The formula resulting after combining the identities above is exactly as claimed in (2). □

3   Partition Querying

In this section we study partition solutions for the problem of querying a set of n nodes with k robots. Notably, even when k is a constant, there are exponentially many (in n) possible uniform or pure-partition search strategies. Nevertheless, we show in this section that the best uniform partition and the best pure-partition strategies can both be computed efficiently (the latter, only when k is a constant).

3.1 Uniform-Partition Querying

Recall that in a partition query, the domain $[n]$ is partitioned into a number, say t, of sets $S_1, \ldots, S_t$ and S_i is the query domain of k_i many robots such that $k_1 + \cdots + k_t = k$. Whenever $t|k$ and $t|n$, in every t-partition query, the query domains are of size n/t, and exactly k/t robots search in each domain. We show in the next theorem that the expected search time in a t-uniform query is independent of the partitioning, for any fixed t.

Theorem 3. *The expected search time of querying n nodes with k robots using any t-uniform query strategy is*

$$E_t[T] = \frac{1}{n} \sum_{i=1}^{n} \frac{1}{1 - (1 - tp_i/n)^{\frac{k}{t}}}. \tag{3}$$

Moreover, $E_t[T]$ is strictly decreasing with respect to t.

Proof (Theorem 3). Consider a t-uniform querying of the n nodes induced by the partition $S_1, \ldots, S_t$. recall that $|S_i| = \frac{n}{t}$, for $i = 1, \ldots, t$. Given that the treasure is in S_i, and by Formula (2), the expected termination time is $E[T_i] = \frac{t}{n} \sum_{r \in S_i} \frac{1}{1 - (1 - tp_r/n)^{\frac{k}{t}}}$. The treasure is uniformly placed at one of the nodes, and thus the probability that the treasure is in S_i is $\frac{1}{t}$. Therefore

$$E_t[T] = \sum_{i=1}^{t} \frac{1}{t} E[T_i] = \sum_{i=1}^{t} \frac{1}{t} \frac{t}{n} \sum_{r \in S_i} \frac{1}{1 - (1 - \frac{tp_r}{n})^{\frac{k}{t}}} = \frac{1}{n} \sum_{r=1}^{n} \frac{1}{1 - (1 - \frac{tp_r}{n})^{\frac{k}{t}}},$$

where the last equality follows since the S_i's form a partition of $[n]$.

To prove the monotonicity of $E_t[T]$ we show that each of the summands is decreasing in t, where $1 \le t \le n$. Note that each of the summands is a function of the form $f(t) := \frac{1}{1 - (1 - at)^{b/t}}$, where a, b are positive constants (depended on the index of each summand), independent of t. We have $\frac{df(t)}{dt} = \frac{(1-at)^{b/t}\left(-\frac{b \log(1-at)}{t^2} - \frac{ab}{t(1-at)}\right)}{\left(1 - (1-at)^{b/t}\right)^2}$. For the i-th summand we have $a = p_i/n$, and since $1 \le t \le n$, it follows that $(1 - at)^{b/t}$ above is positive. Hence, the sign of $\frac{df(t)}{dt}$ is fully determined by $-\frac{\log(1-at)}{t} - \frac{a}{1-at}$, which we show shortly that is negative. Indeed, consider function $g(x) = \ln(1 - x) + \frac{x}{1-x}$ for $x \in (0, 1)$. The derivative of $g'(x) = -\frac{1}{1-x} + \frac{1}{(1-x)^2}$, which is positive since $0 < x < 1$. Therefore for all $x \in (0, 1)$, we have $g(x) > \lim_{x \to 0} g(x) = 0$. For the i-th summand we have $a = p_i/n$, and since $1 \le t \le n$ we see that $ax \in (0, 1)$. But then $g(ax) > 0$, which is equivalent to the assertion $-\frac{\log(1-at)}{t} - \frac{a}{1-at} < 0$, as promised. $\square$

There are two immediate implications of Theorem 3.

Corollary 1. *The optimal uniform querying strategy of n nodes with k-robots is a t-uniform querying strategy where $t = \gcd(n, k)$. Hence, the optimal uniform querying strategy can be found in time $\mathcal{O}(\log n)$, and the optimal expected querying time is given by (3).*

Corollary 2. *Consider an optimal partition querying strategy of n nodes with k robots, where the search domains are $S_1, \ldots, S_t$, and $k_1, \ldots, k_t$ many robots are assigned to each domain, respectively. Then for each $i = 1, \ldots, t$, we have that k_i does not divide $|S_i|$, unless $k_i = 1$.*

3.2 Pure-Partition Querying

For any fixed k, we show in this section how to find the optimal pure-partition query strategy of n nodes and k robots. Consider some pure-partition strategy $\{S_i\}_{i=1,\ldots,k}$. Let the random variable X count the number of steps until the treasure is found. Further, let B_i denote the event that the treasure is in one of the vertices of some set S_i, and so $\Pr[B_i] = \frac{|S_i|}{n}$. Using Theorem 3 and since $E[X] = \sum_{i=1}^{k} \Pr[B_i] E[X|B_i]$, the best pure-partition strategy is the one with query domains $S_1, \ldots, S_k$ that minimizes

$$E[X] = \sum_{i=1}^{k} \frac{|S_i|}{n} \sum_{j \in S_i} \frac{1}{p_j} \tag{4}$$

In the language of [9], we just showed that determining the optimal pure-partition query is a *set partition problem with an additive objective (SPAO)*. In such problems a set $[n]$ has to be partitioned into k (fixed) many sets S_i so as to minimize the expression $\sum_{i=1}^{k} g(S_i)$, given some cost-function $g : 2^n \mapsto \mathbb{R}$.

A special family of cost-functions g will play a critical role in our arguments. Suppose that the elements in $[n]$ are each associated with real numbers r_i, and assume w.l.o.g. that $r_1 < r_2 < \ldots < r_n$.

Definition 1. *Cost-function $g : 2^n \mapsto \mathbb{R}$ is called* concave in the subset sum for fixed cardinality of the subset *if there are concave functions $f_1, \ldots, f_n$ such that for any $B \subset A$ with $|B| = i$ we have $g(B) = f_i\left(\sum_{j \in B} r_j\right)$.*

Clearly, by Equation (4), we see that determining the optimal pure-partition query strategy is a SPAO with a cost-function that is concave in the subset sum for fixed cardinality of the subset. In fact, our functions are linear in the sum of $r_j = 1/p_j$. We invoke the following known result.

Theorem 4 [9]. *In every SPAO with cost-function that is concave in the subset sum for fixed cardinality of the subset, there is an optimal partition $S_1, \ldots, S_k$ in which the elements of each S_i are consecutive integers.*

We are now ready to present an efficient algorithm for determining optimal pure-partition strategies.

Theorem 5. *Consider the problem of querying n nodes with k robots. An optimal pure-partition strategy $S_1, \ldots, S_k$ can be found in time $n^{\mathcal{O}(k)}$, and the optimal expected querying time is given by Eq. (4).*

Proof. By Eq. (4), determining the optimal pure-partition query is a SPAO with a cost-function that is concave in the subset sum for fixed cardinality of the subset. Relabel nodes $[n]$ so that $p_1 \geq p_2 \geq \cdots \geq p_n$. According to Theorem 5, there is an optimal partition $S_1, \ldots, S_k$ in which each $S_i \subseteq [n]$ is composed of consecutive integers. Such "consecutive" partitions are as many as the number of positive solutions to the linear Diophantine equation $x_1 + \cdots + x_k = n$, which is $\binom{n-1}{k-1}$. By exhaustively checking them all, we can find the one minimizing the Formula in (4). Note that once we fix the partition, we can arbitrarily assign each domain to any of the robots. $\square$

As an immediate corollary, we see how to solve the general partition query problem of even many nodes and 2 robots.

Theorem 6. *Consider the problem of querying $2n$ nodes with 2 robots. An optimal partition strategy S_1, S_2 can be found in time $\mathcal{O}(n^2)$, and the optimal expected querying time is given by Eq. (4).*

Proof. In partition querying, sets S_1, S_2 either coincide (unanimous querying) or they form a partition of $[2n]$. By Corollary 1, the 2-uniform querying strategy is better than the unanimous. But the 2-uniform querying strategy is just a pure-partition strategy, and using Theorem 5 we can determine the best among them in time $\mathcal{O}(n^2)$.

4   Overlap Querying with 2 Robots

In this section we consider the problem of querying n nodes with two robots using overlap query strategies and we show the following result.

Theorem 7. *An optimal overlap query strategy of n nodes with 2 robots can be found in $\mathcal{O}(n^2)$ steps.*

Assume that nodes $[n]$ are relabeled so that $p_1 \leq p_2 \leq \ldots \leq p_n$. The main idea behind the proof of Theorem 7 is to show that there is always an optimal strategy S_1, S_2 with two special properties; (a) either S_1, S_2 are disjoint or $S_1 \cap S_2 = \{1, \ldots, s\}$, for some $1 \leq s \leq n$ (Lemma 2) and (b) any nodes explored by any of the robots but not the other have consecutive indices (Lemma 3).

Lemma 2. *Consider an optimal overlap query S_1, S_2 of n nodes with two robots. If $s \in S_1 \cap S_2$, then $\{1, \ldots, s\} \subseteq S_1 \cap S_2$.*

Proof (Lemma 2). Suppose that S_1, S_2 is an optimal overlapping query with 2 robots, where $n_1 = |S_1|$ and $n_2 = |S_2|$. Then the expected search time is

$$E[T] = \frac{1}{n} \left[\sum_{i \in S_1} E[T|i] + \sum_{i \in S_2} E[T|i] + \sum_{i \in S_1 \cap S_2} E[T|i] \right], \tag{5}$$

where $E[T|i]$ is the expected search time given the treasure is at the node i. For nodes $i \in (S_1 \cup S_2) \setminus (S_1 \cap S_2)$ only one robot queries the node i. Therefore for $i \in S_j \setminus (S_1 \cap S_2)$, $j \in \{1,2\}$, we have

$$E[T|i] = \frac{1}{1 - (1 - \frac{p_i}{n_j})}. \tag{6}$$

The nodes $i \in S_1 \cap S_2$ are queried by both robots, and so for the nodes inside the overlap we have $E[T|i] = \frac{1}{1-(1-\frac{p_i}{n_1})(1-\frac{p_i}{n_2})}$.

Clearly, if $S_1 = S_2$ or if $S_1 \cap S_2 = \emptyset$, the statement of the Lemma is true. So, we may assume that $S_2 \setminus S_1 \neq \emptyset$ and that S_1, S_2 are not disjoint. Our goal will be to compare two query strategies which differ only in nodes x, y for which $p_x \leq p_y$. More specifically, we show that if $y \in S_1 \cap S_2$ and $x \in S_2 \setminus S_1$, then by switching x, y, the expected termination time does not increase.

So let $E_y[T]$ be the expected search time when $y \in S_1 \cap S_2$ and $x \in S_2 \setminus S_1$, and $E_x[t]$ be the expected search time when $x \in S_1 \cap S_2$ and $y \in S_2 \setminus S_1$. Then

$$E_x[T] - E_y[T] = \frac{1}{n} \left[\frac{1}{1 - (1 - \frac{p_y}{n_2})} + \frac{1}{1 - (1 - \frac{p_x}{n_1})(1 - \frac{p_x}{n_2})} \right.$$
$$\left. - \frac{1}{1 - (1 - \frac{p_x}{n_2})} - \frac{1}{(1 - \frac{p_y}{n_1})(1 - \frac{p_y}{n_2})} \right]$$

Consider the function $f(t) = \frac{1}{1-t} - \frac{1}{1-rt}$, where $0 < r, t < 1$. Then $f'(t) = \frac{1}{(1-t)^2} - \frac{r}{(1-rt)^2}$. Since $1 - t < 1 - rt$ we conclude that $f'(t) > 0$ and thus $f(t)$ is an increasing function. Now note that $0 < \left(1 - \frac{p_i}{n_j}\right) < 1$ for $1 \leq i \leq n$ and $1 \leq j \leq 2$. This together with the fact that $f(t)$ is an increasing function and $p_x \leq p_y$ implies that $E_x[T] - E_y[T] \leq 0$. $\qquad\square$

Lemma 3. *Consider the problem of querying n nodes with two robots. Let the nodes $1, \dots, n$ be labeled such that $p_1 < \cdots < p_n$. Suppose that S_1, S_2 is an optimal solution among overlapping and partitioning queries. Then $S_1 \cap S_2$, $S_1 \setminus (S_1 \cap S_2)$ and $S_2 \setminus (S_1 \cap S_2)$ are consecutive.*

Proof (Lemma 3). If the optimal solution is a partition solution, i.e. $S_1 \cap S_2 = \emptyset$ then, by Theorem 4, we know that S_1 and S_2 are consecutive. So suppose that the optimal solution is an overlapping solution i.e. $S_1 \cap S_2 \neq \emptyset$. Then by Lemma 2 we know that the nodes inside the overlap, i.e. $S_1 \cap S_2$, are the nodes with the smaller probability of success. Therefore the nodes inside the overlap are consecutive nodes. The formula for the expected search time of overlapping strategy is given by (5).

Consider the nodes outside of the overlapping set i.e. $B = (S_1 \cup S_2) \setminus (S_1 \cap S_2)$. We want to find a partitioning query of B with two robots whose expected search time is minimum. If in an optimal solution, one of the sets $S_1 \setminus (S_1 \cap S_2)$ and $S_2 \setminus (S_1 \cap S_2)$ is empty we are done. Otherwise, the optimal solution is obtained by partitioning the nodes $[n] \setminus (S_1 \cap S_2)$ into two non-empty sets, and assigning each of them to some robot.

We think of the following artificial optimization problem. Assuming that robots will share some fixed collection of nodes $[n] \setminus B$, partition the remaining nodes B so as to minimize the expected termination time. Conditioning on that the treasure is in B, the contribution to the cost is the same as if the two robots were performing pure-partition strategies, where the size of each of their domains is still $|S_1|$ and $|S_2|$ see (5) and (6). In particular, this artificial optimization problem is still a SPAO with a cost-function that is concave in the subset sum for fixed cardinality of the subset. So by Theorem 4, there is an optimal partitioning B into two consecutive sets minimizing the expected search time.   □

We are now ready to prove Theorem 7.

Proof (Theorem 7). Let S_1, S_2 presents an optimal query among partitioning and overlapping solutions. Suppose that the nodes $1, \ldots, n$ are ordered such that $p_1 < \cdots < p_n$. By Lemma 3 we know that $S_1 \cap S_2$, $S_1 \setminus (S_1 \cap S_2)$ and $S_2 \setminus (S_1 \cap S_2)$ are consecutive sets. Therefore there are $\binom{n+1}{2}$ ways to choose $S_1 \cap S_2$, $S_1 \setminus (S_1 \cap S_2)$ and $S_2 \setminus (S_1 \cap S_2)$. By exhaustively checking all of them, we can determine the optimal solution in $\mathcal{O}(n^2)$ many steps.

5 Brief Discussion on Generalizations

Notably, Corollary 2 leaves open the possibility that either unanimous or pure-partition strategies are optimal strategies for the partition query problem. Indeed, when all $n = 3$ nodes are associated with the same probability p, then the expected termination time of unanimous querying is, by Theorem 3, equal to $\frac{1}{1-(1-p/3)^2}$. On the other hand, the other possible partition strategy in which a robot searches two nodes, and the other robot searches the remaining node has cost $\frac{5}{3p}$ (see (4)). Which of the two is better is dependent on the value of p, indicating that the structure of optimal solutions in partition querying might be in general challenging to determine.

Some of our results of Sect. 4 also generalize to querying with 3 (or more) robots. We can show that in an optimal solution, the nodes (if any) that are searched by all robots form a set of consecutive integers, and again this set should include the nodes with the smallest probabilities p_i. However, it is not true necessarily that indices of nodes searched by r robots are smaller than indices of nodes searched by l robots, whenever $r > l$. This is another indication that optimal solutions to overlap querying might be challenging to obtain.

6 Conclusion

In this paper we introduced a new search problem about querying with uncertainty; k robots try to locate a hidden treasure that is placed uniformly at random in one of n locations, each associated with a probability p_i, $i = 1, \ldots, n$. A query by a robot in location i may succeed only if the treasure is located in location i, in which case the treasure will be revealed with probability p_i.

Based on a new closed formula which we proved for the expected number of steps until the treasure is found by a robot in unanimous querying, we were also able to analyze more complex querying such as partition and overlapping and establish the expected number of steps until the treasure is found by a robot in these cases as well. We analyzed in detail querying by two robots, but the general case of overlapping queries with multiple robots still remains open. Further, throughout our analysis we made the assumption that the underlying graph is the complete graph K_n and there was no cost associated either with the movement of the robots or the underlying graph. An interesting variant of our model is considering other graphs, e.g., lines, rings, trees, as well as when the robots incur costs in their movement from node to node, e.g., when there is an underlying, possibly either edge or vertex weighted graph.

References

1. Agarwal, P.K., Aronov, B., Har-Peled, S., Phillips, J.M., Yi, K., Zhang, W.: Nearest neighbor searching under uncertainty ii. In: Proceedings of the 32nd ACM SIGMOD-SIGACT-SIGAI Symposium on Principles of Database Systems, pp. 115–126. ACM (2013)
2. Ahlswede, R., Wegener, I.: Search Problems. Wiley-Interscience, Hoboken (1987)
3. Alpern, S., Gal, S.: The Theory of Search Games and Rendezvous, vol. 55. Kluwer Academic Publishers, Dordrecht (2002)
4. Alpern, S., Gal, S.: The Theory of Search Games and Rendezvous. Springer, Heidelberg (2003). https://doi.org/10.1007/b100809
5. Baeza Yates, R., Culberson, J., Rawlins, G.: Searching in the plane. Inf. Comput. **106**(2), 234–252 (1993)
6. Beck, A.: On the linear search problem. Isr. J. Math. **2**(4), 221–228 (1964)
7. Beck, A., Warren, P.: The return of the linear search problem. Isr. J. Math. **14**(2), 169–183 (1973)
8. Bellman, R.: An optimal search. SIAM Rev. **5**(3), 274 (1963)
9. Chakravarty, A.K., Orlin, J.B., Rothblum, U.G.: Technical note–a partitioning problem with additive objective with an application to optimal inventory groupings for joint replenishment. Oper. Res. **30**(5), 1018–1022 (1982)
10. Chrobak, M., Gąsieniec, L., Gorry, T., Martin, R.: Group search on the line. In: Italiano, G.F., Margaria-Steffen, T., Pokorný, J., Quisquater, J.-J., Wattenhofer, R. (eds.) SOFSEM 2015. LNCS, vol. 8939, pp. 164–176. Springer, Heidelberg (2015). https://doi.org/10.1007/978-3-662-46078-8_14
11. Czyzowicz, J., Georgiou, K., Kranakis, E., Krizanc, D., Narayanan, L., Opatrny, J., Shende, S.: Search on a line with byzantine robots. In: ISAAC. LIPCS (2016)
12. Czyzowicz, J., Kranakis, E., Krizanc, D., Narayanan, L., Opatrny, J.: Search on a line with faulty robots. In: Proceedings of the 2016 ACM Symposium on Principles of Distributed Computing, pp. 405–414. ACM (2016)
13. Deng, X., Kameda, T., Papadimitriou, C.: How to learn an unknown environment. In: FOCS, pp. 298–303. IEEE (1991)
14. Feinerman, O., Korman, A.: Memory lower bounds for randomized collaborative search and implications for biology. In: Aguilera, M.K. (ed.) DISC 2012. LNCS, vol. 7611, pp. 61–75. Springer, Heidelberg (2012). https://doi.org/10.1007/978-3-642-33651-5_5

15. Fomin, F.V., Thilikos, D.M.: An annotated bibliography on guaranteed graph searching. Theoret. Comput. Sci. **399**(3), 236–245 (2008)
16. Hanusse, N., Ilcinkas, D., Kosowski, A., Nisse, N.: Locating a target with an agent guided by unreliable local advice: how to beat the random walk when you have a clock? In: Proceedings of the 29th ACM SIGACT-SIGOPS Symposium on Principles of Distributed Computing, pp. 355–364. ACM (2010)
17. Hanusse, N., Kavvadias, D.J., Kranakis, E., Krizanc, D.: Memoryless search algorithms in a network with faulty advice. TCS **402**(2–3), 190–198 (2008)
18. Kao, M.-Y., Ma, Y., Sipser, M., Yin, Y.: Optimal constructions of hybrid algorithms. J. Algorithms **29**(1), 142–164 (1998)
19. Kao, M.-Y., Reif, J.H., Tate, S.R.: Searching in an unknown environment: an optimal randomized algorithm for the cow-path problem. Inf. Comput. **131**(1), 63–79 (1996)
20. Kranakis, E., Krizanc, D.: Searching with uncertainty. In: 6th International Colloquium on Structural Information & Communication Complexity, SIROCCO 1999, Lacanau-Ocean, France, 1–3 July 1999, pp. 194–203 (1999)
21. Stone, L.: Theory of Optimal Search. Academic Press, New York (1975)

Energy-Optimal Broadcast in a Tree with Mobile Agents

Jerzy Czyzowicz[2(✉)], Krzysztof Diks[1], Jean Moussi[2], and Wojciech Rytter[1]

[1] Département d'informatique, Université du Québec en Outaouais,
Gatineau, QC, Canada
{diks,rytter}@mimuw.edu.pl
[2] Faculty of Mathematics, Informatics and Mechanics,
University of Warsaw, Warsaw, Poland
{jurek,Jean.Moussi}@uqo.ca

Abstract. A set of k mobile agents is deployed at the root r of a weighted, n-node tree T. The weight of each tree edge represents the distance between the corresponding nodes along the edge. One node of the tree, the source s, possesses a piece of information which has to be communicated (broadcasted) to all other nodes using mobile agents. An agent visiting a node, which already possesses the information, automatically acquires it and communicates it to all nodes subsequently visited by this agent. The process finishes when the information is transferred to all nodes of the tree.

The agents spend energy proportionally to the distance traversed. The problem considered in this paper consists in finding the minimal total energy, used by all agents, needed to complete the broadcasting. We give an $O(n \log n)$ time algorithm solving the problem. If the number of agents is sufficiently large (at least equal to the number of leaves of T), then our approach results in an $O(n)$ time algorithm.

When the source of information s is initially at the root r, our algorithm solves the problem of searching the tree (exploring it) by a set of agents using minimal energy. It is known that, even if the tree is a line, the broadcasting problem and the search problem are NP-complete when the agents may be initially placed at possibly many distinct arbitrary positions.

Keywords: Mobile agents · Tree · Data delivery · Broadcast · Search

1 Introduction

A packet of information available at the source node of a network must be disseminated to all other nodes. The task needs to be performed by a collection of mobile agents. Given a network, what is the minimal amount of energy needed for the information to be delivered to all nodes. If the agents are initially distributed

---

J. Czyzowicz—Supported by NSERC grant of Canada.

K. Diks and W. Rytter—Supported by the grant NCN2014/13/B/ST6/00770 of the Polish Science Center.

© Springer International Publishing AG 2017
A. Fernández Anta et al. (Eds.): ALGOSENSORS 2017, LNCS 10718, pp. 98–113, 2017.
https://doi.org/10.1007/978-3-319-72751-6_8

along the network, this problem turns out to be NP-complete, even if the original network is a line (see [13]). In this paper we solve the problem for tree networks, when all agents start at the same node.

Similar problems, studied in operations research (particularly in vehicle routing), sometimes assume limited capacities of robots and quantities of product to be transported. In our case, the amount of product to be carried by a robot is irrelevant. Consequently, similarly to [13], we categorize our problem as *data delivery* or, more exactly as *broadcasting* using mobile agents.

Moreover, in the special case when the data source and the starting position of the robots are the same, our solution produces the optimal search schedule - the movement of the collection of agents searching all nodes of the tree using minimal total energy. It is a folklore knowledge that the same problem when the time of the schedule is to be optimized (i.e. the time of arrival to its destination of the last robot), the problem is NP-complete.

1.1   Preliminaries and the Problem Statement

A set of k mobile agents is placed at the root r of an edge-weighted tree T, where edge weight represents distance between edge endpoints. One node of the tree, that we call *source node*, initially possesses a packet of information, which eventually needs to reach every other node of the tree. The information is transported by mobile agents, which use energy proportionally to the distance travelled. We conservatively assume, that if an agent previously acquired the source information packet, then if such agent is visiting a node, it implicitly leaves a copy of the packet at that node. Any agent later visiting such node automatically copies the packet to its memory and it may then distribute it to other, subsequently visited nodes. We consider the following problem:

Data broadcasting:

> Let T be an edge-weighted tree with two distinguished nodes s, r. The node s is a source node and r is the initial location of k mobile agents. The source node possesses a packet of information, which needs to be transported to all other nodes by mobile agents.
> What is the minimal amount of energy, denoted by $MinCost(T, k)$, that the agents need for their travel so that the source packet is successfully distributed?

By a (global) schedule we mean a set of functions $f_1, f_2, \ldots, f_k$, such that $f_i(t)$ is a position at time t of agent i in the tree. The agent i knows the data packet at time t^*, if $f_i(t) = s$ for some $t \leq t^*$, or $f_i(t_1) = v$, $f_j(t_2) = v$ and $t_2 \leq t_1 \leq t^*$ for some agent j, which knew the packet at time t_2.

We call a schedule *optimal* if it results in the smallest possible usage of energy by mobile agents.

1.2   Our Results

We present almost linear-time, greedy algorithms solving data broadcasting. Our problem can be solved in $O(n \log n)$ time, independently of the number k of available agents. This complexity is reduced to $O(n)$ time in case of unlimited number of agents, or when the number of agents is at least equal to the number of leaves of T.

In the special case when the root, from which all agents start, is also the source node, our approach solves the search problem, when the collection of agents need to search the tree optimally, i.e. using the smallest total energy. Surprisingly, according to our knowledge, this natural setting of the search/exploration problem has not been studied before.

The missing proofs will appear in the full version of the paper.

1.3   Related Work

Recent development of the network technology fuelled the research in mobile agents computing. Several applications involve physical mobile devices, software agents, migrating in a network, or living beings: humans (e.g. soldiers or disaster relief personnel) or animals (e.g. ants). Most important questions for mobile agents concern environment search or exploration (cf. [3,9,15–17]). Some questions involving mobile agents are related to problems from operations research, especially vehicle routing (e.g., see [19]).

Searching and exploration have been extensively investigated in numerous settings. Using collections of mobile agents for the tree environment, the previous papers attempted to optimize the usage of various resources, e.g., the time for exploration (the maximal time used among all agents) [6,16], memory used by agents [4], the number of agents [15], etc. Many papers assumed no knowledge of the tree, which leads to distribute algorithms optimizing the competitive ratio (e.g. [15,17]). In the centralized setting, the minimization of maximal time used by mobile agents is an NP-hard problem, even if the tree is a star and the collection contains only two robots (cf. [6,16]).

The task of broadcast is useful, e.g., when a designated leader needs to share its information with collaborating agents in order to perform together some future tasks. The broadcast problem for stationary processors has been extensively studied both in the case of the message passing model, (e.g. [7]), and for the wireless model, (see [10]).

The question of energy awareness has been investigated in different contexts. Paper [2] studied power management of (not necessarily mobile) devices. Several methods have been proposed to reduce energy consumption of computer systems including power-down strategies (see [2,5,18]) or speed scaling (cf. [20]). However, most research related to energy efficiency attempts to optimize the total power used in the entire system. When the power assignments concern the individual system components (as is the case of our model), the related optimization questions (e.g. see [8]) have a flavour of load balancing.

The communication problem by mobile agents has been studied in [1]. The agents of [1] perform efficient *convergecast* and *broadcast* in line networks. All agents of [1] have energy sources of the same size, allowing to travel the same distance. However, in the case of tree networks, the problems of convergecast and broadcast are proven to be strongly NP-complete in [1].

In the case, where agents may have different initial energy levels, [13] investigated a simpler communication problem of *data delivery*, when the information has to be transmitted between two given network nodes. This problem is proven to be NP-complete in [13] already for line networks. [11] showed that the situation is quite different when the agents are required to return to original locations. [11] gave a polynomial solution for data delivery in trees by returning agents. The problem of energy-efficient data delivery between given set of pairs of graph nodes was investigated in [12]. The data delivery and convergecast for trees with energy-exchanging agents were studied in [14], where linear-time solutions for both problems were proposed.

2 Agents Starting from the Source Node

We start with an easier case when the root r is the same as the source node s, which contains the initial data packet. Observe that, even if we have unlimited number of agents in r, the problem is nontrivial. In the proposed solution, every agent i initially takes an exact amount of energy needed to traverse some subtree $T(i)$ and its traversal of $T(i)$ must be optimal. The union of all subtrees must sum up to the entire tree T and the choice of the subtrees must minimize the total energy needed to traverse them.

We consider separately cases of limited and unlimited number of agents. We will show that not all the agents are always activated, i.e. in some cases making walk too many agents would result in a suboptimal algorithm. We say that an agent is *activated* if it is used for walking (consumes a non-zero amount of energy), copies a data packet to its memory, when arriving to a node at which a copy of data packet is present, and subsequently disseminates it to all nodes visited afterwards.

Denote by T_v the subtree of T rooted at v.

Lemma 1. *Suppose that the source node s is the same as the root r. In every optimal broadcast algorithm, each activated agent should terminate its walk at a leaf of T.*

Proof. The proof goes by contradiction. Suppose that, in an optimal broadcast, some agent i terminates its walk in a non-leaf node v, traversing some edge (w, v) as the last edge of its route. Two cases are possible:

Case 1: The traversal of the last edge (w, v) does not coincide with the first visit of node v by agent i. In this case we can remove the traversal of the last edge (w, v) from the route of agent i and the tree explored by agent i remains the same. However, such shortening of the route of agent i reduces its energy cost by $weight(w, v)$, which contradicts the optimality of the original traversal.

Case 2: The traversal of the last edge (w, v) by agent i coincides with its first visit of node v. In such a case, agent i could not previously enter the subtree T_v (otherwise this would imply the second visit of v). Consequently, as T_v contains at least one leaf, unvisited by agent i, it must be visited by some other agent j. However, to reach any leaf of T_v from the starting position r, agent j must visit v on its route. As v does not need to be visited by two different agents, we can then again shorten the route of i by the last edge (w, v), reducing its cost. This contradicts optimality of its original route.   □

The subset of leaves of T, at which the activated agents of an optimal algorithm terminate their paths, are called *critical leaves*. Each path from root r to a critical leaf is called a *critical path*. The union of all critical paths forms a tree, rooted at r, that we call the *frame* of the algorithm.

2.1   Scheduling Agent Movements When Critical Leaves are Known

We start with the presentation of an algorithm which designs the movements of the agents once the set of critical leaves has been obtained. As agents possess the information about the packet from the start, it is sufficient to generate the trajectories of all robots, disregarding synchronization between actual movements of different agents.

Consider a subset $\mathcal{L}$ of critical leaves of tree T. Define $frame(\mathcal{L})$ as the union of all critical paths, i.e. the subtree of T induced by $\mathcal{L}$ and all its ancestors, see Fig. 1. By $|frame(\mathcal{L})|$ we understand the sum of weights of all edges of $frame(\mathcal{L})$. Observe that the edges $T \setminus frame(\mathcal{L})$ form a set of subtrees rooted at the nodes of $frame(\mathcal{L})$. We call them *hanging subtrees* and we denote the set of hanging subtrees by $H(\mathcal{L})$.

Once we know the optimal set of critical leaves $\mathcal{L}$, then an optimal schedule is easy to construct. Below we give the algorithm ConstructSchedule, which constructs the optimal schedule for the given set of k agents. In fact we concentrate later only on computing the optimal $\mathcal{L}$ (needed in line 1 of the algorithm ConstructSchedule). Our main result is the computation of minimum cost in almost linear time, which also implies computing the optimal set $\mathcal{L}$.

Algorithm ConstructSchedule(k);

1. Compute the set of critical leaves $\mathcal{L}$, such that $|\mathcal{L}| \leq k$, which maximizes $\Delta(\mathcal{L})$.

2. Assign to every critical leaf l_i a different agent i which will terminate its walk at l_i.

3. Assign arbitrarily each subtree $T' \in H(\mathcal{L})$ to a single critical leaf $L(T')$, such that T' has the root on the critical path from r to $L(T')$.

4. **for each** leaf $l_i \in \mathcal{L}$ **do**

   4.1. Agent i follows the critical path from r to the critical leaf l_i,

   4.2. On the way to its assigned critical leaf l_i the agent i makes a full DFS traversal of each hanging subtree $T' \in H(\mathcal{L})$ such that $L(T') = l_i$.

Observe that the total number of edge traversals, generated by the algorithm ConstructSchedule, can be quadratic.

Denote by $path(u, v)$ the set of nodes on the simple path between u and v (including u, v) and let $|path(u, w)|$ denote the distance (sum of edge weights) from node u to w in tree T. We denote also $depth(w) = |path(r, w)|$.

We define below a function $\Delta(\mathcal{L})$ which measures the efficiency of the broadcasting algorithm having $\mathcal{L}$ as its critical leaves.

$$\Delta(\mathcal{L}) = 2|frame(\mathcal{L})| - \sum_{w \in \mathcal{L}} depth(w) \qquad (1)$$

The following lemma shows what is the value of $MinCost(T, k)$ - the energy cost of the schedule produced by the algorithm ConstructSchedule for k agents starting at the root of tree T. The energy depends on the choice of the set of critical leaves $\mathcal{L}$. The construction of the set $\mathcal{L}$ minimizing the energy cost will be discussed in the subsequent sections.

Lemma 2. *Assume k agents are placed initially in the source node $r = s$ of T. Then*

$$MinCost(T, k) = 2|E| - \Delta(\mathcal{L}),$$

where $\mathcal{L}$ is a subset of leaves maximizing $\Delta(\mathcal{L})$ over $|\mathcal{L}| \leq k$.

Proof. The algorithm has enough agents, so that to every critical leaf l_i corresponds a different agent i, which terminates its walk at l_i. The edges of all hanging subtrees $H(\mathcal{L})$, i.e. all edges of $T \setminus frame(\mathcal{L})$, are traversed twice in step 4.2 of the algorithm. Moreover each edge of a critical path is traversed in step 4.1 as many times as there are critical paths containing this edge. Consequently, the total cost of such traversal of T is twice the sum of lengths of edges belonging to $T \setminus frame(\mathcal{L})$, and the sum of the critical path lengths of the $frame(\mathcal{L})$. Hence the total cost equals

$$2|T \setminus frame(\mathcal{L})| + \sum_{w \in \mathcal{L}} depth(w) = 2|E| - (2 \cdot |frame(\mathcal{L})| - \sum_{w \in \mathcal{L}} depth(w)) = 2|E| - \Delta(\mathcal{L})$$

$$(2)$$

By Lemma 1, each optimal algorithm using at most k agents, corresponds to $frame(\mathcal{L})$ for some $\mathcal{L}$. Therefore, the cost represented in Eq. 2 is minimized for maximal $\Delta(\mathcal{L})$. □

Consequently, the broadcasting problem reduces in this case to the computation of $\mathcal{L}$ which maximizes $\Delta(\mathcal{L})$ with $|\mathcal{L}| \leq k$. The set $\mathcal{L}$ will be computed incrementally and in a greedy way. We conclude this section with some observations needed for the incremental construction of the optimal set of critical leaves.

Assume $\mathcal{L}$ is a set of leaves and consider a leaf $w \notin \mathcal{L}$. Denote by $LCA(w, \mathcal{L})$ the lowest common ancestor of w and some leaf from $\mathcal{L}$ (i.e. the lowest node belonging to $path(w, r)$ and $frame(\mathcal{L})$). Define $LCA(w, \emptyset) = r$. Let

$$\delta(w, \mathcal{L}) = |path(u, w)| - |path(r, u)|, \tag{3}$$

where $u = LCA(w, \mathcal{L})$. Equivalently we have

$$\delta(w, \mathcal{L}) = depth(w) - 2 \cdot depth(LCA(w, \mathcal{L})) \tag{4}$$

Observation 1. *For $\mathcal{L}_1, \mathcal{L}_2$ such that $\mathcal{L}_1 \subseteq \mathcal{L}_2$ and for any leaf w we have $\delta(w, \mathcal{L}_1) \geq \delta(w, \mathcal{L}_2)$.*

Indeed the statement of the Observation 1 follows from the fact that $\mathcal{L}_1 \subseteq \mathcal{L}_2$ implies $depth(LCA(w, \mathcal{L}_1)) \leq depth(LCA(w, \mathcal{L}_2))$.

Lemma 3. *For a given subset of leaves $\mathcal{L}$ and a leaf $\tilde{w} \notin \mathcal{L}$:*

$$\Delta(\mathcal{L} \cup \{\tilde{w}\}) = \Delta(\mathcal{L}) + \delta(\tilde{w}, \mathcal{L}). \tag{5}$$

Proof. If we add $\tilde{w}$ to $\mathcal{L}$, the new path between $LCA(\tilde{w}, \mathcal{L})$ and $\tilde{w}$ is added to $frame(\mathcal{L})$ (cf. Fig. 1). Hence, according to formula 1 we have

$$\Delta(\mathcal{L} \cup \{\tilde{w}\}) - \Delta(\mathcal{L}) = 2|frame(\mathcal{L} \cup \{\tilde{w}\})| - 2|frame(\mathcal{L})|$$
$$- \sum_{w \in (\mathcal{L} \cup \{\tilde{w}\})} depth(w) + \sum_{w \in \mathcal{L}} depth(w)$$
$$= 2|path(LCA(\tilde{w}, \mathcal{L}), \tilde{w})| - depth(\tilde{w}) = depth(\tilde{w})$$
$$- 2 \cdot depth(LCA(\tilde{w}, \mathcal{L}) = \delta(\tilde{w}, \mathcal{L})$$

$\square$

2.2 A Schematic Algorithm Computing the Minimal Cost

The formula (5) from Lemma 3 is used to design our algorithm Schematic-MinCost. The idea of the algorithm may be viewed as an incremental, greedy construction of the optimal set of critical leaves $\mathcal{L}$, by adding them, one by one. At each step we have a current version of the frame, which is augmented by a new subpath when a new leaf is added to $\mathcal{L}$. Consider $frame(\mathcal{L})$ obtained from the leaves $\mathcal{L}$ assigned to the first $i - 1$ agents (that terminate their paths at $i - 1$ leaves of $\mathcal{L}$). In the i-th iteration of the main loop we try to decide what is the best use of the next available agent. The i-th agent will terminate its traversal at some leaf w_i of T, not yet present in $frame(\mathcal{L})$. Therefore, $frame(\mathcal{L} \cup \{w_i\})$ will contain some new subpath, disjoint with $frame(\mathcal{L})$, starting at some vertex of $LCA(w_i, \mathcal{L})$ and ending at w_i. Observe, that the usage of agent i, permits the subpath from $LCA(w_i, \mathcal{L})$ to w_i to be traversed once (by a new agent) rather than twice (by some other agent which would need to perform a complete traversal of some subtree containing this path), which results in some energy gain. However such energy benefit is at the expense of bringing the agent from the root r to $LCA(w_i, \mathcal{L})$. The main loop executions continue as long as such gain is possible (i.e. benefit minus expense is positive) and there are still available agents to be used. Such benefit is represented by the function $\delta(w_i, \mathcal{L})$ and our algorithm chooses the leaf offering the largest benefit. We prove later that this greedy approach results in construction of the best possible set of critical leaves.

Algorithm Schematic-MinCost(T, k);

1. $\mathcal{L} := \emptyset$;

2. **while** $|\mathcal{L}| \leq k$ and $\exists (w \notin \mathcal{L}) \, \delta(w, \mathcal{L}) > 0$ **do**

3.   choose a leaf $w \notin \mathcal{L}$ with maximum $\delta(w, \mathcal{L})$;

4. $\mathcal{L} := \mathcal{L} \cup \{w\}$;

5. **return** $|2E| - \Delta(\mathcal{L})$;

Example 1. Figure 1 illustrates the execution of one step of the algorithm. The set $\mathcal{L}$ contains leaves w_1, w_2. The value of $\Delta(\mathcal{L}) = 36 - 19 = 17$, cf. formula 1. Among the remaining leaves, the maximal benefit is obtained by including w_4 in the set of critical leaves as $\delta(w_4, \mathcal{L}) = 3$. Then $\Delta(\{w_1, w_2, w_4\}) = 20$. As for the remaining leaves the values of δ are not positive, only three agents are activated (even if more are available) and, by Lemma 2, the cost of the optimal algorithm equals

$$2|E| - \Delta(\{w_1, w_2, w_4\}) = 56 - 20 = 36.$$

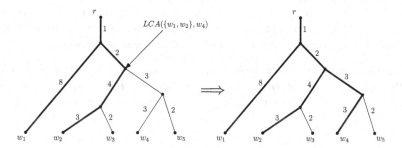

Fig. 1. The iteration which starts with a set of leaves $\mathcal{L} = \{w_1, w_2\}$, then w_4 is added to $\mathcal{L}$. Bold edges belong to the frames (subtrees $frame(\mathcal{L})$ of paths from the root to set $\mathcal{L}$ of critical leaves) before and after inclusion of w_4. We have $\Delta(\{w_1, w_2, w_4\}) = \Delta(\{w_1, w_2\}) + \delta(w_4, \{w_1, w_2\})$

Observe that algorithm Schematic-MinCost is in fact non-deterministic as it is possible that more than one leaf having the same value of δ may be chosen in line 3. Moreover, among optimal broadcasting algorithms, it is possible that the number of agents used may be different. This is possible if we activate an agent terminating at a leaf w for which $\delta(w, \mathcal{L}) = 0$.

Lemma 4. *Assume that in the algorithm Schematic-MinCost we insert the sequence $w_1, w_2, \ldots w_m$ of leaves into $\mathcal{L}$ and there is a set $\mathcal{L}'$ maximizing $\Delta(\mathcal{L}')$ such that $\{w_1, w_2, \ldots w_t\} \subseteq \mathcal{L}', t < m$. Then there exists a set $\mathcal{L}''$, also maximizing $\Delta(\mathcal{L}'')$, which contains $\{w_1, w_2, \ldots w_t, w_{t+1}\}$.*

Sketch of the Proof. The idea of the proof is to show that the set $\mathcal{L}'$ must contain some leaf w^*, such that $w^* \neq w_i$, for $i = 1, 2, \ldots, t+1$ and that exchanging w^* by w_{t+1} in the set $\mathcal{L}$" will not increase the cost of the corresponding broadcasting algorithm.

Lemma 5.

(a) *The set $\mathcal{L}$ computed by the algorithm Schematic-MinCost maximizes $\Delta(\mathcal{L})$.*
(b) *The value $2|E| - \Delta(\mathcal{L})$, output by the algorithm Schematic-MinCost, is the minimum amount of energy needed for broadcasting using k agents initially placed in the source $r = s$.*

Proof.

(a) Using inductively Lemma 4 we prove that the entire set $w_1, w_2, \ldots w_m$ belongs to a set of critical leaves used by an optimal algorithm. By the exit condition of the while loop at line 2 of the algorithm Schematic-MinCost, there is no other leaf which may be added to such critical set of leaves improving the cost of the algorithm.
(b) This point follows directly from (a) and Lemma 2.   □

Observe that every agent possesses the information about the packet at the very beginning of the algorithm. Then, as observed before, once the trajectories of each activated agent are determined, the timing of the travel of each agent is independent of the timing of the travel of any other agent. We conclude then by the following observation, which will be useful in the next section.

Observation 2. *Any energy-optimal schedule may be designed in such a way that the time intervals, during which agents perform their travel, are pairwise disjoint. In particular, we can choose any agent and make this agent complete its walk before any other agent starts walking.*

2.3   Efficient Implementation of Algorithm Schematic-MinCost

Efficiency of Schematic-MinCost depends on the cost of computing *on-line* the best $\delta(w, \mathcal{L})$. We replace it by introducing a more efficient function Gain(v) which does not depend on $\mathcal{L}$ and can be computed *off-line* in linear time. The algorithm Schematic-MinCost subsequently adds leaves to the set $\mathcal{L}$, each time choosing the leaf w offering the largest gain, i.e. the largest reduction $\delta(w, \mathcal{L})$ in the cost of the broadcasting schedule. The values of function δ for any leaf w, which does not yet belongs to $\mathcal{L}$, may change with subsequent modifications of $frame(\mathcal{L})$. In order to avoid recalculations of the function δ we propose the following solution.

Consider the moment when the leaf w is being added to the current set $\mathcal{L}$. Let v be a child of $LCA(w, \mathcal{L})$, which belongs to the path from $LCA(w, \mathcal{L})$ to w. Let $maxpath(v)$ be the longest path starting at v (and going away from the root). If there is more than one such path, we choose any one of them arbitrarily. We denote by leaf($maxpath(v)$) the last node on such path. Observe that, at the moment when w is being added to $\mathcal{L}$, we have $|maxpath(v)| = |path(v, w)|$. For any node $v \neq r$ we define

$$\text{Gain}(v) = |maxpath(v)| + \text{weight}(\text{parent}(v), v) - |path(r, \text{parent}(v))| \qquad (6)$$

By convention, we also set $\text{Gain}(r) = |maxpath(r)|$.

Observation 3. *Assume $\mathcal{L}$ is a set of leaves. It follows from Eq. 3, that*

$$\max\{\text{Gain}(v) : v \notin frame(\mathcal{L})\} = \max\{\delta(w, \mathcal{L}) : w \notin \mathcal{L}\}$$

Following the above Observation, in our algorithm we will be looking for nodes v, which are not in the current $frame(\mathcal{L})$.

Algorithm MinCostLimited(T, k);

1. $X := \{v \in V \ : \ \text{Gain}(v) > 0\}$;

2.   Sort X with respect to $\text{Gain}(v)$ in non-increasing order;

3. $\varDelta := 0 \ ; \ \mathcal{L} := \emptyset$;

4.   **while** $X \neq \emptyset$ and $|\mathcal{L}| \leq k$ **do**

5.   choose $v \in X$ with maximum $\text{Gain}(v)$;

6. $\varDelta := \varDelta + \text{Gain}(v)$;

7.   remove from X all nodes belonging to $maxpath(v)$;

8. $\mathcal{L} := \mathcal{L} \cup \text{leaf}(maxpath(v))$;

9.   **return** $2 \cdot |E| - \varDelta$

/* $|\mathcal{L}|$ equals the number of *activated* agents */

Theorem 1. *The algorithm MinCostLimited(T, k) correctly computes in $O(n \log n)$ time the minimal amount of energy, which is needed to perform the data broadcast by k agents.*

Proof. We prove, by induction on the iteration of the while loop from line 4, that the node v chosen in line 5 does not belong to the current $frame(\mathcal{L})$. Indeed, in the first iteration of the while loop from line 4, $frame(\mathcal{L})$ is empty. In every other iteration, because of the leaf added to $\mathcal{L}$ in line 8, $frame(\mathcal{L})$ is augmented by the nodes of $maxpath(v)$, but all these nodes are then removed from set X in line 7. Therefore, in each execution of line 5 no node of X belongs to $frame(\mathcal{L})$.

Consequently, by Observation 3, every value of Gain chosen in line 5 of algorithm MinCostLimited is the same as the value of δ from the corresponding iteration of line 3 of algorithm Schematic-MinCost. Moreover, the same leaf is added to the set of critical set of leafs $\mathcal{L}$ in the corresponding iterations of both algorithms. The final critical set of leaves is then the same for both algorithms.

In the variable $\varDelta$ is accumulated the sum of the values of function Gain for all nodes chosen in all iterations of the while loop. By Observation 3, after exiting the while loop, $\varDelta$ equals the sum of values of function δ for all leafs from the final critical set $\mathcal{L}$. By Lemma 3, this sum equals $\Delta(\mathcal{L})$ and the final value of the computed cost equals $2|E| - \Delta(\mathcal{L})$. By Lemma 5 this proves the correctness of algorithm MinCostLimited.

We consider now the time efficiency of the algorithm. Observe first, that in the preprocessing, the values of $Gain(v)$ can be computed in linear time. Recall that, by formula 6, we need to compute the values of $|maxpath(v)|$, weight$(parent(v), v)$ and $|path(r, parent(v))|$ Observe, that all these values may be computed using depth-first-search traversal (DFS) of T. Indeed weight$(parent(v), v)$ and $|path(r, parent(v))|$ may be obtained when DFS enters node v from its parent. On the other hand, $|maxpath(v)|$ is obtained when DFS visits v for the last time (arriving from its last child).

The amortized complexity of line 7 is also linear. Assume that X is implemented as a bidirectional list and each node v of the tree T contains a pointer to the element of X corresponding to $Gain(v)$. Then the removal operation in line 7 takes constant time for each considered node v, hence the $O(n)$ time overall. As each other instruction inside the while loop takes constant time, the complexity of all lines of the algorithm, except line 2, is $O(n)$. The overall complexity is then dominated by the $O(n \log n)$ sorting in line 2. $\square$

2.4   Unlimited Number of Agents in the Source

For a set of nodes Y denote by children(Y) the set of all children of nodes in Y.

Algorithm MinCostUnlimited(T);

/* The number of agents is unlimited */

1. $X := \{r\}$; $\Delta := 0$; $\mathcal{L} := \emptyset$;

2. **while** $X \neq \emptyset$ **do**

3. $v := $ Extract any element of X;

4.   Add to X each $x \notin maxpath(v)$ such that
           $parent(x) \in maxpath(v)$ and $Gain(x) > 0$;

5. $\Delta := \Delta + Gain(v)$; $\mathcal{L} := \mathcal{L} \cup \{leaf(maxpath(v))\}$;

6.   / * $X = \{v \in children(frame(\mathcal{L})) : v \notin frame(\mathcal{L}) \}$ */

7.   **return** $2 \cdot |E| - \Delta$

Theorem 2. *Assume that the number of agents initially placed at the source node is at least equal to the number of leaves of T. Then algorithm MinCostUnlimited(T) correctly computes in $O(n)$ time the minimal amount of energy needed for the broadcast in T.*

The idea of the proof (see the Appendix) is based on the fact that the set X is now restricted only to the children of the current frame and we choose each of them at some time. Since any two nodes, which are present in X at a same moment, never interfere (i.e. choosing one of them does never affect the Gain function of the other one), they may be treated in any order (as the number of available agents is sufficient for taking each of them at some time). This allows to avoid sorting and the time of the entire treatment is proportional to the number of edges of T.

3   All Agents Start from the Same Node r Different from the Source s

In this section we extend the consideration to the case when the initial position of the packet is not at the root of the tree. We show that this setting may be reduced to the case studied in the previous sections.

It would be helpful if we design the schedule, so that a robot moves along its trajectory independently from the timing of the motion of any other robot. By Observation 2, this was possible when the robots were initially placed at the source node s. However, in the current setting, at every time moment the robots executing an optimal schedule can be divided into two categories: the robots which already know the packet and the robots that do not. Clearly, the former category of robots are not restricted by their movement. On the other hand, the robots not knowing the packet might need to delay their movement as they may have to visit a node after the packet is deposited there.

The path between the root r and the source s we call the *backbone* of tree T and we denote it by B. We start with the following lemma.

Lemma 6. *There exists an optimal broadcasting algorithm in which the first activated agent starts moving towards the source node s, eventually returning to r, before any other agent is activated.*

Proof. The packet initially present at the source node s needs to be transported to all other nodes of the tree, including root r. Therefore, there must exist an agent which travels from r to s to pick up the packet. After that, a copy of the packet must be transported along the backbone B, starting at s and ending at r. During this travel of the packet along B, it may be transported by divers agents. However, when the packet is left by some agent i at a point p of B and picked later by some agent j, we can make agent i wait at point p until the arrival of agent j. At that moment, as agents are identical, we could exchange the roles of agents i and j and it is still agent i which continues to transport the packet. We conclude, by induction, that the packet is transported all the way by the same agent.

Observe as well, that the remaining agents that were exchanging roles with the agent i, in fact, do not need to start their travel before agent i reaches r. Indeed, they may wait at r until the packet is brought there by agent i and start their respective routes afterwards.   □

We now construct the reduction from the setting where $r \neq s$ to the case $r = s$. For every instance I of the problem for $r \neq s$ we create an instance I'

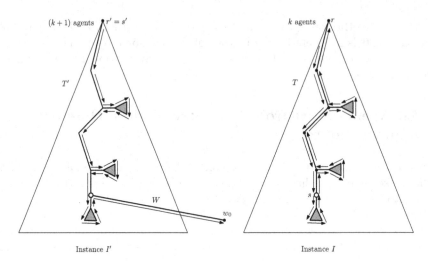

Fig. 2. On the left: the trajectory of the first activated agent, when the algorithm MinCostLimited is run for instance I'. The first agent terminates its walk in w_0. On the right: trajectory of the first activated agent for instance I. The first agent traverses the same nodes (except w_0) but returns to r to be reused later.

of the problem where $r = s$. We show that to solve I it is sufficient to solve I', where we use the results from the previous sections.

Let I be a given instance of the broadcast problem, in which we have tree T with k agents $1, 2, \ldots, k$ initially placed at root r and the source $s \neq r$.

We describe an instance I' of the broadcast problem with $s = r$ and $k + 1$ agents. We construct the tree T' by adding to T one extra leaf w_0 and an edge from node s to w_0 of weight W, where W equals the sum of weights of all edges of T. We define its root $r' = r$ in which we place $k' = k + 1$ mobile agents, represented by the integers $0, 1, \ldots, k$. We also set the source $s' = r'$ (see Fig. 2).

Lemma 7 [Reduction-Lemma]. *If $s \neq r$ in T then*

$$MinCost(T, k) = MinCost(T', k + 1) + |path(r, s)| - W,$$

where in T' the source node s' equals the initial location of $k + 1$ agents.

Proof. Consider first an optimal solution to the instance I' produced by algorithm ConstructSchedule. The weight of the edge incoming to node w_0 is so large that the leaf w_0 must belong to the set of critical leaves $\mathcal{L}$ and some agent 0 must terminate its walk in w_0. By Observation 2, we can suppose that agent 0 is the very first agent activated and the remaining agents didn't start before agent 0 reaches node w_0. Denote by T_H the set of all edges, traversed by agent 0, outside the simple path from r to w_0. By algorithm ConstructSchedule, T_H forms a subset of hanging subtrees.

Consider now an optimal solution to the instance I, which verifies Lemma 6. In this solution, the first activated agent 1, starting at r, travels along the backbone B to the source s (without any detour) and then continues its walk, eventually returning to r (bringing the packet), before any other agent starts moving. Obviously, on its way back along the backbone (i.e. from s to r) agent 1 may visit some nodes outside the backbone before returning to r.

Assume then, that agent 1, during its return from s to r along the backbone, traverses exactly the subtrees formed by the edges T_H (cf. Fig. 2). Consider the time moment in instance I' where agent 0 arrives at leaf w_0 and the time moment when in instance I agent 1 returns to root r. In both cases we have k agents and the packet available at the root r and the part of the tree that still needs to be explored equal to $T \setminus (T_H \cup B)$. Therefore, if we use the trajectories of the remaining k agents $1, 2, \ldots, k$ from the optimal solution of instance I' to complete the instance I the obtained solution of I is also optimal.

Observe that the assumption that agent 1 visited the subtrees formed by the edges of T_H may be dropped. Indeed, all subtrees of T_H are the hanging subtrees of the optimal solution and each of them is DFS traversed by some agent. Assigning any such subtree to agent 1 or any other agent visiting its root does not change the cost of the solution (recall line 3 of algorithm ConstructSchedule).

As from the moments when the situations in instances I' and I are identical all agents walk along the same trajectories in T and T', respectively, the cost of the solution of instance I differs from the solution of instance I' by the difference in the amounts of energy spent by the first agents of each instance, respectively. As this difference is $W - |path(r, s)|$, we have

$$MinCost(T, k) = MinCost(T', k + 1) + |path(r, s)| - W$$

□

Theorem 3. *Suppose that in the tree T the root r is different from the source s. We can solve the limited broadcast problem in $O(n \log n)$ time. If k is at least equal to the number of leaves in T we solve the broadcast problem in $O(n)$ time.*

Proof. Due to Lemma 7 limited broadcast reduces in linear time to the case when the source is the same as starting location of agents. In the unlimited case we can use Lemma 7 with $k = n$.

Hence the time complexity is asymptotically of the same order as that of the algorithm MinCostLimited(T, k), which is $O(n \log n)$. The case of unlimited broadcast can be done similarly in $O(n)$ time, by reduction to the algorithm MinCostUnlimited. □

4 Final Remarks

There are several open questions related to the communication problems and data delivery for mobile agents. One possible extension is to try to perform broadcast from a single source having mobile agents initially distributed in some nodes of the tree.

When the initial amounts of energy are a priori assigned to the agents, most communication problems for mobile agents are shown to be NP-complete (cf. [1,13]). However, when the assignment of energy levels to the agents is left to the algorithm, minimization of total energy used for the communication problems remains open.

Another variation consists in broadcasting from a set of source nodes rather than a single one. If such set of sources involve all the tree nodes the problem becomes gossiping, in which the union of initial information of all nodes must reach every node of the network.

More general open question, which generalizes all communication protocols, concern the delivery of the union of information of a given set of nodes to another set of nodes. Finally, we can consider delivery from a set of specific sources to the respective specific target nodes. We believe that some variants of the question will lead to NP-hard problems for trees.

References

1. Anaya, J., Chalopin, J., Czyzowicz, J., Labourel, A., Pelc, A., Vaxès, Y.: Collecting information by power-aware mobile agents. In: Aguilera, M.K. (ed.) DISC 2012. LNCS, vol. 7611, pp. 46–60. Springer, Heidelberg (2012). https://doi.org/10.1007/978-3-642-33651-5_4
2. Albers, S.: Energy-efficient algorithms. Commun. ACM **53**(5), 86–96 (2010)
3. Albers, S., Henzinger, M.R.: Exploring unknown environments. SIAM J. Comput. **29**(4), 1164–1188 (2000)
4. Ambühl, C., Gasieniec, L., Pelc, A., Radzik, T., Zhang, X.: Tree exploration with logarithmic memory. ACM Trans. Algorithms **7**(4), 1–21 (2011)
5. Augustine, J., Irani, S., Swamy, C.: Optimal powerdown strategies. SIAM J. Comput. **37**, 1499–1516 (2008)
6. Averbakh, I., Berman, O.: A heuristic with worst-case analysis for minimax routing of two traveling salesmen on a tree. Discret. Appl. Math. **68**, 17–32 (1996)
7. Awerbuch, B., Goldreich, O., Peleg, D., Vainish, R.: A trade-off between information and communication in broadcast protocols. J. ACM **37**(2), 238–256 (1990)
8. Azar, Y.: On-line load balancing. In: Fiat, A., Woeginger, G.J. (eds.) Online Algorithms. LNCS, vol. 1442, pp. 178–195. Springer, Heidelberg (1998). https://doi.org/10.1007/BFb0029569
9. Baeza-Yates, R.A., Schott, R.: Parallel searching in the plane. Comput. Geom. **5**, 143–154 (1995)
10. Bar-Yehuda, R., Goldreich, O., Itai, A.: On the time-complexity of broadcast in multi-hop radio networks: an exponential gap between determinism and randomization. J. Comput. Syst. Sci. **45**(1), 104–126 (1992)
11. Bärtschi, A., Chalopin, J., Das, S., Disser, Y., Geissmann, B., Graf, D., Labourel, A., Mihalák, M.: Collaborative delivery with energy-constrained mobile robots. In: Suomela, J. (ed.) SIROCCO 2016. LNCS, vol. 9988, pp. 258–274. Springer, Cham (2016). https://doi.org/10.1007/978-3-319-48314-6_17
12. Bärtschi, A., Chalopin, J., Das, S., Disser, Y., Graf, D., Hackfeld, J., Penna, P.: Energy-efficient delivery by heterogeneous mobile agents. In: Proceedings of STACS, pp. 10:1–10:14 (2017)

13. Chalopin, J., Jacob, R., Mihalák, M., Widmayer, P.: Data delivery by energy-constrained mobile agents on a line. In: Esparza, J., Fraigniaud, P., Husfeldt, T., Koutsoupias, E. (eds.) ICALP 2014. LNCS, vol. 8573, pp. 423–434. Springer, Heidelberg (2014). https://doi.org/10.1007/978-3-662-43951-7_36

14. Czyzowicz, J., Diks, K., Moussi, J., Rytter, W.: Communication problems for mobile agents exchanging energy. In: Suomela, J. (ed.) SIROCCO 2016. LNCS, vol. 9988, pp. 275–288. Springer, Cham (2016). https://doi.org/10.1007/978-3-319-48314-6_18

15. Das, S., Dereniowski, D., Karousatou, C.: Collaborative exploration by energy-constrained mobile robots. In: Scheideler, C. (ed.) Structural Information and Communication Complexity. LNCS, vol. 9439, pp. 357–369. Springer, Cham (2015). https://doi.org/10.1007/978-3-319-25258-2_25

16. Dynia, M., Korzeniowski, M., Schindelhauer, C.: Power-aware collective tree exploration. In: Grass, W., Sick, B., Waldschmidt, K. (eds.) ARCS 2006. LNCS, vol. 3894, pp. 341–351. Springer, Heidelberg (2006). https://doi.org/10.1007/11682127_24

17. Fraigniaud, P., Gasieniec, L., Kowalski, D.R., Pelc, A.: Collective tree exploration. In: Farach-Colton, M. (ed.) LATIN 2004. LNCS, vol. 2976, pp. 141–151. Springer, Heidelberg (2004). https://doi.org/10.1007/978-3-540-24698-5_18

18. Irani, S., Shukla, S.K., Gupta, R.: Algorithms for power savings. ACM Trans. Algorithms 3(4), 41 (2007)

19. Toth, P., Vigo, D.: Vehicle routing: problems, methods, and applications. SIAM (2014)

20. Yao, F.F., Demers, A.J., Shenker, S.: A scheduling model for reduced CPU energy. In: Proceedings of 36th FOCS, pp. 374–382 (1995)

Searching for a Non-adversarial, Uncooperative Agent on a Cycle

Jurek Czyzowicz[1], Stefan Dobrev[2], Maxime Godon[1], Evangelos Kranakis[3(✉)],
Toshinori Sakai[4], and Jorge Urrutia[5]

[1] Dép. d'informatique, Université du Québec en Outaouais, Gatineau, Canada
[2] Institute of Mathematics, Slovak Academy of Sciences, Bratislava, Slovak Republic
[3] School of Computer Science, Carleton University, Ottawa, ON, Canada
`kranakis@scs.carleton.ca`
[4] Research Institute of Educational Development, Tokai University, Tokyo, Japan
[5] Instituto de Matematicas, UNAM, 04510 Mexico D.F., Mexico

Abstract. Assume k robots are placed on a cycle–the perimeter of a unit (radius) disk–at a position of our choosing and can move on the cycle with maximum speed 1. A non-adversarial, uncooperative agent, called *bus*, is moving with constant speed s along the perimeter of the cycle. The robots are searching for the moving bus but do not know its exact location; during the search they can move anywhere on the perimeter of the cycle. We give algorithms which minimize the worst-case search time required for at least one of the robots to find the bus.

The following results are obtained for one robot. (1) If the robot knows the speed s of the bus but does not know its direction of movement then the optimal search time is shown to be exactly (1a) $2\pi/s$, if $s \geq 1$, (1b) $4\pi/(s+1)$, if $1/3 \leq s \leq 1$, and (1c) $2\pi/(1-s)$, if $s \leq 1/3$. (2) If the robot does not know neither the speed nor the direction of movement of the bus then the optimal search time is shown to be $2\pi(1 + \frac{1}{s+1})$. Moreover, for all $\epsilon > 0$ there exists a speed s such that any algorithm knowing neither the bus speed nor its direction will need time at least $4\pi - \epsilon$ to meet the bus.

These results are also generalized to $k \geq 2$ robots and analogous tight upper and lower bounds are proved depending on the knowledge the robots have about the speed and direction of movement of the bus.

Keywords: Cycle · Direction of movement · Moving bus
Robot · Search · Speed

1 Introduction

Due to their fundamental nature, the problems of searching and exploration have been investigated in many areas of mathematics and computer science, e.g., by

J. Czyzowicz and E. Kranakis—Research supported in part by NSERC Discovery grant.

A. Fernández Anta et al. (Eds.): ALGOSENSORS 2017, LNCS 10718, pp. 114–126, 2017.
https://doi.org/10.1007/978-3-319-72751-6_9

addressing fundamental perspectives in robotics and autonomous mobile agent computing. The robots move with certain speeds (not necessarily the same) and the objective of the search is to find a (usually static) target placed at an unknown location of the domain in a (provably) optimal time. This search problem was first proposed by Bellman [5] and independently by Beck [4].

In this paper we consider a similar search problem concerning k mobile autonomous robots which are initially located on a cycle–the perimeter of a unit disk–and which can move on this cycle with maximum speed 1. Unlike previous research which considers a static target, in our work the robots are aware that a bus (non-adversarial, uncooperative agent) is moving with constant speed, say s, along this cycle but do not know its exact location and may or may not know its direction of movement. We assume that during their search the robots can move in any direction anywhere on this cycle.

More specifically, we are interested in investigating the following search problem: Give an algorithm which places the robots on the perimeter of the cycle and minimizes the search time, i.e. the time it takes for at least one of the robots to "catch the bus". By the "robot catching the bus" we mean that the robot and the bus are at the same location at the same time.

1.1 Preliminaries and Model of Computation

The robots are assumed to traverse a cycle (always assumed to be the perimeter of a disk of unit radius). Furthermore, there is a bus which is traveling around the cycle at constant linear speed s and its location is unknown to the robots. The robots can move at speed at most 1 on the perimeter of the disk and can change direction at will at any time during the search depending on the specifications of the algorithm. An algorithm specifies the initial position and trajectories of the robots. For k robots, their movement is specified by a k-tuple $(f_1(t), f_2(t), \ldots, f_k(t))$ of k functions such that $f_i(t)$ gives the precise location of the i-th robot on the cycle at time t, where $i = 1, 2, \ldots, k$. Without loss of generality we may assume that the robots start at the same time while the bus is always in motion around the cycle. The search algorithm succeeds when at least one of the k robots catches the bus. Note that in our model no wireless communication between the robots and the bus is required, rather the robots can recognize the bus when they cross its path. Throughout we use the abbreviations CW and CCW to denote clockwise and counter clockwise movement, respectively.

1.2 Related Work

Our problem can be seen as a rendezvous/meeting problem with an uncooperative, but not adversarial agent, a middle case between rendezvous and cops and robbers. In the standard rendezvous model, all agents fully cooperate to the common meeting goal. Indeed, this is the case in the related paper [11] on rendezvous of two robots with different speeds in a cycle; our problem is different in that one of the two vehicles—namely the bus—has a fixed speed and cannot change direction. At the other extreme, in cops and robbers problems (e.g., see [6]), the cops have the same goal of meeting the robber, but the robber is adversarial and actively tries to avoid meeting. Here (at least for search), we are also trying to meet with an agent. However, that agent does not cooperate, but goes doing its own business, not caring whether it is met or not.

The underlying domain which is traversed by the robots is a continuous curve (in our case the perimeter of a disk of unit radius). In this setting, in addition to the rendezvous paper [11], related to our rendezvous problem is the work on probabilistic rendezvous on a cycle for robots with different speeds [15], rendezvous on a cycle for multiple robots with different speeds in [13], and rendezvous in arbitrary graphs for two robots with different speeds [16].

Related is also the literature on search on a line involving a robot and a static exit in the seminal papers [3–5] as well as extensive discussions and models in the books on search problems [1], on the theory of rendezvous games [2], and on the game of cops and robbers [6]. More recently, there is research on robot evacuation which is like search but measures the quality of search by the time it takes the last robot to find an exit; this has been investigated in the wireless model as well as in the face-to-face model [8]. Related papers on robot evacuation include two robots in the face-to-face model [7,9] and [10] in the wireless model when the underlying domain is a triangle or a square.

There is also related work on gathering a collection of identical memoryless, mobile robots in one node of an anonymous unoriented ring. Robots start from different nodes of the ring and operate in Look-Compute-Move cycles and have to end up in the same node [14], as well as oblivious mobile robots in the same location of the plane when the robots have limited visibility [12].

1.3   Outline and Results of the Paper

In this paper we determine the search time when the robots (1) do not know the direction of movement but know the speed of the bus, and (2) know neither the direction of movement nor the speed of the bus. We note that if the robot knows the direction of movement of the bus, then it follows from the proof in [11] that $2\pi/(s+1)$ is a tight bound on the search time. In Sect. 2 we provide tight upper

Table 1. Optimal search time for a single robot of maximum speed 1. The column "Speed" refers to what the robot knows about the speed of the bus, while the last column indicates the theorem where the result is proved.

Direction	Speed	Optimal search time	Theorem
Known	s	$2\pi/(s+1)$	Theorem 1
Unknown	$s \geq 1$	$2\pi/s$	Theorem 2
Unknown	$1/3 \leq s \leq 1$	$4\pi/(s+1)$	Theorem 2
Unknown	$s \leq 1/3$	$2\pi/(1-s)$	Theorem 2
Unknown	Unknown	4π	Theorem 3

and lower bounds for single robot search, while in Sect. 3 tight upper and lower bounds for multiple robot search. In both sections we consider the impact of knowing the direction of movement of the bus. Table 1 summarizes the results of Sect. 2 for a single robot and Table 2 the results of Sect. 3 concerning multiple robots.

Table 2. Optimal search time for k robots of maximum speed 1. The column "Speed" refers to what the robots know about the speed of the bus, while the last column indicates the theorem where the result is proved (if not OPEN).

Direction	Speed	k	Search time (U.B.)	Lower bound	Theorem	Note
Known	s	all	$2\pi/k(s+1)$	$2\pi/k(s+1)$	Theorem 4	optimal
Unknown	$s \geq 1$	all	$2\pi/ks$	$2\pi/ks$	Theorem 5	optimal
Unknown	$s \leq 1$	even	$2\pi/k$	$2\pi/k$	Theorem 5	optimal
Unknown	$s \leq 1$	odd	$\min(2\pi/ks, 2\pi/(k-s))$	$2\pi/k$	Theorem 5	OPEN
Unknown	Unknown	even	$2\pi/k$	$2\pi/k$	Theorem 6	optimal
Unknown	Unknown	odd	$2\pi/(k-1)$	$2\pi/k$	Theorem 6	OPEN

2 One Robot

Consider the case of a single robot R and let B denote the bus, and P the path followed by the robot. Throughout this section we assume that the bus is moving at constant speed s and cannot change direction, while the robot is moving with speed 1. Our analysis is divided into three subsections depending on the knowledge the robot has about the bus. In Subsect. 2.1 we assume only that the robot knows the direction of movement of the bus, in Subsect. 2.2 the robot does not know the direction of movement of the bus but knows its speed, while in Subsect. 2.3 the robot knows neither the direction nor the speed s of the bus. Table 1 summarizes the results of Sect. 2 for a single robot.

2.1 Known Direction of Movement of the Bus

Assume that the robot knows only the direction of movement of the bus. The following theorem was first proved in [11] but we state it here for completeness.

Theorem 1 [11]. *If the robot knows the direction of movement of the bus then $\frac{2\pi}{s+1}$ is the worst-case optimal search time.* □

Observe that the upper bound stated in Theorem 1 is obvious.

2.2 Unknown Direction of Movement but Known Speed of the Bus

Assume that the robot does not know the direction of movement but knows the speed s of the bus. We show that for small bus speeds the robot should walk at its maximal speed, eventually catching the bus (head on or from behind). For large bus speeds, the best strategy for the robot is to wait until the bus arrives. Finally for intermediate bus speeds, the robot should walk at full speed in one direction, then reverse the direction at some point and continue until the bus is met.

Theorem 2. *If the robot knows the speed s of the bus but does not know its direction of movement then the optimal search time is exactly*

1. $2\pi/s$ *if $s \geq 1$.*
2. $4\pi/(s+1)$ *if $\frac{1}{3} \leq s \leq 1$.*
3. $2\pi/(1-s)$ *if $s \leq \frac{1}{3}$.*

Proof (Theorem 2, upper bounds). The upper bounds are relatively simple; we present below three algorithms, one for each case.

To prove Statement 1 assume that $s \geq 1$. In the search algorithm below, the robot stays put and waits for the bus to arrive.

Algorithm Wait (Direction Unknown: $s \geq 1$).

1. Robot waits for the bus to arrive.

The upper bound $\frac{2\pi}{s}$ in Statement 1 is immediate since the bus travels with speed s and the robot is at distance at most 2π from the bus. To prove Statement 2, assume that $\frac{1}{3} \leq s \leq 1$ and consider the following search algorithm.

Algorithm ZigZag (Direction Unknown: $1/3 \leq s \leq 1$).

1. Robot chooses a direction and walks at full speed for time $\frac{2\pi}{s+1}$;
2. If bus not found then changes direction and walks until bus is met;

The upper bound $\frac{4\pi}{s+1}$ in Statement 2 is easy since in the first part of the algorithm the robot walks for time $\frac{2\pi}{s+1}$. If it did not meet the bus by this time it is because the bus is moving in the same direction as the robot. Therefore at the moment the robot changes direction, in the second part of the algorithm, it is certain that it is moving against the bus. Therefore it will meet the bus in additional time $\frac{2\pi}{s+1}$. To prove Statement 3, assume that $s \leq \frac{1}{3}$ and consider the following search algorithm.

Algorithm GoStraight (Direction Unknown: $s \leq 1/3$).

1. Robot chooses an arbitrary direction and walks until bus is found;

The upper bound $\frac{2\pi}{1-s}$ in Statement 3 is easy since in the worst case the bus and the robot are moving in the same direction with initial distance less than 2π. This proves the upper bounds.   □

Lower bounds. Let us introduce a visualization that will be used in the other lower bounds as well. The x-axis represents time and the y-axis represents positions in the circle. The bus trajectory is then represented by a line passing though the initial position of the bus, with the slope determined by the speed and direction of the bus (let down-slope mean counterclockwise direction).

The robot's trajectory from time 0 until time T will be represented by a contiguous curve P (possibly consisting of straight line segments) in this time-space diagram. Let p_s and p_e denote the start and end-points of P. In order for the robot to catch the bus, its trajectory will have to cross the bus lines corresponding to all possible initial positions, directions (and possibly speeds, if the speed is unknown) of the bus. Such a robot trajectory will be called a *valid* one. Note that the validity of the trajectory depends on the assumptions/knowledge about the bus's speed s, i.e. a trajectory valid for a given s might not be valid for different s.

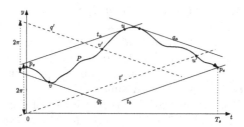

Fig. 1. Trajectory and other concepts: u, v, u' and v' are *key points*. t_a, t', q_q and q' are *support lines*.

Consider first the case where the speed of the bus is known to be s, but its direction is unknown. For a fixed P, let t_a^s and t_b^s denote tangents of slope s touching P from above and below, respectively. Since here we deal with fixed s, we will omit superscripts, and use shorthands q_a and q_b for t_a^{-s} and t_b^{-s}, respectively (refer to Fig. 1).

Let $z(x)$ denote the y-coordinate of line z at time x. Hence, $t_a(0)$ and $t_b(0)$ represent the starting positions of the buses moving at speed s that touch the trajectory of the robot from above and from below.

Lemma 1. *If the bus speed s is known but its direction is unknown, then P is valid if and only if $t_a(0) - t_b(0) \geq 2\pi$ and $q_a(0) - q_b(0) \geq 2\pi$*

Proof (Lemma 1). As the y axis represents the unfolded perimeter of the cycle, every point of the cycle is represented in any segment of y-axis of length at least 2π. In particular, the segment $\langle t_b(0), t_a(0) \rangle$. Note that since P is contiguous, every line of slope s lying between t_b and t_a represents a bus starting at this segment intersecting P. As the same argument holds for direction $-s$, P is valid.

If (without loss of generality) $t_a(0) - t_b(0) < 2\pi$, there is a point in the circle not covered by the segment $\langle t_b(0), t_a(0) \rangle$. A bus line of slope s crossing this point does not intersect P, therefore P is not valid.   □

Suppose that P is valid. Let u be the earliest (in time) of the intersections of t_a or t_b with P. Let t' be a line of slope s at vertical distance exactly 2π from u and lying between t_a and t_b. Let u' be the earliest intersection of t' with P. Note that since t' is between t_a and t_b, it is ensured to intersect P, hence u' is well defined. Define v and v' for the slope $-s$ analogously (see Fig. 1). Let us call points u, v, u', v' the *key* points of P.

Let P' be a trajectory starting at the earliest of the key points, following P and finishing at the latest key point (v to u' in Fig. 1). We will call such trajectory a *pruned* trajectory. By Lemma 1 and its construction, P' is also a valid trajectory for s. (Note that translating a trajectory does not change its validity status, as translations correspond to different starting time and position of the robot.)

We are now ready to prove Statement 1 of Theorem 2:

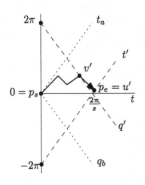

Fig. 2. The case $s \geq 1$.

Proof (Theorem 2, Statement 1 lower bound). Since $s \geq 1$, we immediately get that $u = v = p_s$. By Lemma 1, in order for P to be valid, it must touch both t' and q' (see Fig. 2). Note that the robot can do so in time $2\pi/s$ by simply waiting at the starting location. Assume, on the contrary, that it moves and first touches (without loss of generality) q' at point v'. Then the earliest it can touch t' is by moving towards it (counterclockwise). However, it will not be able to reach t' sooner than q' (which is also moving counterclockwise, but at speed $s \geq 1$), which reaches t' at time exactly $2\pi/s$. □

Consider now the case $s < 1$. The span of a robot trajectory is simply the difference between its starting time and its end time. We know that we can prune the trajectory without increasing its span or breaking its validity. In fact, the following Lemma tells us that it is sufficient to consider only trajectories with robot moving at full speed at all time:

Lemma 2. *Let $s < 1$ and let P' be a valid pruned trajectory of span T. Then there exists a valid trajectory P'' of span at most T in which the robot always travels at full speed.*

Proof (Lemma 2). Let a, b, c and d be the key points of P', ordered by time. Let r_a, r_b, r_c and r_d be the corresponding support lines. P' will be modified as follows (see Fig. 3):

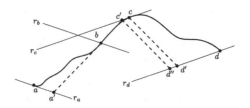

Fig. 3. Speeding up pruned trajectory.

- consider the robot moving from c to r_d at full speed and let d' be the point where it reaches r_d
- consider the robot moving from b to r_a at full speed back in time and let a' be the point where it reaches r_a (alternatively, a' is the point on r_a from where a robot moving at full speed towards r_b reaches b)
- finally, let c' be the point where the robot moving at full speed from b to r_c reaches r_c and let d'' be the point where the robot moving from c' at full speed towards r_d reaches it

As the support lines did not change, the trajectory $a'bc'd''$ is still valid. Note that a' is not earlier than a and d' is not later than d. Also, as c' is not later than c, then d'' is not later than d' and hence the resulting span $a'd''$, as depicted in Fig. 3, has not increased.   □

Hence, it is sufficient to consider only trajectories consisting of at most three line segments of slope 1 and -1. In fact, it is sufficient to consider only one- and two-segment trajectories:

Lemma 3. *Let P be a valid pruned full-speed trajectory consisting of three segments of alternating directions. Then there is a valid pruned full-speed trajectory of at most two segments with smaller span.*

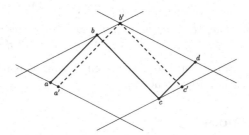

Fig. 4. Optimizing a three-segment trajectory.

Proof (Lemma 3). As P has three segments, it must change direction in each of its interior key points and each key point touches exactly one support line. Without loss of generality the first segment is of speed 1. As P changes direction whenever it touches a support line, it can't leave the quadrangle into which it started. Since P is valid, it touches all four support lines, hence the only possibility is shown in Fig. 4.

Consider now the trajectory $a'b'c'$. It touches all four support lines and hence is valid, however its span is smaller than then the span of P : $|bc| = |b'c'|$, but $|ab| + |cd| > |a'b'|$, because $s < 1$ and hence the distance of c from the support line $b'd$ is greater than the distance of b from it.   □

Now we are ready to prove Statements 2 and 3.

Proof (Theorem 2, Statements 2 and 3 lower bounds). By Lemma 3, we need to consider only two strategies:

- **One segment:** robot travels in one direction until it meets the bus, or
- **Two segments:** robot travels in one direction for distance $2\pi/(1 + s)$ (any two segment trajectory needs to cross from the starting support line to the opposing one; if it reverses sooner, it can only achieve the needed 2π separation between support lines by travelling $2\pi/(1 - s)$ since reversal, at which point one segment strategy is better), then reverses and travels until it meets the bus

In the one segment strategy, the worst case time to meet the bus is $2\pi/(1 - s)$, in the two segments strategy, it is $4\pi(1 + s)$. The first is better for $s < \frac{1}{3}$, the latter for $s \in \langle \frac{1}{3}, 1 \rangle$, which proves Statements 2 and 3. This completes the lower bound in all three cases and proves Theorem 2. □

2.3   Robot Knows Neither the Direction nor the Speed of the Bus

Now assume the robot knows neither the direction of movement nor the speed of the bus.

Theorem 3. *If the robot knows neither the speed s of the bus being used nor its direction of movement then the search time is* $2\pi \left(1 + \frac{1}{s+1}\right)$. *Moreover, for all $\epsilon > 0$ there exists a bus speed s such that any algorithm knowing neither the bus speed nor its direction will need time at least $4\pi - \epsilon$ to meet the bus.*

Observe that the upper and the lower bounds from Theorem 3 converge to the same value 4π when s approaches zero.

3   Multiple Robots

In this section we consider the case of a collection of $k > 1$ robots. Throughout we assume that the bus is moving at constant speed s and cannot change direction, while the robots are moving with speed 1. Our analysis is divided into three subsections depending on the knowledge the robots have about the bus. In Subsect. 3.1 we assume only that the robots know the direction of movement of the bus, in Subsect. 3.2 the robots do not know the direction of movement of the bus but know its speed, while in Subsect. 3.3 the robots know neither the direction nor the speed s of the bus.

3.1   Known Direction of Movement of the Bus

Now assume the robots know the direction of movement of the bus.

Theorem 4. *If the robots know the direction of movement of the bus, then the optimal search time is* $\frac{2\pi}{k(s+1)}$.

3.2   Unknown Direction of Movement but Known Speed of the Bus

Now assume the robots do not know the direction of movement but know the speed s of the bus.

Theorem 5. *If the robots know the speed s of the bus but do not know its direction of movement then the search time is the following:*

1. *If $s \geq 1$, then $2\pi/ks$ is the lower and the upper bound.*
2. *If $s \leq 1$ and k is even, then $2\pi/k$ is the lower and the upper bound.*
3. *Furthermore, if $s \leq 1$ and k is odd, then $2\pi/k$ is the lower bound and the upper bound is*
   (a) $\frac{2\pi}{ks}$ *for* $s \in \left(\frac{k}{k+1}, 1 \right)$
   (b) $\frac{2\pi}{k-s}$ *for* $s \leq \frac{k}{k+1}$

Proof (Theorem 5). We prove separately the upper and lower bounds.
   Upper bounds. For Statement 1, assume $s \geq 1$.

Search Algorithm (Direction Unknown, Speed Known $s \geq 1$).

---
1. The robots are initially placed on the perimeter of the cycle at distance $\frac{2\pi}{k}$ from each other and wait motionless for the bus to arrive.

It is clear that the bus will meet one of the robots in time at most $\frac{2\pi}{ks}$. This upper bound is valid regardless of the parity of k.

Next consider Statement 2. Recall that in this case k is even. Assume $s \leq 1$.

Search Algorithm (Direction Unknown, Speed Known $s \leq 1$: k even).

---
1. The robots are initially placed in pairs on the perimeter of the cycle at distance $\frac{2\pi}{k}$ from each other;
2. The robots in each pair move in opposite direction

For k even, the resulting distance between pairs is exactly $\frac{2\pi}{k/2} = \frac{4\pi}{k}$. Observe that the bus is located between two robots moving against each other. Since these two robots will meet no later than in time $\frac{4\pi}{2k} = \frac{2\pi}{k}$, the resulting running time for this algorithm will be $\frac{2\pi}{k}$.

Finally, consider the case of k odd. We evaluate two algorithms and choose the best, depending on s. The first option is to use the algorithms for speed larger than 1, i.e. spread the agents evenly and wait until the bus meets a robot. In this case, the meeting time is $2\pi/ks$.

The second option is to use the following algorithm:

Search Algorithm (Direction Unknown, Speed Known $s \leq 1$: k odd).

---
1. Let $X = \frac{2\pi(1-s)}{k-s}$ and $Y = \frac{2(2\pi - X)}{k-1}$
2. $k + 1$ robots are initially placed in pairs on the perimeter of the cycle at distance Y from each other; One robot at a node u neighbouring a segment of length X is then removed to bring down the number of robots used to k (see Fig. 5)
3. The robots in each pair move in opposite direction, the lone robot moves away from the X segment

Observe that if the bus started in a Y segment, two robots will be traveling towards each other from the opposite ends of this segment and will meet it at time at most $Y/2$. If the bus started in the X segment, the lone robot crossing the X segment will catch it in time at most $\frac{X}{1-s}$. X was selected so that these two

times are equal: $T = \frac{Y}{2} = \frac{2\pi - X}{k-1} = \frac{\frac{2\pi(k-s)-2\pi(1-s)}{k-s}}{k-1} = \frac{2\pi(k-1)}{(k-s)(k-1)} = \frac{2\pi}{k-s} = \frac{X}{1-s}$

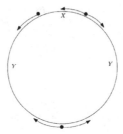

Fig. 5. The algorithm for odd number of agents. Here $k = 5$.

For $s > \frac{k}{k+1}$, the bound $2\pi/k$ of the waiting algorithm is better, while for $s < \frac{k}{k+1}$, this algorithm yields a better bound of $\frac{2\pi}{k-s}$.

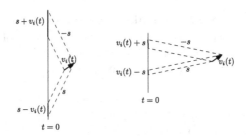

Fig. 6. Sizes of excluded regions. Left: case $s \geq 1$. Right: case $s < 1$.

Lower bounds. Let $v_i(t)$ denote the speed of the i-th robot at time t as it is searching for the bus. Further, consider the movement of the robot at an infinitesimal time dt. Let us express how much of the possible initial bus positions can the robot exclude (cover) in the time interval dt: These are the initial bus positions from which the bus crosses the trajectory of the robot in the interval dt, i.e. if the robot does not meet the bus at time interval dt, then the bus could not have started at those positions (see Fig. 6). As dt is infinitesimal, the robot's trajectory can be seen as a line segment of horizontal size dt and vertical size $v_i(t)dt$. Hence, in the interval dt it covers a segment of starting bus positions of size $(s + v_i(t))dt$ among the busses with opposite direction as the robot, and of size $|s - v_i(t)|dt$ (where $||$ here means absolute value) among the buses with the same direction as the robot. Let us define $f(t) = s + v_t(t) + |s - v_i(t)|$.

Let us first consider the lower bound in Statement 1, i.e. $s \geq 1 \geq v_i(t)$. In this case, $s + v_i(t)$ and $s - v_i(t)$ are always positive, hence we can ignore the absolute value and sum up the regions for both directions directly: $f(t) = (s + v_i(t)) + (s - v_i(t)) = 2s$.

Let T be the time it takes for at least one of the robots to find the bus according to the execution of an optimal search algorithm. Therefore in time T, the i-th robot can cover at most (if at every time moment it excluded different regions) length $\int_0^T f(t)dt = 2sdt = 2sT$. Thus, all k robots taken together can cover at most a length of $2Tks$, and only if all of them cover different areas. However, this last quantity must be at least 4π (2π for clockwise and another 2π for counterclockwise bus directions), otherwise there is a non-excluded starting bus position and direction which would escape the robot's search. It follows that $2ksT \geq 4\pi$, which yields $T \geq 2\pi/ks$. This proves the lower bound in Statement 1.

Now consider the lower bound in Statement 2. i.e. $s \leq 1$. If $s \geq v_i(t)$, then $f(t) = 2s$, as shown before. Otherwise $|s - v_i(t)| = v_i(t) - s$ and $f(t) = (s + v_i(t)) + (v_i(t) - s) = 2v_i(t)$. In any case, as both s and $v_i(t)$ do not exceed 1, we have $f(t) \geq 2$. Hence, in time T, the i-th robot covers length at most $\int_0^T 2f(t)dt \leq \int_0^T 2dt = 2T$. It follows that k robots can cover a length of at most $2kT$. However, this last quantity must be at least 4π (otherwise there is a trajectory of the bus which will escape the robots' search). Therefore, $2kT \geq 4\pi$, which yields $T \geq 2\pi/k$. This proves the lower bound in Statements 2 and 3, as the proof does not rely on the parity of k. □

3.3   Robots Know Neither the Direction nor the Speed of the Bus

Assume the robots know neither direction of movement nor speed s of the bus.

Theorem 6. *If the robots know neither direction of movement nor speed of the bus then the search can be completed in time $\frac{2\pi}{k}$ for k even, and $\frac{2\pi}{k-1}$ for k odd. Moreover, the lower bound of $\frac{2\pi}{k}$ is valid regardless of the parity of k.*

The exact answer is not known for k odd.

4   Conclusion

Several interesting questions arise from our research. One possibility is to look more closely to the case of non-constant speed (i.e. known only upper and lower bound). In fact, one can consider also the case of known movement function (e.g. $f(t)$, where $f(t)$ gives the speed of the bus at time t), but unknown initial location and direction (determined by the sign at $f(t)$), and perhaps also the time shift (e.g. knowing that the agent moves according to $f(t + t_0)$, for some t_0, but not knowing t_0). Similarly one could consider the case of robots with different possibly non-constant speeds. The possibility of exploring other domains, like trees or arbitrary graphs is quite challenging, and also communication models like wireless (e.g., limited visibility, or face-to-face) could be considered.

References

1. Ahlswede, R., Wegener, I.: Search Problems. Wiley-Interscience, Hoboken (1987)
2. Alpern, S., Gal, S.: The Theory of Search Games and Rendezvous. Springer, Boston (2003). https://doi.org/10.1007/b100809
3. Baeza Yates, R., Culberson, J., Rawlins, G.: Searching in the plane. Inf. Comput. **106**(2), 234–252 (1993)
4. Beck, A.: On the linear search problem. Isr. J. Math. **2**(4), 221–228 (1964)
5. Bellman, R.: An optimal search. Siam Rev. **5**(3), 274 (1963)
6. Bonato, A., Nowakowski, R.: The Game of Cops and Robbers on Graphs. AMS, Providence (2011)
7. Brandt, S., Laufenberg, F., Lv, Y., Stolz, D., Wattenhofer, R.: Collaboration without communication: evacuating two robots from a disk. In: Fotakis, D., Pagourtzis, A., Paschos, V.T. (eds.) CIAC 2017. LNCS, vol. 10236, pp. 104–115. Springer, Cham (2017). https://doi.org/10.1007/978-3-319-57586-5_10
8. Czyzowicz, J., Gąsieniec, L., Gorry, T., Kranakis, E., Martin, R., Pajak, D.: Evacuating robots via unknown exit in a disk. In: Kuhn, F. (ed.) DISC 2014. LNCS, vol. 8784, pp. 122–136. Springer, Heidelberg (2014). https://doi.org/10.1007/978-3-662-45174-8_9
9. Czyzowicz, J., Georgiou, K., Kranakis, E., Narayanan, L., Opatrny, J., Vogtenhuber, B.: Evacuating robots from a disk using face-to-face communication (Extended Abstract). In: Paschos, V.T., Widmayer, P. (eds.) CIAC 2015. LNCS, vol. 9079, pp. 140–152. Springer, Cham (2015). https://doi.org/10.1007/978-3-319-18173-8_10. CoRR abs/1501.04985
10. Czyzowicz, J., Kranakis, E., Krizanc, D., Narayanan, L., Opatrny, J., Shende, S.: Wireless autonomous robot evacuation from equilateral triangles and squares. In: Papavassiliou, S., Ruehrup, S. (eds.) ADHOC-NOW 2015. LNCS, vol. 9143, pp. 181–194. Springer, Cham (2015). https://doi.org/10.1007/978-3-319-19662-6_13
11. Feinerman, O., Korman, A., Kutten, S., Rodeh, Y.: Fast rendezvous on a cycle by agents with different speeds. In: Chatterjee, M., Cao, J., Kothapalli, K., Rajsbaum, S. (eds.) ICDCN 2014. LNCS, vol. 8314, pp. 1–13. Springer, Heidelberg (2014). https://doi.org/10.1007/978-3-642-45249-9_1
12. Flocchini, P., Prencipe, G., Santoro, N., Widmayer, P.: Gathering of asynchronous robots with limited visibility. Theor. Comput. Sci. **337**(1), 147–168 (2005)
13. Huus, E., Kranakis, E.: Rendezvous of many agents with different speeds in a cycle. In: Papavassiliou, S., Ruehrup, S. (eds.) ADHOC-NOW 2015. LNCS, vol. 9143, pp. 195–209. Springer, Cham (2015). https://doi.org/10.1007/978-3-319-19662-6_14
14. Klasing, R., Markou, E., Pelc, A.: Gathering asynchronous oblivious mobile robots in a ring. Theor. Comput. Sci. **390**(1), 27–39 (2008)
15. Kranakis, E., Krizanc, D., MacQuarrie, F., Shende, S.: Randomized rendezvous algorithms for agents on a ring with different speeds. In: ICDCN, Goa, India, 4–7 January, pp. 9:1–9:10 (2015)
16. Kranakis, E., Krizanc, D., Markou, E., Pagourtzis, A., Ramírez, F.: Different speeds suffice for rendezvous of two agents on arbitrary graphs. In: Steffen, B., Baier, C., van den Brand, M., Eder, J., Hinchey, M., Margaria, T. (eds.) SOFSEM 2017. LNCS, vol. 10139, pp. 79–90. Springer, Cham (2017). https://doi.org/10.1007/978-3-319-51963-0_7

Improved Leader Election for Self-organizing Programmable Matter

Joshua J. Daymude[1]([✉]), Robert Gmyr[2], Andréa W. Richa[1],
Christian Scheideler[2], and Thim Strothmann[2]

[1] Computer Science, CIDSE, Arizona State University, Tempe, AZ, USA
{jdaymude,aricha}@asu.edu
[2] Department of Computer Science, Paderborn University, Paderborn, Germany
{gmyr,scheidel,thim}@mail.upb.de

Abstract. We consider programmable matter that consists of computationally limited devices (called *particles*) that are able to self-organize in order to achieve some collective goal without the need for central control or external intervention. We use the geometric amoebot model to describe such self-organizing particle systems, which defines how particles can actively move and communicate with one another. In this paper, we present an efficient local-control algorithm which solves the leader election problem in $\mathcal{O}(n)$ asynchronous rounds with high probability, where n is the number of particles in the system. Our algorithm relies only on local information — particles do not have unique identifiers, any knowledge of n, or any sort of global coordinate system — and requires only constant memory per particle.

1 Introduction

The vision for *programmable matter* is to create some material or substance that can change its physical properties like shape, density, conductivity, or color in a programmable fashion based on either user input or autonomous sensing of its environment. Many realizations of programmable matter have been proposed — including DNA tiles, shape-changing molecules, synthetic cells, and reconfiguring modular robots — each of which is pursuing solutions applicable to its own situation, subject to domain-specific capabilities and constraints. We envision programmable matter as a more abstract system of computationally limited devices (which we refer to as *particles*) which can move, bond, and exchange information in order to collectively reach a given goal without any outside intervention. *Leader election* is a central and classical problem in distributed computing that is very interesting for programmable matter; e.g., most known shape formation techniques for programmable matter suppose the existence of

J. J. Daymude and A. W. Richa—Supported in part by NSF awards CCF-1422603 and CCF-1637393.

R. Gmyr, C. Scheideler and T. Strothmann—Supported in part by DFG grant SCHE 1592/3-1.

© Springer International Publishing AG 2017
A. Fernández Anta et al. (Eds.): ALGOSENSORS 2017, LNCS 10718, pp. 127–140, 2017.
https://doi.org/10.1007/978-3-319-72751-6_10

a leader/seed particle (examples can be found in [23] for the nubot model, [20] for the abstract tile self assembly model and [12,13] for the amoebot model).

In this paper, we present a fully asynchronous local-control protocol for the leader election problem, improving our previous algorithm for leader election in [14] which was only described at a high level, lacking specific rules for each particle's execution. Moreover, while the analysis in [14] used a simplified, synchronous setting and only achieved its linear runtime bound in expectation, here we prove with high probability[1] correctness and runtime guarantees for the full local-control protocol[2]. Finally, as this algorithm is both conceptually simpler than that of [14] and presented directly from the point-of-view of an individual particle, it is more easily understood and implemented.

1.1   Amoebot Model

We represent any structure the particle system can form as a subgraph of the infinite graph $G = (V, E)$, where V represents all possible positions the particles can occupy relative to their structure, and E represents all possible atomic movements a particle can perform as well as all places where neighboring particles can bond to each other. In the *geometric amoebot model*, we assume that $G = G_{eqt}$, where G_{eqt} is the infinite regular triangular grid graph. We recall the properties of the geometric amoebot model necessary for this algorithm; a full description can be found in [14].

Each particle occupies either a single node (i.e., it is *contracted*) or a pair of adjacent nodes in G_{eqt} (i.e., it is *expanded*), and every node can be occupied by at most one particle. Particles move through *expansions* and *contractions*; however, as our leader election algorithm does not require particles to move, we omit a detailed description of these movement mechanisms.

Particles are *anonymous*; they have no unique identifiers. Instead, each particle has a collection of *ports* — one for each edge incident to the node(s) the particle occupies — that have unique labels from the particle's local perspective. We assume that the particles have a common *chirality* (i.e., a shared notion of clockwise direction), which allows each particle to label its ports in clockwise order. However, particles do not have a common sense of global orientation and may have different offsets for their port labels.

Two particles occupying adjacent nodes are connected by a *bond*, and we refer to such particles as *neighbors*. Neighboring particles establish bonds via the ports facing each other. The bonds not only ensure that the particle system forms a connected structure, but also are used for exchanging information. Each particle has a constant-size local memory that can be read and written to by any neighboring particle. Particles exchange information with their neighbors by

---

[1] An event occurs *with high probability* (*w.h.p.*), if the probability of success is at least $1 - n^{-c}$, where $c > 1$ is a constant; in our context, n is the number of particles.

[2] An astute reader may note that a w.h.p. guarantee on correctness is weaker than the absolute guarantee given for the algorithm in [14], but the latter was given without considering the necessary particle-level execution details.

simply writing into their memory. Due to the constant-size memory constraint, particles know neither the total number of particles in the system nor any estimate of this number.

We assume the standard asynchronous model, wherein particles execute an algorithm concurrently and no assumptions are made about individual particles' activation rates or computation speeds. A classical result under this model is that for any asynchronous concurrent execution of atomic particle activations, there exists a sequential ordering of the activations which produces the same end configuration, provided conflicts which arise from the concurrent execution are resolved (namely, only conflicts of shared memory writes can happen in our algorithm; we simply assume that an arbitrary particle wins). Thus, it suffices to view particle system progress as a sequence of *particle activations*; i.e., only one particle is active at a time. Whenever a particle is activated, it can perform an arbitrary, bounded amount of computation involving its local memory and the memories of its neighbors and can perform at most one movement. We define an *asynchronous round* to be complete once each particle has been activated at least once.

1.2 Related Work

A variety of work related to programmable matter has recently been proposed and investigated. One can distinguish between active and passive systems. In passive systems, the computational units either have no intelligence (moving and bonding is based only on their structural properties or interactions with their environment), or have limited computational capabilities but cannot control their movements. Examples of research on *passive systems* are DNA computing [1,3, 7,21], tile self-assembly systems (e.g., the surveys in [15,19,22]), and population protocols [2]. We will not describe these models in detail as they are of little relevance to our approach. *Active systems*, on the other hand, are composed of computational units which can control the way they act and move in order to solve a specific task. Prominent examples of active systems are *swarm robotics* (see, e.g., [17,18]), *modular self-reconfigurable robotic systems* (e.g., [16,24]) — especially *metamorphic robots* [8] — and the *nubot* model [5,6,23] by Woods et al. For an in depth discussion of these models and how they relate to our amoebot model, we refer the reader to the full version of this paper [9].

The *amoebot* model [10] is a model for self-organizing programmable matter that aims to provide a framework for rigorous algorithmic research for nanoscale systems. In [14], the authors describe a leader election algorithm for an abstract (synchronous) version of the amoebot model that decides the problem in expected linear time. Recently, a universal shape formation algorithm [13], a universal coating algorithm [11] and a Markov chain algorithm for the compression problem [4] were introduced, showing that there is potential to investigate a wide variety of problems under this model.

1.3 Problem Description

We consider the classical problem of *leader election*. An algorithm is said to solve the leader election problem if for any connected particle system of initially contracted particles with empty memories, eventually a single particle *irreversibly* declares itself the *leader* (e.g., by setting a dedicated bit in its memory) and no other particle ever declares itself to be the leader. We define the running time of a leader election algorithm to be the number of asynchronous rounds until a leader is declared. Note that we do not require the algorithm to terminate for particles other than the leader.

2 Algorithm

Before we describe the leader election algorithm in detail, we give a short high-level overview. The algorithm consists of six *phases*. These phases are not strictly synchronized among each other, i.e., at any point in time, different parts of the particle system may execute different phases. Furthermore, a particle can be involved in the execution of multiple phases at the same time. The first phase is *boundary setup* (Sect. 2.1). In this phase, each particle locally checks whether it is part of a *boundary* of the particle system. Only the particles on a boundary participate in the leader election. Particles occupying a common boundary organize themselves into a directed cycle. The remaining phases operate on each boundary independently. In the *segment setup* phase (Sect. 2.2), the boundaries are subdivided into *segments*: each particle flips a fair coin. Particles that flip heads become *candidates* and compete for leadership whereas particles that flip tails become *non-candidates* and assist the candidates in their competition. A segment consists of a candidate and all subsequent non-candidates along the boundary up to the next candidate. The *identifier setup* phase (Sect. 2.3) assigns a random identifier to each candidate. The identifier of a candidate is stored distributively among the particles of its segment. In the *identifier comparison* phase (Sect. 2.4), the candidates compete for leadership by comparing their identifiers using a token passing scheme. Whenever a candidate sees an identifier that is higher than its own, it revokes its candidacy. Whenever a candidate sees its own identifier, the *solitude verification* phase (Sect. 2.5) is triggered. In this phase, the candidate checks whether it is the last remaining candidate on the boundary. If so, it initiates the *boundary identification* phase (Sect. 2.6) to determine whether it occupies the unique *outer boundary* of the system. In that case, it becomes the leader; otherwise, it revokes its candidacy.

2.1 Boundary Setup

The boundary setup phase organizes the particle system into a set of *boundaries*. This approach is directly adopted from [14], but we give a full description here to introduce important notation. Let $A \subset V$ be the set of nodes in $G_{eqt} = (V, E)$ that are occupied by particles. According to the problem definition, the subgraph

$G_{\text{eqt}}|_A$ of G_{eqt} induced by A is connected. Consider the graph $G_{\text{eqt}}|_{V \setminus A}$ induced by the unoccupied nodes in G_{eqt}. We call a connected component R of $G_{\text{eqt}}|_{V \setminus A}$ an *empty region*. Let $N(R)$ be the neighborhood of an empty region R in G_{eqt}; that is, $N(R) = \{u \in V \setminus R : \exists v \in R \text{ such that } (u,v) \in E\}$. Note that by definition, all nodes in $N(R)$ are occupied by particles. We refer to $N(R)$ as the *boundary* of the particle system corresponding to R. Since $G_{\text{eqt}}|_A$ is a finite graph, exactly one empty region has infinite size while the remaining empty regions have finite size. We define the boundary corresponding to the infinite empty region to be the unique *outer boundary* and refer to a boundary that corresponds to a finite empty region as an *inner boundary*.

For each boundary of the particle system, we organize the particles occupying that boundary into a directed cycle. Upon first activation, each particle p instantly determines its place in these cycles using only local information as follows. First, p checks for two special cases. If p has no neighbors, it must be the only particle in the particle system since the particle system is connected. Thus, it immediately declares itself the leader and terminates. If all neighboring nodes of p are occupied, p is not part of any boundary and terminates without participating in the leader election process any further.

If these special cases do not apply, then p has at least one occupied node and one unoccupied node in its neighborhood. Interpret the neighborhood of p as a directed ring of six nodes that is oriented clockwise around p. Consider all maximal sequences of unoccupied nodes $(v_1, \ldots v_k)$ in this ring; call such a sequence an *empty sequence*. Such a sequence is part of some empty region and hence corresponds to a boundary that includes p. Let v_0 be the node before v_1 and let v_{k+1} be the node after v_k in the ring. Note that we might have $v_0 = v_{k+1}$. By definition, v_0 and v_{k+1} are occupied. Particle p implicitly arranges itself as part of a directed cycle spanning the aforementioned boundary by considering the particle occupying v_0 to be its *predecessor* and the particle occupying v_{k+1} to be its *successor* on that boundary. It repeats this process for each empty sequence in its neighborhood.

A particle can have up to three empty sequences in its neighborhood, and consequently can be part of up to three distinct boundaries. However, a particle cannot locally decide whether two distinct empty sequences belong to two distinct empty regions or to the same empty region. To guarantee that the executions on distinct boundaries are isolated, we let the particles treat each empty sequence as a distinct empty region. For each such sequence, a particle acts as a distinct *agent* which executes an independent instance of the algorithm encompassing the remaining five phases of the leader election algorithm. Whenever a particle is activated, it sequentially executes the independent instances of the algorithm for each of its agents in an arbitrary order, i.e., whenever a particle is activated also its agents are activated. Each agent a is assigned the predecessor and successor — denoted $a.\text{pred}$ and $a.\text{succ}$, respectively — that was determined by the particle for its corresponding empty sequence. This organizes the set of all agents into disjoint cycles spanning the boundaries of the particle system (see Fig. 1). As consequence of this approach, a particle can occur up to three times on the same boundary as different agents. While we can ignore this property for most of the remaining phases, it will remain a cause for special consideration in the solitude verification phase (Sect. 2.5).

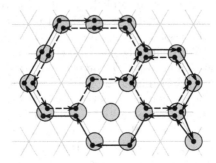

Fig. 1. Boundaries and agents. Particles are depicted as gray circles and the agents of a particle are depicted as black dots inside of the corresponding circle. After the boundary setup phase, the agents form disjoint cycles that span the boundaries of the particle system. The solid arrows represent the unique outer boundary and the dashed arrows represent the two inner boundaries.

2.2   Segment Setup

All remaining phases (including this one) operate exclusively on boundaries, and furthermore execute on each boundary independently. Therefore, we only consider a single boundary for the remainder of the algorithm description. The goal of the segment setup phase is to divide the boundary into disjoint "segments". Each agent flips a fair coin. The agents which flip heads become *candidates* and the agents which flip tails become *non-candidates*. In the following phases, candidates compete for leadership while non-candidates assist the candidates in their competition. A *segment* is a maximal sequence of agents $(a_1, a_2, \ldots, a_k)$ such that a_1 is a candidate, a_i is a non-candidate for $i > 1$, and $a_i = a_{i-1}$.succ for $i > 1$. Note that the maximality condition implies that the successor of a_k is a candidate. We refer to the segment starting at a candidate c as c.seg and call it the segment of c. In the following phases, each candidate uses its segment as a distributed memory.

2.3   Identifier Setup

After the segments have been set up, each candidate generates a random *identifier* by assigning a random digit to each agent in its segment. The candidates use these identifiers in the next phase to engage in a competition in which all but one candidate on the boundary are eliminated. Note that the term identifier is slightly misleading in that two distinct candidates can have the same identifier.

Nevertheless, we hope that the reader agrees that the way these values are used makes this term an appropriate choice.

To generate a random identifier, a candidate c sends a *token* along its segment in the direction of the boundary. A token is simply a constant-size piece of information that is passed from one agent to the next by writing it to the memory of a neighboring particle. While the token traverses the segment, it assigns a value chosen uniformly at random from $[0, r - 1]$ to each visited agent where r is a constant that is fixed in the analysis. The identifier generated in this way is a number with radix r consisting of $|c.\text{seg}|$ digits where c holds the most significant digit and the last agent of $c.\text{seg}$ holds the least significant digit. We refer to the identifier of a candidate c as $c.\text{id}$. The competition in the next phase of the algorithm is based on comparing identifiers. When comparing identifiers of different lengths, we define the shorter identifer to be lower than the longer identifier.

After generating its random identifier, each candidate creates a copy of its identifier that is stored in *reversed digit order* in its segment. This step is required as a preparation for the next phase. To achieve this, we use a single token that moves back and forth along the segment and copies one digit at a time. More specifically, we reuse the token described above that generated the random identifier. Once this token reaches the end of the segment, it starts copying the identifier by reading the digit of the last agent of the segment and moving to the beginning of the segment. There, it stores a copy of that digit in the candidate c. It then reads the digit of c and moves back to the end of the segment where it stores a copy of that digit in the last agent of the segment. It proceeds in a similar way with the second and the second to last agent and so on until the identifier is completely copied. Afterwards, the token moves back to c to inform the candidate that the identifier setup is complete.

Note that for ease of presentation we deliberately opted for simplicity over speed when creating a reversed copy of the identifier. As we will show in Sect. 3.2, the running time of this simple algorithm is dominated by the running time of the next phase so that the overall asymptotic running time of the leader election algorithm does not suffer.

2.4   Identifier Comparison

During the identifier comparison phase the agents use their identifiers to compete with each other. Each candidate compares its own identifier with the identifier of every other candidate on the boundary. A candidate with the highest identifier eventually progresses to the solitude verification phase, described in the next section, while any candidate with a lower identifier withdraws its candidacy. To achieve the comparison, the non-reversed copies of the identifiers remain stored in their respective segments while the reversed copies move backwards along the boundary as a sequence of tokens. More specifically, a *digit token* is created for each digit of a reversed identifier. A digit token created by the last agent of a segment is marked as a *delimiter token*. Once created, the digit tokens traverse the boundary against the direction of the cycle spanning it. Each agent is allowed to hold at most two tokens at a time, which gives the tokens some space to move along the boundary. The tokens are not allowed to overtake each other, so whenever an agent stores two tokens, it keeps track of the order they were received in and forwards them accordingly. An agent forwards at most one token per activation. Furthermore, an agent can only receive a token after it creates its own digit token. We define the *token sequence* of a candidate c as the sequence of digit tokens created by the agents in c.seg. Note that according to the rules for forwarding tokens, the token sequences of distinct candidates remain separated and the tokens within a token sequence maintain their relative order along the boundary.

Whenever a token sequence traverses a segment c.seg of a candidate c, the agents in c.seg cooperate with the tokens of the token sequence to compare the identifier c.id with the identifier stored in the token sequence. This comparison has three possible outcomes: (i) the token sequence is longer than c.seg or the lengths are equal and the token sequence stores an identifier that is strictly greater than c.id, (ii) the token sequence is shorter than c.seg or the lengths are equal and the token sequence stores an identifier that is strictly smaller than c.id, or (iii) the lengths are equal and the identifiers are equal. In the first case, c does not have the highest identifier and withdraws its candidacy. In the second case, c might be a candidate with the highest identifier and therefore remains a candidate. Finally, in the third case, c initiates the solitude verification phase, which is then executed in parallel to the identifier comparison phase. Solitude verification might be triggered quite frequently, especially for candidates with short segments; we describe how this is handled in the next section. Due to space constraints, we omit the exact token passing scheme for identifier comparison and refer to the full version of this paper [9].

2.5   Solitude Verification

The goal of the solitude verification phase is for a candidate c to check whether it is the last remaining candidate on its boundary. Solitude verification is triggered during the identifier comparison phase whenever a candidate detects equality between its own identifier and the identifier of a token sequence that traversed its segment. Note that such a token sequence can either be the token sequence created by c itself or the token sequence created by some other candidate that generated the same identifier. Once the solitude verification phase is started, it runs in parallel to the identifier comparison phase and does not interfere with it. This phase is based on the idea of solitude verification given in [14], but greatly simplifies many of the original ideas to obtain a more easily understood protocol.

A candidate c can check whether it is the last remaining candidate on its boundary by determining whether or not the next candidate in direction of the cycle is c itself. To achieve this, the solitude verification phase has to span not only c.seg but also all subsequent segments of former candidates that already withdrew their candidacy during the identifier comparison phase. We refer to the union of these segments as the *extended segment* of c. The basic idea of the algorithm is the following. We treat the edges that connect the agents on the boundary as vectors in the two-dimensional Euclidean plane. If c is the last remaining candidate on its boundary, the vectors corresponding to the directed edges of the boundary cycle in the extended segment of c and the next edge (connecting the extended segment of c to the next candidate) sum to the zero vector, implying that the next candidate and c occupy the same node. To perform this summation in a local manner, c locally defines a two-dimensional coordinate system (e.g., by choosing two consecutive ports as the x and y axes, respectively) and uses two token passing schemes to generate and sum the x and y coordinates of these vectors in parallel. Again, due to space constraints, the details of this token passing scheme for summing x or y vector coordinates is detailed in the full version of this paper [9].

Using the token passing scheme, a candidate c can decide whether the next candidate along the boundary is itself. However, this is not sufficient to decide whether c is the last remaining candidate on the boundary. As described in Sect. 2.1, a particle can occur up to three times as different agents on the same boundary. Therefore, there can be distinct agents on the same boundary that occupy the same node of G_{eqt}. If an extended segment reaches from one of these agents to another, the vectors induced by the extended segment sum up to the zero vector even though there are at least two agents left on the boundary. To handle this case, each particle assigns a locally unique agent identifier from $\{1, 2, 3\}$ to each of its agents in an arbitrary way. The token passing scheme then additionally checks that the agent identifier of the last agent in the extended segment matches that of c, ensuring that c is the last remaining candidate on its boundary.

Finally, we must address the interaction between the solitude verification phase and the identifier comparison phase. As noted in the previous section, solitude verification may be triggered quite frequently. Therefore, it may occur that solitude verification is triggered for a candidate c while c is still performing a previously triggered execution of solitude verification. In this case, c simply continues with the already ongoing execution and ignores the request for another execution. Furthermore, c might be eliminated by the identifier comparison phase while it is performing solitude verification. In this case, c waits for the ongoing solitude verification to finish and only then withdraws its candidacy.

2.6 Boundary Identification

Once a candidate c determines that it is the only remaining candidate on its boundary, it initiates the boundary identification phase to check whether or not it lies on the unique outer boundary of the particle system. If it lies on the outer boundary, the particle acting as candidate agent c declares itself the leader. Otherwise, c revokes its candidacy. To achieve this, we make use of the observation that the outer boundary is oriented clockwise while an inner boundary is oriented counter-clockwise (see Fig. 1), a property resulting directly from the way the an agent's predecessor and successor are defined in Sect. 2.1.

A candidate c can distinguish between clockwise and counter-clockwise oriented boundaries using a simple token passing scheme introduced in [14]. It sends a token along the boundary that sums up the angles of the turns it takes according to Fig. 2, storing the results in a counter α. When the token returns to c, the absolute value $|\alpha|$ represents the external angle of the polygon induced by the boundary. It is well known that the external angle of a polygon in the Euclidean plane is $|\alpha| = 360°$. Since the outer boundary is oriented clockwise and an inner boundary is oriented counter-clockwise, we have $\alpha = 360°$ for the outer boundary and $\alpha = -360°$ for an inner boundary. The token can encode α as an integer k such that $\alpha = k \cdot 60°$. To distinguish the two possible final values of k it is sufficient to store k modulo 5 so that we have $k = 1$ for the outer boundary and $k = 4$ for an inner boundary. Therefore, the token only needs three bits of memory.

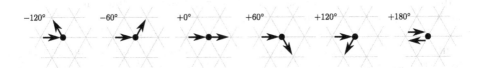

Fig. 2. Determining the external angle α. The incoming and outgoing arrows represent the directions in which the token enters and leaves an agent, respectively. Only the angle between the arrows is relevant; the absolute global direction of the arrows cannot be detected by the agents since they do not posses a common compass.

3 Analysis

We now turn to the analysis of the leader election algorithm. We first show its correctness in Sect. 3.1 and then analyze its running time in Sect. 3.2. Due to space constraints, some of the supporting lemmas and their proofs are omitted; they can be found in the analysis section of the full version of this paper [9].

3.1 Correctness

To show the correctness of the algorithm we must prove that eventually a single particle irreversibly declares itself to be the leader of the particle system and no other particle ever declares itself to be the leader. Any agent on an inner boundary can never cause its particle to become the leader; even if the algorithm reaches the point at which there is exactly one candidate c on some inner boundary, c will withdraw its candidacy in the boundary identification phase. Therefore, we can focus exclusively on the behavior of the algorithm on the unique outer boundary. We focus only on the major theorem here.

Theorem 1. *The algorithm solves the leader election problem, w.h.p.*

Proof. We must show that eventually a single particle irreversibly declares itself to be the leader of the particle system and no other particle ever declares itself to be the leader. Again, we consider only the agents on the outer boundary as agents on an inner boundary will never cause their particles to declare themselves as leaders. Once every particle has finished the boundary setup phase, every agent has finished the segment setup phase, and every candidate has finished the identifier setup phase, with high probability[3] there is a unique candidate c^* that has the highest identifier on the outer boundary. Since c^* has the highest identifier, it does not withdraw its candidacy during the identifier comparison phase. In contrast, every other candidate $c \neq c^*$ eventually withdraws its candidacy because the token sequence of c^* eventually traverses c.seg. Therefore, such an agent c cannot cause its particle to become the leader. Once c^* is the last remaining candidate on the outer boundary, it eventually triggers the solitude verification phase because the token sequence of c^* eventually traverses c^*.seg while c^* is not already performing solitude verification. After verifying that it is the last remaining candidate, c^* executes the boundary identification phase and determines that it lies on the outer boundary. It then instructs its particle to declare itself the leader of the particle system.   □

3.2   Running Time

Recall from Sect. 1.3 that the running time of an algorithm for leader election is defined as the number of asynchronous rounds until a leader is declared. Since the given algorithm always establishes a leader on the outer boundary, we can limit our attention to that boundary. Let n be the number of particles in the system and L be the number of agents on the outer boundary.

The first two phases of the algorithm, namely the boundary setup and segment setup phases, consist entirely of computations based on local neighborhood information. Therefore, these phases can be completed instantly by each particle upon its first activation. Since each particle is activated at least once in every round, every particle completes these first two phases after a single round. When an agent becomes a candidate, it initiates the identifier setup phase. We have the following lemma.

Lemma 1. *All candidates on the outer boundary complete the identifier setup phase after $\mathcal{O}(\log^2 n)$ rounds, w.h.p.*

After the identifiers have been generated, they are compared in the identifier comparison phase. In this phase, a set of digit tokens, one for each agent on the boundary, traverses the boundary against the direction of the cycle spanning it. Each agent can store at most two tokens. The tokens are not allowed to overtake each other, so agents maintain the order of the tokens when forwarding them. Note that a token is never delayed unless it is blocked by tokens in front of it.

---

[3] This w.h.p. guarantee results from there being a small but nonzero probability that either (a) all agents flip tails and become non-candidates in the segment setup phase, or (b) more than one candidate generates the same highest identifier in the identifier setup phase. See [9] for more details.

Therefore, an agent a forwards a token whenever a.pred can hold an additional token. Finally, an agent forwards at most one token for each activation.

We define the number of *steps* a token has taken as the number of times it's been forwarded from one agent to the next since its creation. Let T be the earliest round such that at its beginning every agent on the outer boundary has created its digit token. We have the following lemma.

Lemma 2. *At the beginning of round $T + i$ for $i \in \mathbb{N}$, each digit token on the outer boundary has taken at least i steps.*

Next, the following lemma provides an upper bound on the running time of the solitude verification phase.

Lemma 3. *For an extended segment of length ℓ, the solitude verification phase takes $\mathcal{O}(\ell)$ rounds.*

The boundary identification phase is only executed once a candidate determines that it is the last remaining candidate on the boundary. The following lemma provides an upper bound for the running time of this phase.

Lemma 4. *The boundary identification phase on the outer boundary takes $\mathcal{O}(L)$ rounds.*

Finally, we can show the following runtime bound.

Theorem 2. *The algorithm solves the leader election problem in $\mathcal{O}(L)$ rounds, w.h.p.*

Theorem 2 specifies the running time of the leader election algorithm in terms of the number of *agents* on the outer boundary. Let C be the number of *particles* on the outer boundary. Since each particle on the outer boundary corresponds to at most three agents on the outer boundary, we have that the algorithm solves the leader election problem in $\mathcal{O}(C)$ rounds, w.h.p.. Moreover, the number of particles on the outer boundary is obviously at most n; thus, the runtime bound can also be formulated as $\mathcal{O}(n)$ rounds, w.h.p.. Note that compared to the $\mathcal{O}(C)$ bound, the $\mathcal{O}(n)$ bound is quite pessimistic since the number of particles on the outer boundary can much lower than n. For example, a solid square of n particles only has $C = \mathcal{O}(\sqrt{n})$ particles on its outer boundary.

4   Conclusion

In this paper we presented a randomized leader election algorithm for programmable matter which requires $\mathcal{O}(n)$ asynchronous rounds with high probability. The main idea of this algorithm is to use coin flips to set up random identifiers for each leader candidate in such a way that at least one candidate has an identifier of logarithmic length, leading to a unique leader w.h.p.. In the full version of this paper [9], we consider several variants of the leader election problem and detail how our algorithm can be modified to solve them. These variants include allowing particle systems to contain both expanded and contracted particles, enforcing that all particles terminate their executions of the algorithm (instead of requiring only the leader to terminate), and improving the with high probability guarantee on electing a leader to a with probability 1 guarantee without changing the $\mathcal{O}(L)$, w.h.p. runtime bound.

References

1. Adleman, L.M.: Molecular computation of solutions to combinatorial problems. Science **266**(11), 1021–1024 (1994)
2. Angluin, D., Aspnes, J., Diamadi, Z., Fischer, M.J., Peralta, R.: Computation in networks of passively mobile finite-state sensors. Distrib. Comput. **18**(4), 235–253 (2006)
3. Boneh, D., Dunworth, C., Lipton, R.J., Sgall, J.: On the computational power of DNA. Discrete Appl. Math. **71**, 79–94 (1996)
4. Cannon, S., Daymude, J.J., Randall, D., Richa, A.W.: A Markov chain algorithm for compression in self-organizing particle systems. In: Proceedings of the 2016 ACM Symposium on Principles of Distributed Computing, PODC 2016, Chicago, IL, USA, 25–28 July 2016, pp. 279–288 (2016)
5. Chen, H.-L., Doty, D., Holden, D., Thachuk, C., Woods, D., Yang, C.-T.: Fast algorithmic self-assembly of simple shapes using random agitation. In: Murata, S., Kobayashi, S. (eds.) DNA 2014. LNCS, vol. 8727, pp. 20–36. Springer, Cham (2014). https://doi.org/10.1007/978-3-319-11295-4_2
6. Chen, M., Xin, D., Woods, D.: Parallel computation using active self-assembly. In: Soloveichik, D., Yurke, B. (eds.) DNA 2013. LNCS, vol. 8141, pp. 16–30. Springer, Cham (2013). https://doi.org/10.1007/978-3-319-01928-4_2
7. Cheung, K.C., Demaine, E.D., Bachrach, J.R., Griffith, S.: Programmable assembly with universally foldable strings (moteins). IEEE Trans. Rob. **27**(4), 718–729 (2011)
8. Chirikjian, G.: Kinematics of a metamorphic robotic system. In: Proceedings of the 1994 IEEE International Conference on Robotics and Automation, IRCA 1994, vol. 1, pp. 449–455 (1994)
9. Daymude, J.J., Gmyr, R., Richa, A.W., Scheideler, C., Strothmann, T.: Improved leader election for self-organizing programmable matter. CoRR, abs/1701.03616 (2017)
10. Derakhshandeh, Z., Dolev, S., Gmyr, R., Richa, A.W., Scheideler, C., Strothmann, T.: Brief announcement: amoebot - a new model for programmable matter. In: 26th ACM Symposium on Parallelism in Algorithms and Architectures, SPAA 2014, Prague, Czech Republic, 23–25 June 2014, pp. 220–222 (2014)
11. Derakhshandeh, Z., Gmyr, R., Porter, A., Richa, A.W., Scheideler, C., Strothmann, T.: On the runtime of universal coating for programmable matter. In: Rondelez, Y., Woods, D. (eds.) DNA 2016. LNCS, vol. 9818, pp. 148–164. Springer, Cham (2016). https://doi.org/10.1007/978-3-319-43994-5_10
12. Derakhshandeh, Z., Gmyr, R., Richa, A.W., Scheideler, C., Strothmann, T.: An algorithmic framework for shape formation problems in self-organizing particle systems. In: Proceedings of the Second Annual International Conference on Nanoscale Computing and Communication, NANOCOM 2015, Boston, MA, USA, 21–22 September 2015, pp. 21:1–21:2 (2015)

13. Derakhshandeh, Z., Gmyr, R., Richa, A.W., Scheideler, C., Strothmann, T.: Universal shape formation for programmable matter. In: Proceedings of the 28th ACM Symposium on Parallelism in Algorithms and Architectures, SPAA 2016, Asilomar State Beach/Pacific Grove, CA, USA, 11–13 July 2016, pp. 289–299 (2016)
14. Derakhshandeh, Z., Gmyr, R., Strothmann, T., Bazzi, R., Richa, A.W., Scheideler, C.: Leader election and shape formation with self-organizing programmable matter. In: Phillips, A., Yin, P. (eds.) DNA 2015. LNCS, vol. 9211, pp. 117–132. Springer, Cham (2015). https://doi.org/10.1007/978-3-319-21999-8_8
15. Doty, D.: Theory of algorithmic self-assembly. Commun. ACM **55**(12), 78–88 (2012)
16. Fukuda, T., Nakagawa, S., Kawauchi, Y., Buss, M.: Self organizing robots based on cell structures - CEBOT. In: Proceedings of the 1988 IEEE International Conference on Intelligent Robots and Systems, IROS 1988, pp. 145–150 (1988)
17. Kernbach, S. (ed.): Handbook of Collective Robotics - Fundamentals and Challanges. Pan Stanford Publishing, Singapore (2012)
18. McLurkin, J.: Analysis and implementation of distributed algorithms for multi-robot systems. Ph.D. thesis, Massachusetts Institute of Technology (2008)
19. Patitz, M.J.: An introduction to tile-based self-assembly and a survey of recent results. Nat. Comput. **13**(2), 195–224 (2014)
20. Rothemund, P.W.K., Winfree, E.: The program-size complexity of self-assembled squares (extended abstract). In: Proceedings of the Thirty-Second Annual ACM Symposium on Theory of Computing, Portland, OR, USA, 21–23 May 2000, pp. 459–468 (2000)
21. Winfree, E., Liu, F., Wenzler, L.A., Seeman, N.C.: Design and self-assembly of two-dimensional DNA crystals. Nature **394**(6693), 539–544 (1998)
22. Woods, D.: Intrinsic universality and the computational power of self-assembly. In: Proceedings of MCU 2013, pp. 16–22 (2013)
23. Woods, D., Chen, H.-L., Goodfriend, S., Dabby, N., Winfree, E., Yin, P.: Active self-assembly of algorithmic shapes and patterns in polylogarithmic time. In: Proceedings of the 4th Conference on Innovations in Theoretical Computer Science, ITCS 2013, pp. 353–354 (2013)
24. Yim, M., Shen, W.-M., Salemi, B., Rus, D., Moll, M., Lipson, H., Klavins, E., Chirikjian, G.S.: Modular self-reconfigurable robot systems. IEEE Robot. Autom. Mag. **14**(1), 43–52 (2007)

Conflict-Free Data Aggregation on a Square Grid When Transmission Distance is Not Less Than 3

Adil Erzin[1,2(✉)] and Roman Plotnikov[1]

[1] Sobolev Institute of Mathematics, Novosibirsk, Russia
{adilerzin,prv}@math.nsc.ru
[2] Novosibirsk State University, Novosibirsk, Russia

Abstract. In this paper a Convergecast Scheduling Problem on a unit square grid, in each node of which there is a sensor with transmission distance d which is not less than 3, is considered. For the cases $d = 1$ and $d = 2$, polynomial algorithms, which construct the optimal solution to the problem, are known. For an arbitrary d, an approximate algorithm is proposed, the application of which gives an upper bound on the length of the conflict-free data aggregation schedule, depending on d. We conducted a priori and a posteriori analysis of the accuracy of this algorithm for various d comparing either with the optimal length of the schedule, or with a lower bound, the value of which we improved.

Keywords: Wireless sensor networks · Data aggregation
Min-length conflict-free scheduling · Convergecast Scheduling Problem
Grid graph

1 Introduction

The elements of the wireless sensor network (WSN) must regularly transmit the collected data to the base station (BS). If the collected data is insignificantly related to a specific sensor, then it is enough to deliver *aggregated* information to the BS (e.g. min, max, mean, etc.). It is often assumed that the time of packet transfer along the edge of the communication graph is one time slot. Sensors, as usual, share common radio frequency to transmit messages. Therefore, if more than one transmitter is operating in the receiving area, the receiver cannot get the data packet intended for it because of such a phenomenon as *interference*. This situation is called a *conflict* or *collision*. Moreover, in *half-duplex* communication systems, the sensor cannot receive and transmit, and also receive more than one packet at a time. And, finally, for reason of energy saving, each sensor transmits a data packet only once during the data aggregation session. This means that in the communication network it is necessary to define the spanning *aggregation tree* (AT) with the root in the BS along the edges of which the packets are transmitted [6]. Obviously, any tree node cannot send a packet before receiving messages from all of its children.

In the *Convergecast Scheduling Problem* (CSP), it is required to find the AT, as well as the min-length schedule for the conflict-free aggregation of data [2, 7].

© Springer International Publishing AG 2017
A. Fernández Anta et al. (Eds.): ALGOSENSORS 2017, LNCS 10718, pp. 141–154, 2017.
https://doi.org/10.1007/978-3-319-72751-6_11

CSP is NP-hard even in the case of a given AT [3]. However, in the case when the communication graph is represented by a square unit grid, in each node of which there is a sensor (this is a unit disk graph in the L_1 metric, which naturally arises in the construction of *regular* covers, which use square *tiles* [9]), and the transmission distance is 1, the problem is polynomially solvable [5]. In [3] the case when the transmission distance is greater than 1 is considered and an approximate algorithm is proposed to construct a feasible conflict-free schedule. Later it was proved that when a transmission range is 2, the proposed in [3] algorithm constructs the *optimal* solution to the CSP on the grid graph [4].

In this paper, we use the Integer Programming Problem (IPP) from [8] to construct optimal schedules for the conflict-free aggregation of data on a grid with an arbitrary transmission distance and compare it with solutions constructed by the approximate algorithm. Unfortunately, the IPP can only be solved with small dimensions, so new lower bounds on the length of the schedule were found.

The rest of the paper is organized as follows. The mathematical formulation of the problem is given in Sect. 2. In Sect. 3, a linear graph is considered as a communication network. A CSP on a square grid is considered in Sect. 4. Simulation results are presented in Sect. 5, and the paper is concluded in Sect. 6.

2   Problem Formulation

We consider a WSN consisting of stationary sensor nodes with one sink – base station. We assume correct reception of a message if and only if there is no simultaneous transmission within proximity of the receiver. We assume that the interference range is equal to the transmission range. Then the WSN with sink node 0 can be represented as a graph $G = (V, E)$, where V denotes all the sensor nodes, and edge $(i, j) \in E$ if and only if the distance between the nodes i and j is within the transmission range.

The CSP considered in this paper is defined as follows. Given a connected undirected graph $G = (V, E)$ and a sink node $0 \in V$, find the minimum length conflict-free schedule of data aggregation from all the vertices of $V \setminus \{0\}$ to 0 under the following conditions:

- at the same time slot any vertex can either receive or send a message;
- each vertex can receive at most one message during one time slot;
- each vertex can send a message only once.

In the next sections we will specify the communication graph as linear graph or grid graph.

3   Linear Graph

In a linear graph the vertices $V = \{0, 1, \ldots, n\}$ are on the line and the distance between the neighbouring vertices i and $i + 1$, $i = 0, \ldots, n - 1$, is 1. The edge between vertices i and j exists if and only if the distance $|i - j|$ between them does not exceed d, where $d \geq 3$ is the transmission distance.

Suppose first that $n = Nd$, where N is a positive integer. We call the vertices with numbers d multiple *red*, and all others *blue*. The idea of the algorithm is as follows. At each time slot, if at least the two most remote vertices are red, then we send the packet from the outermost red vertex to the left by a distance d, if this does not lead to a conflict. Then we look for the most remote blue vertex and send the packet from it to the most remote vertex, the transmission to which does not lead to a conflict (taking into account the transmissions on the right). We repeat the last procedure as long as possible. If there are no valid transmissions, then delete all the vertices from which the packets were sent, increase the time and repeat the procedure for the remaining vertices. The pseudocode of the algorithm is given below.

Algorithm A_l.
Step 0. Set $t = 0$ and $W = \{1, \ldots, n\}$.
Step 1. Set $t = t + 1$ and $S = \emptyset$. (% S is the set of transmitting vertices)
    Step 1.1. If there are no blue vertices left between the two outermost red vertices of W, then the packet transfer from the outermost red vertex i to the left by a distance d (to the vertex $i - d$). Set $W = W \setminus \{i\}$ and $S = S \cup \{i\}$.
    Step 1.2. Find the outermost blue vertex $j \in W$ transmission from which does not lead to conflicts taking into account transfers from vertices of the set S. Send the packet from j to the most remote vertex from W, if this does not lead to a conflict taking into account transfers from the vertices of the set S. If such node j exists, then set $W = W \setminus \{j\}$ and $S = S \cup \{j\}$. If there is no such vertex j and $W \neq \emptyset$, then go to Step 1.
Step 2. Aggregation time is $T_h(n, d) = t$. Stop.

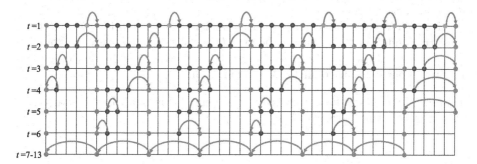

Fig. 1. Example of conflict-free data aggregation in the linear graph ($d = 5$). (Color figure online)

Lemma 1. *If $n = Nd + r$, $0 \leq r \leq d - 1$, then for $N \geq 2$ algorithm A_l constructs a feasible schedule of length at most $N + d + r$.*

Proof. Let $n = Nd$. After the time slot $d - 1$, one red vertex has already sent the packet and the outermost red vertex has the number $(N - 1)d$. It cannot send a packet at time $d - 1$. One more additional time slot is needed to provide the possibility of transfer to a distance d. We will show that $d + 1$ time slots are sufficient for all blue vertices to transmit packets to the nearest red vertices.

In algorithm A_l at the first time slot, blue vertices, distances between which are not less than d, transmit their packets. Assume that the distance between blue vertices with numbers less than $(N - 1)d$, that transmit packets during the first time round (let's call these vertices *green*), is $d + 2$ or (if the next vertex is red) $d + 3$ (Fig. 1). Such a change in the set of vertices that transmit at the first time slot does not reduce the length of the schedule. Therefore, if we prove a statement for such a slightly modified algorithm, then the statement is true for the original algorithm.

The elements of each group of blue vertices located between adjacent green vertices are transmitted in turn, beginning with the rightmost vertex. In each group there are d vertices. They transmit at $2, 3, \ldots, d + 1, d$ moments of time. Therefore, after the $(d + 1)$-th moment, all the blue vertices will send packets.

If $r > 0$, then it suffices to add r time slots to obtain the situation described above.

Lemma 2. *If $n = Nd$, then the length of the conflict-free aggregation schedule in the linear graph with n nodes cannot be less than $N + d - 1$, where d is the transmission distance.*

Proof. Assume the opposite, i.e. there is a schedule whose length is less than $N + d - 1$. Let D be the distance from vertex 0 to the most remote vertex that did not transmit data during the first d moments in such a schedule.

If $D \geq (N - 1)d$, then the vertex D needs at least $N - 1$ time slots for transmission, and therefore the length of the schedule cannot be less than $N + d - 1$. A contradiction.

If $D < (N - 1)d$, then during d time slots, all $d + 1$ most remote vertices must transmit. Therefore, during the first d time slots, the two vertices from the last $d + 1$ must transmit the packets simultaneously. But such transfer cannot be carried out without conflicts. A contradiction.

Remark 1. The lower bound proved in the Lemma 2 is tight. In order to verify this, it is sufficient to refer to Fig. 2.

4   Grid Graph

Suppose that communication graph is given as a square $(n + 1) \times (m + 1)$ grid with unit distance between the neighbour nodes. At each node of grid (x, y), $x = 0, 1, \ldots, n$, $y = 0, 1, \ldots, m$, except the origin $(0, 0)$, there is a sensor with the transmission distance $d \geq 3$ in L_1 metric.

In this section we describe the novel algorithm, which constructs an approximate solution to the considered problem on a grid. Also we give an upper bound on the length of a schedule constructed by the algorithm.

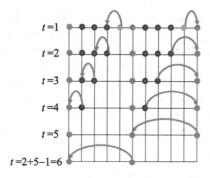

Fig. 2. An example of the tightness of the lower bound ($d = 5$). (Color figure online)

To further clarity the exposition, let us color the nodes, whose ordinate is multiple of d, in red, and let us color the other nodes in blue. Let $M = \lfloor m/d \rfloor, r_y = m - Md$. Let the i-th *row* be the set of nodes which ordinate equals i, where $i = 0, \ldots, m$. Let the j-th *strip* stands for the set of rows $(j-1)d + 1, \ldots, jd, j = 1, \ldots, \lfloor m/d \rfloor$. Let us call the *distance* between rows i and j, $i, j = 0, \ldots, m, i \neq j$ the value of $|i - j|$.

The algorithm A_{sq} of aggregation on square grid consists of two sequential stages: *vertical aggregation* and *horizontal aggregation*. At the first stage, in vertical aggregation, the data transmission is performed only upward or downward until the moment when the data from all vertices is transmitted to the elements of 0-th row. After that, at the second stage, the horizontal aggregation is performed with algorithm A_l described above. The algorithm A_v, which performs vertical aggregation, is described below.

For each time slot the algorithm A_v chooses the pairs of rows, each pair contains the row of sending vertices (*a row-transmitter*) and the row of receiving vertices (*a row-receiver*). The rows are chosen sequently from top to bottom. Note, that vertical transmission of a row at larger distance is more preferable, because in this case more vertices from row-transmitter are able to send simultaneously their packets without conflicts. Each time, the highest allowable blue row is selected as row-transmitter and then the row at largest distance not exceeding d from it is selected as row-receiver. If there are two candidates for a row-receiver, then the highest one is chosen. The transmission between the chosen pair of rows should not lead to a conflict with previously defined transmissions at the same time slot. The red row is chosen as a row-transmitter only in a case when all vertices above and $d-1$ blue rows below have already sent their packets. Given a row-transmitter and a row-receiver the method *TransmitRow* chooses a subset of vertices to transmit without conflicts. The pseudocodes of the algorithms A_v and *TransmitRow* (due to lack of space) are presented in Appendix section.

Lemma 3. *If $r_x \leq d-1, r_y \leq d-1, N \geq 2$, and $M \geq 4$ all positive integers, then algorithm A_{sq} constructs a feasible schedule on a grid $(Nd+r_x+1) \times (Md+r_y+1)$ of length at most $M + N + r_x + r_y + \lfloor d^2/4 \rfloor + 2d + \max\{1, \lfloor (d-1)/2 \rfloor\} - 2$.*

Proof. The length of a schedule constructed by the algorithm A_{sq} on a grid $(Nd + r_x + 1) \times (Md + r_y + 1)$ is equal to a sum of the length of a schedule of vertical aggregation obtained by the A_v on a grid $(Nd + r_x + 1) \times (Md + r_y + 1)$ and the length of a schedule obtained by the A_l on a linear graph with $Nd + r_x + 1$ nodes. As it follows from Lemma 1, the number of time slots of horizontal aggregation does not exceed $N + r_x + d$. According to the algorithm A_v, in each of the first r_y time slots the highest row from those, which did not transmit data yet, transfers data downwards by d units. It remains to prove that $M + \lfloor d^2/4 \rfloor + d + \max\{1, \lfloor (d-1)/2 \rfloor\} - 2$ time slots are sufficient to transfer the data to the vertices on the 0-th row from all other vertices.

First, we will prove four propositions. Then we will show that the trueness of Lemma 3 follows from these propositions.

Proposition 1 [3]. *All blue vertices of the M-th strip transmit the data during the first $d + \lfloor d^2/4 \rfloor - 1$ time slots.*

Proposition 2. *All blue vertices of the $(M-1)$-th strip transmit the data during the first $d - 1$ time slots.*

Proof. The blue vertices of the $(M-1)$-th strip are able to transmit the packets downwards at a distance d without conflicts, while the blue vertices of the M-th strip transmit their packets up to the Md-th row. According to the Step 1.3 of A_v, the blue vertices of the M-th strip transmit upwards during the first $\sum_{t=2}^{\lceil (d+1)/2 \rceil} t$ time slots. Let us show that $d - 1 \leq \sum_{t=2}^{\lceil (d+1)/2 \rceil} t$ for any integer $d > 0$. Consider two cases:

1. $d = 2l - 1$:
$\sum_{t=2}^{\lceil (d+1)/2 \rceil} t - (d - 1) = \sum_{t=2}^{l} i - 2l + 2 = 0.5(l - 1)(l + 2) - 2l + 2 = 0.2l^2 - 1.5l + 1 \geq 0 \; \forall l \in (-\infty; 1] \cup [2; +\infty)$;
2. $d = 2l$:
$\sum_{t=2}^{\lceil (d+1)/2 \rceil} t - (d-1) = \sum_{t=2}^{l+1} i - 2l + 1 = 0.5l(l+3) - 2l + 1 = 0.5l^2 - 0.5l + 1 > 0 \; \forall l \in (-\infty; +\infty)$.

In both cases, the inequality $d - 1 \leq \sum_{t=2}^{\lceil (d+1)/2 \rceil} t$ holds for any integer value of d.

Proposition 3. *All blue vertices of the $(M-2)$-th strip transmit the data during the first $d + \lfloor d^2/4 \rfloor + \lfloor (d - 1)/2 \rfloor - 2$ time slots.*

Proof. According to Proposition 2, during the first $d-1$ time slots all blue vertices of the $(M - 1)$-th strip transmit their packets downwards to the blue vertices of the $(M - 2)$-th strip at the distance d. At the time slot t ($1 \leq t \leq d - 1$) the $((M - 2)d - t)$-th row of the $(M - 2)$-th strip receive packets from the $((M - 1)d - t)$-th row of the $(M - 1)$-th strip. At the same time slot either $((M - 2)d - t - 1)$-th row of the $(M - 3)$-th strip (when $t < d - 1$) or the $((M - 4)d - 1)$-th row of the $(M - 4)$-th strip (when $t = d - 1$) transmits its packets downwards. This process is shown in Fig. 3 (the vertices, which already sent their packets, are marked grey).

After the end of the $(d-1)$-th time slot all blue vertices of the $(M-1)$-th strip and all blue vertices of the $(M-3)$-th strip, except the row $((M-3)d-1)$, have already sent their data, unlike the vertices of $(M-2)$-th strip and the $((M-3)d-1)$-th row. As it is shown in Fig. 4, during the next $\sum_{t=1}^{\lceil (d-1)/2 \rceil} t$ time slots the highest $\lceil (d-1)/2 \rceil$ rows of the $(M-2)$-th strip transmit their packets downwards to the row $((M-3)d-1)$. After that the remained $\lfloor (d-1)/2 \rfloor$ rows of the $(M-2)$-th strip transmit their packets upwards to the $(M-2)$-th row during the next $\sum_{t=2}^{\lfloor (d-1)/2 \rfloor +1} t$ time slots. In total, all blue vertices of the $(M-2)$-th strip transmit their packets during the first $d-1+\sum_{t=1}^{\lceil (d-1)/2 \rceil} t+\sum_{t=2}^{\lfloor (d-1)/2 \rfloor +1} t = d + \lfloor d^2/4 \rfloor + \lfloor (d-1)/2 \rfloor - 2$ time slots.

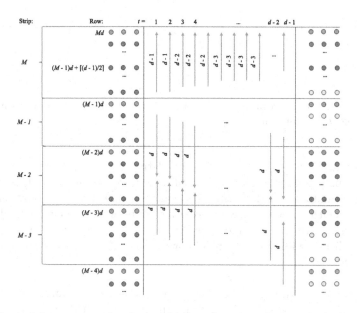

Fig. 3. Vertical data aggregation during the first $d-1$ time slots on the highest 4 strips in the grid graph with transmission range d. (Color figure online)

Proposition 4. *All blue vertices below the $(M-2)$-th strip transmit the data during the first $d + \lfloor d^2/4 \rfloor + \lceil (d-1)/2 \rceil + 1$ time slots.*

Proof. In order to facilitate understanding of this proof, the reader is referred to the Fig. 5 where the illustration of vertical aggregation on a grid with $d = 4$ and $M = 8$ is presented. As it is shown above, all the vertices of $(M-3)$-th row transmit their packets earlier than the vertices of $(M-2)$-th row, i.e., within $d + \lfloor d^2/4 \rfloor + \lfloor (d-1)/2 \rfloor - 2$ time slots. Let us prove that the vertices below the $(M-3)$-th strip transmit the data during the first $d + \lfloor d^2/4 \rfloor + \lceil (d-1)/2 \rceil + 1$ time slots. At the first time slot the data transmission below the $(M-3)$-th strip is performed in the following regular way. The row i transmits the packets downwards to the row $i-d$, the row $j = i-2d-1$ transmits the packets upwards

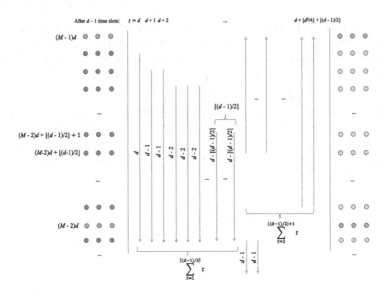

Fig. 4. Vertical data aggregation on the $(M-2)$-th strip of the grid graph with transmission range d. (Color figure online)

at the distance d (or $j = i - 2d - 2$ if $i - 2d - 1$-th row is red), the highest blue row below j transmits downwards at the distance d and so on. Note, that all rows transmitting data at the first time slot can be divided into pairs in such way that there is either one red row or no any rows between the elements of a pair, and there is exactly $2d - 2$ blue rows and 2 or 3 red rows between neighboring pairs. The distance between two neighboring transmitting pairs of rows is either $2d + 1$ or $2d + 2$.

The highest pair of rows, which transmit the packets at the first time slot below the $(M - 3)$-th strip, consists of rows $(M - 3)d - 2$ and $(M - 3)d - 3$, because the row $(M - 1)d - 1$ transmits the packets to the row $(M - 2)d - 1$. If $d > 3$, then at the second time slot the two highest blue rows below the $((M - 3)d - 3)$-th row transmit their packets at the distance d (one — upwards, another — downwards), and all the rows below (maybe except one row in the first strip) transmit the data by pairs at the distance d. If $d > 5$ then the data transmission at the third time slot is performed in a similar way again. Note, that the rows below the $(M - 3)$-th strip transmit the data in mentioned way during the first $\lceil (d - 1)/2 \rceil$ time slots, because in that period there exist a guaranteed receivers at the distance d for all transmitting vertices. Notice that after the $\lceil (d - 1)/2 \rceil$-th time slot all the blue rows, which did not transmit packets yet, can be grouped in such way that each group contains either d or $d + 1$ sequential blue rows (depending on the parity of d), and the distance between two groups is greater than d. Obviously, elements in one group are not able to receive a packet from elements of another group. Next, the highest rows of each group transmit the packets downwards at the distance d while the number of blue rows in group exceeds $d - 1$ — this is done by at maximum 2 time slots.

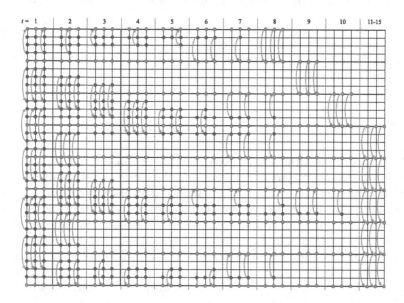

Fig. 5. Example of conflict-free vertical data aggregation in the grid graph ($d = 4$). (Color figure online)

It remains only to show that duration of data transmission of the remaining $d - 1$ rows in a group does not exceed $d + \lfloor d^2/4 \rfloor - 1$ time slots. In a worst case, at first, the highest $\lceil (d - 1)/2 \rceil$ blue rows transmit the data downwards to the lowest row of a group. Note that in this case the packet can be sent downwards only if the receiver is red or if there exists one red row between the transmitter and receiver, because otherwise transmission to the red row is preferable. Therefore, the highest $\lceil (d - 1)/2 \rceil$ blue rows transmit their packets not longer than $\sum_{t=2}^{\lceil (d+1)/2 \rceil} t$ time slots. After that the remaining $\lfloor (d - 1)/2 \rfloor$ blue rows sequentially transmit data to the most remote red row. Notice that the conflicts with other groups are not possible at this step. Therefore, the remaining $\lfloor (d - 1)/2 \rfloor$ rows transmit their packets not longer that $\sum_{t=2}^{\lfloor (d+1)/2 \rfloor} t$ time slots. We have $\sum_{t=2}^{\lceil (d+1)/2 \rceil} t + \sum_{t=2}^{\lfloor (d+1)/2 \rfloor} t = d + \lfloor d^2/4 \rfloor - 1$.

In total, the duration of data transmission from all blue rows below the $(M - 2)$-th strip can not exceed $\lceil (d - 1)/2 \rceil + 2 + d + \lfloor d^2/4 \rfloor - 1 = d + \lfloor d^2/4 \rfloor + \lceil (d - 1)/2 \rceil + 1$.

Let us return to the proof of the Lemma 3. Note that each red row transmits data downward at the distance d as soon as all rows above and $d - 1$ blue rows below have already transmitted their packets. Therefore, the upper bound of a schedule length obtained by the algorithm A_v is maximum of the next three values:

(1) $M + d + \lfloor d^2/4 \rfloor - 1$ (this estimate follows from Proposition 1)
(2) $M + d + \lfloor d^2/4 \rfloor + \lfloor (d-1)/2 \rfloor - 2$ (this estimate follows from Proposition 3)
(3) $M + d + \lfloor d^2/4 \rfloor + \lceil (d-1)/2 \rceil - 3$ (this estimate follows from Proposition 4)

Remark 2. In [3] the less accurate upper bound for the minimum length of a convergecast schedule on a square grid is found for the case when $r_x = r_y = 0$. The upper bound obtained above in Lemma 3 is less than the previous upper bound from [3] by $\min\{d, \lceil (d-1)/2 \rceil + 2\}$.

5   Simulation

All the proposed algorithms have been implemented in C++ using the Visual Studio Integrated Development Environment. We have run a simulation in order to compare the solutions obtained by the proposed algorithms with optimal solutions or, in cases when an optimal solution is not known, to check the proximity of the objective to the upper and lower bounds. We used CPLEX with IP formulation from [8] to obtain an optimal solution. In Table 1 the results of the experiment on linear graph are presented. The values marked italic stand for the cases when CPLEX failed to find an optimal solution within 1000 s, and the best feasible solution was taken. The two useful points should be noticed: (a)

Table 1. The length of a schedule obtained by A_l on a linear graph with $n+1$ vertices compared with CPLEX solution, lower and upper bounds.

n	d = 3				d = 4				d = 5				d = 6			
	CPLEX	A_l	LB	UB	CPLEX	A_l	LB	UB	CPLEX	A_l	LB	UB	CPLEX	A_l	LB	UB
2	2	2	-	2	2	2	-	2	2	2	-	2	2	2	-	2
3	3	3	3	4	3	3	-	3	3	3	-	3	3	3	-	3
4	4	4	-	5	4	4	4	5	4	4	-	4	4	4	-	4
5	4	5	-	6	5	5	-	6	5	5	5	6	5	5	-	5
6	4	4	4	5	5	6	-	7	6	6	-	7	6	6	6	7
7	5	5	-	6	5	6	-	8	6	7	-	8	7	7	-	8
8	5	6	-	7	5	5	5	6	6	7	-	9	7	8	-	9
9	6	6	5	6	6	6	-	7	6	7	-	10	7	8	-	10
10	6	6	-	7	6	7	-	8	6	6	6	7	7	8	-	11
11	6	7	-	8	7	7	-	9	7	7	-	8	7	8	-	12
12	7	7	6	7	7	7	6	7	7	8	-	9	7	7	7	8
13	7	7	-	8	7	8	-	8	*8*	8	-	10	8	8	-	9
14	7	8	-	9	*10*	8	-	9	*8*	8	-	11	8	9	-	10
15	8	8	7	8	8	8	-	10	-	8	7	8	*10*	9	-	11
16	8	8	-	9	8	8	7	8	-	9	-	9	*9*	9	-	12
17	*10*	9	-	10	8	9	-	9	-	9	-	10	*14*	9	-	13
18	*13*	9	8	9	9	9	-	10	-	9	-	11	*13*	9	8	9

in cases when $n = Nd$ algorithm A_l always yielded an optimal solution; (b) in cases when $n = Nd$ and $N \leq 2$ the objective of optimal solution equals to the lower bound $N + d - 1$, but in cases when $n = Nd$ and $N > 2$ the objective of optimal solution equals to the upper bound $N + d$. We believe that these two properties hold for any integer $N \geq 1$ and $d \geq 3$, but unfortunately we could not prove this theoretically.

Table 2 represents the results of A_{sq} on a square compared with the upper bound which equals to the sum of the upper bounds of A_l and A_v. Unfortunately the optimal solutions are not known for these cases because CPLEX failed to solve the problem on a square in acceptable time. The rather good lower bounds

Table 2. The length of a schedule obtained by A_{sq} on a grid graph with $(n+1) \times (n+1)$ vertices compared with upper bound.

n		10	11	12	13	14	15	16	17	18	19	20	21	22	23	24	25	26	27	28	29	30
$d = 3$	A_{sq}	15	17	15	16	18	17	18	20	19	20	22	21	22	24	23	24	26	25	26	28	27
	UB	-	-	15	17	19	17	19	21	19	21	23	21	23	25	23	25	27	25	27	29	27
$d = 4$	A_{sq}	20	21	19	20	22	23	19	20	22	23	21	22	24	25	23	24	26	27	25	26	28
	UB	-	-	-	-	-	-	19	21	23	25	21	23	25	27	23	25	27	29	25	27	29
$d = 5$	A_{sq}	23	25	27	29	29	24	25	27	28	29	24	25	27	28	29	26	27	29	30	31	28
	UB	-	-	-	-	-	-	-	-	-	-	24	26	28	30	32	26	28	30	32	34	28
$d = 6$	A_{sq}	27	28	28	30	32	33	34	35	30	31	33	34	35	36	29	30	32	33	34	35	31
	UB	-	-	-	-	-	-	-	-	-	-	-	-	-	-	29	31	33	35	37	39	31

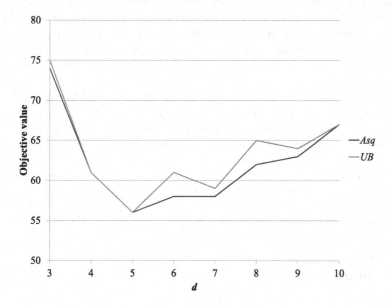

Fig. 6. Objective of solution obtained by the algorithm A_{sq} compared with upper bound for the CSP on a square grid 101×101 with different d. (Color figure online)

are not known as well: the best one is $(n+m)/d+d+1$ which is rather far from the upper bound. The upper bounds were calculated for the cases when $n = Nd + r$ and $N \geq 4$ because of the conditions of Lemma 3. It should be noticed, that the upper bound appeared to be tight for the cases when n is multiple of d. In Fig. 6 the objectives obtained by the algorithm A_{sq} and the upper bounds for a case when $n = m = 100$ in dependency of d are presented. The length of obtained schedule is minimum when $d = 5$. If $n = m = 1000$, then the minimum schedule length is 248, when $d = 13$. In the general case, the function of the upper bound of length of the schedule, depending on d, is convex.

6 Conclusion

In this paper, we present a new algorithm for approximate solution of the conflict-free convergecast scheduling problem on a unit square grid. We found a guaranteed upper and lower bounds on objective value in linear case and an upper bound in general case. The experiment showed, that the proposed algorithm allows to obtain an optimal solution on a linear graph in cases when the number of nodes except the base station is multiple of the transmission distance. Also, according to the experiment results, the obtained upper bound in general case is tight.

In general, CSP is NP-hard even when the aggregation tree is given [3]. We have considered a special case of the problem that is interesting not only from the theoretical point of view, but can also be used to assess the quality of the functioning of the WSN. For example, in regular covers, the sensors are located at the vertices of a regular grid, each cell (tile) of which is either a triangle, a square, or a hexagon [9]. When the sensors are placed arbitrarily, each sensor can be assigned to one of the grid nodes. This will make it possible to select the transmission distance and estimate the data aggregation time in the original sensor network [1].

In future we plan to define a rather accurate lower bound for vertical aggregation in order to estimate quality of the proposed algorithm. Also we plan to implement and run the best of known heuristics for the conflict-free convergecast scheduling problem and compare them with our approach.

Acknowledgments. The research of A. Erzin is partly supported by the Russian Foundation for Basic Research (grants 16-07-00552 and 17-51-45125) and by the Ministry of Science and Education of the Russian Federation under the 5–100 Excellence Programme. The research of R. Plotnikov is partly supported by the Russian Foundation for Basic Research (grant 16-37-60006).

A   Appendix

Algorithm A_v.
Step 0. Set $t = 0$, $W = \{0, \ldots, n\}^m$, $S = \emptyset$. (% W is arrays of vertices which did not transmit data yet, grouped by rows, S is a schedule stored as an array

of arrays of pairs of vertices: each pair contains the sender and the receiver, the number of the array where the pair is placed stands for the time slot)

Step 1. Set $s = max\{i : W[i] \neq \emptyset\}$.

If $s = 0$ then **Stop**. The vertical aggregation schedule is $S = \{S_1, \ldots, S_t\}$. The length of the schedule is t;

Set $t = t + 1$; Set $S_t = \emptyset$;

Step 1.1. If $s > Md$, then set $S_t^s = TransmitRow(W, s, s - d)$, set $S_t = S_t \cup S_t^s$ and go to Step 1. Otherwise, go to Step 1.2.

Step 1.2. If $s > 0$, s is multiple of d and each element from $\{W_{s-d+1}, \ldots, W_{s-1}\}$ is empty, then set $S_t^s = TransmitRow(W, s, s - d)$, set $S_t = S_t \cup S_t^s$ and go to Step 1.4. Otherwise, go to Step 1.3.

Step 1.3. If $s > (M - 1)d$, then do the following.

Set $s = min\{i : i > (M - 1)d$ and $W_i \neq \emptyset\}$;

Set $r = M$ if $s - (M - 1) < \lceil (d - 1)/2 \rceil$ or set $r = M - 1$ otherwise;

Set $S_t^s = TransmitRow(W, s, r)$;

Set $S_t = S_t \cup S_t^s$;

Set $s_{prev} = s$;

Set $s = min\{s - 1, r - d - 1\}$;

Step 1.4. While $s > 0$ do the following.

If the s-th row is red or W_s is empty, then set $s = s - 1$ and go to Step 1.4.

Find the most remote blue row r from the row s, such that $|s - r| < d + 1$, $r < s_{prev} - d$ and W_r is not empty. If there are two appropriate rows, then take the maximum of them. If there is no any appropriate row, then set $s = s - 1$ and go to Step 1.4.

Set $S_t^s = TransmitRow(W, s, s - d)$;

Set $S_t = S_t \cup S_t^s$;

Set $s_{prev} = s$;

Set $s = min\{s - 1, r - d - 1\}$;

Algorithm $TransmitRow(W, s, r)$.

Set $p = 0$; $A = \emptyset$;

while $p < n$ **do**

    Add a pair of vertices $((W_{s,p}, s), (W_{s,p}, r))$ to A;

    Set $p_{next} = p + 1$;

    while $p_{next} < n$ **and** $W_{s,p_{next}} - W_{s,p} < d + 1 - |s - r|$ **do**

        $p = p + 1$;

    Set $W_s = W_s \setminus W_{s,p}$;

    Set $p = p_{next}$;

return A.

References

1. Aldyn-ool, T.A., Erzin, A.I., Zalyubovskiy, V.V.: The coverage of a planar region by randomly deployed sensors. Vestn. Novosib. Gos. Univ. Ser. Mat. Mekh. Inform. **10**(4), 7–25 (2010)
2. De Souza, E., Nikolaidis, I.: An exploration of aggregation convergecast scheduling. Ad Hoc Netw. **11**, 2391–2407 (2013)
3. Erzin, A., Pyatkin, A.: Convergecast scheduling problem in case of given aggregation tree. The complexity status and some special cases. In: 10th International Symposium on Communication Systems, Networks and Digital Signal Processing, article 16, 6 p. IEEE-Xplore, Prague (2016)
4. Erzin, A.: Solution of the convergecast scheduling problem on a square unit grid when the transmission range is 2. In: Battiti, R., Kvasov, D.E., Sergeyev, Y.D. (eds.) LION 2017. LNCS, vol. 10556, pp. 50–63. Springer, Cham (2017). https://doi.org/10.1007/978-3-319-69404-7_4
5. Gagnon, J., Narayanan, L.: Minimum latency aggregation scheduling in wireless sensor networks. In: Gao, J., Efrat, A., Fekete, S.P., Zhang, Y. (eds.) ALGOSENSORS 2014. LNCS, vol. 8847, pp. 152–168. Springer, Heidelberg (2015). https://doi.org/10.1007/978-3-662-46018-4_10
6. Incel, O.D., Ghosh, A., Krishnamachari, B., Chintalapudi, K.: Fast data collection in tree-based wireless sensor networks. IEEE Trans. Mob. Comput. **11**(1), 86–99 (2012)
7. Malhotra, B., Nikolaidis, I., Nascimento, M.A.: Aggregation convergecast scheduling in wireless sensor networks. Wirel. Netw. **17**, 319–335 (2011)
8. Tian, C.: Neither shortest path nor dominating set: aggregation scheduling by greedy growing tree in multihop wireless sensor networks. IEEE Trans. Veh. Technol. **60**(7), 3462–3472 (2011)
9. Zalyubovskiy, V., Erzin, A., Astrakov, S., Choo, H.: Energy-efficient area coverage by sensors with adjustable ranges. Sensors **9**(4), 2446–2460 (2009)

Uniform Dispersal of Robots with Minimum Visibility Range

Attila Hideg[1](✉) and Tamás Lukovszki[2]

[1] Department of Automation and Applied Informatics,
Budapest University of Technology and Economics, Budapest, Hungary
attila.hideg@aut.bme.hu
[2] Faculty of Informatics, Eötvös Loránd University, Budapest, Hungary
lukovszki@inf.elte.hu

Abstract. We consider the filling problem, in which autonomous mobile robots enter a connected orthogonal area from several entry points and have to disperse in order to reach full coverage. The entry points are called doors. The area is decomposed into cells. The robots are autonomous, anonymous, they have a limited visibility range of one unit, and do not use explicit communication. Collision of the robots is not allowed. First we describe an algorithm solving the filling problem for the single door case in $O(n)$ time steps in the synchronous model, where n is the number of cells in the area. This algorithm is optimal in terms of visibility range, and asymptotically optimal in running time and size of persistent memory used by the robots. Moreover, we show that our algorithm solves the multiple door filling problem in $O(n)$ time, as well. For the multiple door case, our algorithm is asymptotically worst-case optimal, and its running time is at most k times the running time of the optimal algorithm for any input, where k is the number of doors.

1 Introduction

In swarm robotics a huge number of simple, cheap, tiny robots can perform complex tasks collectively. The greatest advantages of such systems are scalability, reliability, and fault tolerance. Contrary to a single-robot system, which requires complex, expensive hardware and software components with redundancy, the same attributes can be achieved by simply adding more robots to a multi-robot system. In case of mobile robots, the spatial distribution of the robots has huge advantage in problems related to exploration, coverage, demining, toxic waste cleanup, etc. In this paper we study the *uniform dispersal* (or *filling*) of synchronous robots in an unknown, connected area.

The area is decomposed into cells and the robots are injected one by one into the area through an entry point, which is called the *door*. The robots have to reach full coverage by occupying all cells. This problem is called filling, and was introduced by Hsiang et al. [5]. When more than one door is present in the area the problem is called *multiple door filling* or *k-door filling*.

A. Fernández Anta et al. (Eds.): ALGOSENSORS 2017, LNCS 10718, pp. 155–167, 2017.
https://doi.org/10.1007/978-3-319-72751-6_12

In [5] the goal was to achieve a rapid filling, minimizing the *make-span* (time to reach full coverage) of the algorithm. Their solution required $2n - 1$ cycles to reach full coverage, where n was the number of cells in the area.

Barrameda et al. [1] investigated the minimum hardware requirements and the possibilities of solving the filling problem by robots with constant visibility radius, communication range, and constant number of bits of persistent memory.

The algorithms in [1,2,5] used the *Leader-Follower* method, where one robot is elected as a leader and the rest of the robots follow it until the leader is blocked, then the leadership is transferred to another robot. The leader-follower method results in a DFS-like dispersion in the area. Collisions are prevented through the property that the leader explores new cells (cells that were never visited before) and the followers simply moving towards their predecessor (the robot they are following).

In case of the multiple door filling problem the robots enter through multiple doors and there are several leaders in the area. In [1] the robots were colored according to the door they entered, and the robots required to have their color visible to other robots within the visibility range. In [3] Das et al. showed that allowing visible colors or lights yields a more powerful computational model than allowing infinite visibility range but no lights.

In this paper a fundamental question is, whether it is possible to reduce the hardware requirements of the robots and still fill an unknown connected orthogonal area, even in presence of holes, and maintain $O(n)$ runtime. These hardware requirements are: visibility range, size of persistent memory, and avoidance of explicit communication and the usage of lights.

1.1   Our Contribution

We present a method for filling an unknown, connected orthogonal region S consisting of $|S| = n$ square shaped cells by a set of n autonomous anonymous robots with a visibility radius of 1 hop in $O(n)$ time in the synchronous computational model. The robots require $O(1)$ bits of persistent memory and cannot communicate (they do not use explicit communication, nor colors or lights). The only precondition is that they require a common coordinate system.

First, we consider the single door case, and present an algorithm which solves the problem without collisions and terminates in $O(n)$ time. Then, we show that the presented approach solves both the single door and multiple door filling problem in orthogonal areas.

Regarding this model our algorithm is optimal in terms of visibility range and asymptotically optimal in the size of the memory. Moreover, it is asymptotically optimal in running time in the single door case, asymptotically worst-case optimal in the multiple door case, and its running time is at most k times the running time of the optimal algorithm for any input. The optimality regarding the visibility range follows from the fact that with a visibility range less than 1 the robots cannot even distinguish between occupied and unoccupied neighboring cells. The asymptotic optimality of the memory size $O(1)$ follows from the result by Barrameda et al. [1]; they proved that oblivious (memoryless) robots cannot deterministically solve the problem. The asymptotic optimality of the running time $O(n)$ follows from the fact that we can place one robot per round in the single door case and n robots must be placed. For the asymptotically worst-case optimality for the k-door case, we show inputs where almost all robots must pass through a single cell to get into a large component of the area. This single cell behaves similarly to a single door, and the dispersion must take $\Omega(n)$ time.

Organization: In Sect. 2 we define our model. In Sect. 3 we present previous results on the filling problem and related problems. Then in Sect. 4 we describe and analyze our algorithm for the single door and for the multiple door cases. Finally, Sect. 5 summarizes the paper.

2   Model

We are given an orthogonal area, i.e. polygonal with sides either parallel or perpendicular to one another, which is decomposed into equal sized, square shaped cells (see [2]). The size of each cell allows only one robot to occupy it at any given time. Each cell has at most four adjacent cells in fixed directions: North, South, East, and West.

The robots' actions are divided into three phases: *Look, Compute,* and *Move.* During the Look phase, the robots take a snapshot of their surroundings, in the Compute phase they perform their computations (e.g. which action should they perform), and during the Move phase they move there. This is called the Look-Compute-Move (LCM) model, which is commonly used in distributed robotics.

When the robots perform their LCM cycles at the same time, the model is called fully synchronous (FSYNC). In the FSYNC model, each robot takes snapshots at the same time, compute, and move at the same time based on a global tick.

The robots are *anonymous*, i.e. they cannot distinguish each other, and are equipped with limited hardware. They have a visibility range of 1 hop, i.e. each robot can 'see' only the cells which they are occupying and the cells adjacent to it. In one LCM cycle a robot can move to one of its neighboring cells or stay at place. The robots are *silent*, i.e. they cannot communicate at all. They are *finite-state* robots, i.e. they have a constant number of bits of persistent memory. The robots have a common notion of North, South, East, and West.

The entry points, called *doors* are always occupied by a robot. Whenever a robot moves from a door cell, a new one is placed there. The doors cannot be distinguished from other cells by the robots, moreover, the robots do not know which door they used to enter the area.

3   Related Work

Hsiang et al. [5] investigated the make-span (i.e. the time to reach full coverage) of filling of a connected orthogonal region measured in rounds. They assumed that robots have a limited ability to communicate with nearby robots, i.e. a robot is able to exchange a constant-size message. They proposed two solutions, BFLF and DFLF, both modeling generally known algorithms: BFS and DFS. In DFLF, the method maintained a distance of 2 hops between the robot and its successor. As a consequence the method only required visibility range of 2 hops. However, the robots had to be able to detect the orientation of each other.

Barrameda et al. [1] assumed common top-down and left-right directions for the robots and showed that robots with visibility range of 1 hop and 2 bits of persistent memory can solve the problem in an orthogonal area if the area does not contain holes, without using explicit communication. Holes are cells in the area which cannot be occupied by robots (e.g. obstacles).

In [2] Barrameda et al. presented two methods for filling an unknown orthogonal area in presence of obstacles (holes). Their first method, called **TALK**, requires a visibility range of 1 hop if the robots have explicit communication. The other method, called **MUTE**, do not use explicit communication between the robots, but it requires visibility range of 6. Both methods need $O(1)$ bits of persistent memory.

For the multiple door case, in [1] a method, called **MULTIPLE**, has been presented. It solves the problem for robots with visibility range of 2 hops, no explicit communication and a $O(1)$ bits of persistent memory. In this solution the doors are colored with different colors and the robots are colored according to the door they enter. The color of the robots is visible to other robots within the visibility range.

A summary of these previous results and a comparison to our contribution is presented in Table 1.

Table 1. Summary of the requirements of Filling algorithms.

Method	Visibility range (hops)[a]	Comm. range (hops)[a]	Memory (bits)	Area
DFLF [5]	2	2	2	Arbitrary
TALK [2]	2	2	4	Orthogonal
MUTE [2]	6	0	9	Orthogonal
MULTIPLE [1]	3	0	4	Orthogonal
Our method: (here)				
Single door	1	0	13 bits	Orthogonal
Multiple doors	1	0	13 bits	Orthogonal

[a]In [2], grid cells sharing a common edge or a common corner with the current cell c of the robot are assumed within the visibility range of the robot. In our model a cell sharing only one corner with c has a hop distance of two. Thus, it is outside of the visibility range. Only the cells sharing a common edge have hop distance of one.

In [6] a related problem, the pattern formation problem has been investigated in the FSYNC model. In the pattern formation problem n robots are placed arbitrarily in the 2 dimensional grid and they have to form a given connected pattern F, known for all robots. In [6] a common coordinate system for the robots has been assumed. The presented solution requires no explicit communication, a visibility range of 2 hops, and 2 bits of persistent memory. The robots are gathered in a certain point. They reach this point one-by-one and then start to traverse a spanning tree of F and fill the tree by this traversal. The pattern formation algorithm needs $O(|F| + d)$ rounds, where d is the diameter of the initial configuration. Only considering the filling phase of this algorithm, it needs $O(|F|)$ rounds.

For an excellent overview of distributed mobile robotics, we refer to the book by Flocchini et al. [4].

4 Method

In this section we describe the algorithm for the filling problem, i.e. we define the states and state-transitions of the robots. Then we prove that the algorithm solves both the single door and multiple door filling problem without collisions.

4.1 Concept

Our algorithm is based on the leader-follower approach, which means a certain robot is the leader while the rest are following it. This is a common approach to eliminate collisions, as the leader is the only robot which is capable of moving to cells, that were never occupied before. Such cells are called *free cells*. Each follower has a *predecessor* which is the robot it is following during the dispersion, and each robot, except the one at the door, has a *successor* which follows it.

Each robot stores its state, which can be one of the following: *None, Leader, Follower, Stopped*. The state of the robot is not visible for other robots. The transitions between these states are shown in Fig. 1. The robots placed at the door are initialized with None state. They can switch to either Leader or Follower state. In Follower state, the robot can switch to Leader state if and only if its predecessor was the Leader and that Leader switched to Stopped state. This ensures that the number of Leaders does not increase.

The robots repeatedly perform their LCM cycles. We label each cycle with a direction and call it a *step*. There are four possible directions: North, East, West, South; the steps are labeled correspondingly N-step, E-step, S-step, W-step. In each step the robots can only move toward the cell in the corresponding direction. A *round* is a sequence of four consecutive steps starting with an N-step, followed by an E-step, followed by an S-step, ending with a W-step (see Fig. 2). The rounds and the steps start synchronously at the robots.

After a robot r is placed, in the first round (and then in every odd round) it always observes its surroundings, then it can only move during the second round (and during every even round), irrespective of its current state. Note: odd and even rounds are relative to the starting round for r.

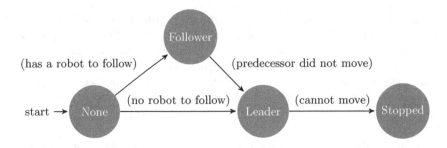

Fig. 1. States and state-transitions of the robots. The edges are labeled with the condition for the transition.

R_1				R_2				R_3			
s_1	s_2	s_3	s_4	s_5	s_6	s_7	s_8	s_9	s_{10}	s_{11}	s_{12}
N-step	E-step	S-step	W-step	N-step	E-step	S-step	W-step	N-step	E-step	S-step	W-step

Fig. 2. The structure of the rounds. Three rounds (denoted by R_1, R_2, R_3), each consists of four consecutive steps in fixed order: step s_1 is an N-step, s_2 is an E-step and so forth.

Our algorithm mimics a DFS exploration of the unknown region simultaneously started from the doors. A current Leader defines a path in the DFS tree from the root (door) it entered, which path is traversed by it. When the Leader switches to Stopped state it is in a leaf of the DFS-tree. The Follower robots will fill the path segment between the last branch vertex and the Leader (in the leaf). Then the leadership transfers backwards until we reach a branch vertex. Then the robot on that branch vertex becomes a new Leader and traverses to the next branch. As the destination selection is deterministic, the robots on branch vertices are checking the neighbors in the same order, even without the explicit

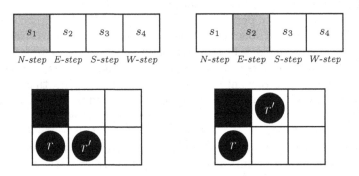

Fig. 3. Robot r following r'.

knowledge of the DFS-trees, or if the predecessor becomes Stopped. Therefore, they will move in the direction of the first free branch.

The main difficulty to realize this concept is that it must be ensured during the algorithm that the Leaders are able to determine which neighboring cells are free and the Followers must know where their predecessors are, despite the fact that the predecessors can be outside of the 1 hop visibility range. Now we describe the exact behavior of the robots in the different states.

Stopped: If a robot r is in Stopped state it does not move anymore. Other robots will treat it like it would be an obstacle.

Follower: If a robot r is a Follower it follows its predecessor r' until r' stops for two consecutive rounds (r' is not able to move in its even round). Then the leadership will be transfered from r' to r. In the odd rounds of r it checks the occupancy of the cell where its predecessor is. When r takes a snapshot in the Look phase in its odd rounds it always sees its predecessor in the N-step. If r' moves in a certain step, then r will not see r' in the next step. E.g., if r' moves to north in the N-step, then after the Look phase of the E-step r does not see r', but it knows that r' moved to the north. Therefore, r knows where r' is, even if the distance between them is 2 and r does not have r' in its visibility range. In the next round r occupies the former cell of r' and r' becomes visible again. In every even round r moves toward its predecessor, occupying the previous cell of r' in the step corresponding to the direction r moves. As a result, the successor of r also knows which direction r moves to and follows r.

There might be a case during the odd round of r, that r' does not move in that round. It is only possible if r' does not have any cells it can move to, either because it is surrounded by obstacles or other robots. It implies that r' switches to Stopped state and r can switch to Leader state after this round (more precisely, after the Look phase of the N-step of the next round, when r recognizes that r' did not move in the odd round of r). We refer to this event as the *transfer of the leadership*. It guarantees that r can only switch to Leader state after its predecessor r' switched to Stopped state. Therefore, the invariant that there can only be one Leader at any time (or k Leaders in the multiply door case) is fulfilled. Note: r switches to Leader state only in the N-step of its even round. Then in that round r performs the actions of robots in Leader state.

Leader: If r is a Leader it moves to a free cell in every second (even) round. In each step of every odd round r checks each of its neighboring cells if it is occupied. Occupied cells can not be free and they can be excluded from the set of potential moving directions in the next (even) round. In the next (even) round in each step r checks the neighboring cell c corresponding to the direction of the step again. If c is unoccupied and it has been unoccupied in the previous (odd) round, r moves to c and r does not check the remaining directions in that round. If r cannot move in any direction it switches to Stopped state and terminates its actions. Note: r can switch to Stopped state in the same round it switched to Leader state.

None: After r has been placed at a door in round R_i, its initial state is None. It only has to know, which step is currently performed by the system, i.e. an N-, E-, S-, or W-step. If the current step is an N-step and no robots are visible, then r is the first robot placed at that door and it has to switch to Leader state. In every other scenario, its predecessor robot r' is in the direction corresponding to the previous step, e.g., if the current step is an E-step then the predecessor r' leaved in the previous step, which is an N-step. Therefore, it must be in the northern neighbor cell. Let $R_{i'}$ be the round in which r' has moved from the door. Note that $i' = i - 1$ if r is placed in an N-step, and $i' = i$ otherwise. If $i' = i$ then round R_i is a shortened round for the newly placed robot r and r does not perform any actions in this round. Then the first round of r starts with the next N-step.

As the predecessor r' of the newly placed robot r has moved during round $R_{i'}$, round $R_{i'+2}$ is the next even round for r' (i.e. the next round r' moves). We have to ensure that $R_{i'+2}$ will be an odd round round for r. Therefore, the only action r performs in $R_{i'+1}$ is switching in Follower state at the end of the round. The first round when r performs its actions is $R_{i'+2}$, which is an odd round for r, and it acts as a Follower. Note: in $R_{i'+3}$, robot r can switch to Leader state if it recognizes that r' is in Stopped state, and r can switch even to Stopped if r cannot find any free cells.

4.2   Single Door Case

First, we show that the described algorithm fills the area without collisions in the case, when the robots are entering in a single door.

Lemma 1. *No collisions can occur during the dispersion.*

Proof. In each step each cell can only be occupied from one sole direction. Even if two robots would move to the same cell during the same round, only one will move there, depending on the direction it wants to enter.   □

Lemma 2. *During each step each Follower knows where its predecessor is.*

Proof. Consider a Follower r. Let r' be its predecessor. Based on the timing of the movement, i.e. in which step r' moved, r knows the target cell of r'. With other words, when a robot r sees its predecessor r' on cell c during step s_i, but if r detects c is unoccupied in s_{i+1}, it implies that r' has moved during s_i. The robot r' can only move to one direction which corresponds to the direction of s_i. Consequently, after the Look phase of s_{i+1} the robot r knows where its predecessor r' is, even if r' is outside of its visibility range. In the next round r moves to c and r' becomes visible again.

The argument above does not apply when r' moves from the door, as r has not been placed yet. However, r knows, in which step it has been placed. Let s_i be the step in which r' has moved from the door and s_{i+1} the step, in which r has been placed. Then r' had to move in the direction corresponding to step s_i. Therefore, r knows where r' is.   □

Lemma 3. *The Leader only moves to free cells.*

Proof. The leader has to move to free cells to increase the coverage with each of its movement and to prevent infinite loops. Based on one single snapshot the Leader r can only distinguish between occupied and unoccupied cells. Let R_i be the round in which r has moved (its even round), then R_{i+1} will be its odd round. In R_{i+1}, if a neighboring cell is occupied at any time during R_{i+1}, r recognizes it as not being free. In the next (even) round R_{i+2} in each step r checks the neighboring cell c corresponding to the direction of the step again. If c is unoccupied and it has been unoccupied in the previous round, r recognizes c as a free cell and moves to c. Then r does not check the remaining directions in that round.

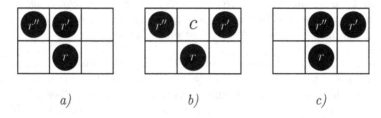

$a)$ $b)$ $c)$

Fig. 4. A non-free cell c and its neighborhood in three consecutive rounds.

Assume r would like to determine whether a neighboring cell c is free in R_i, and assume c is not a free cell. There are two cases: c contains a robot at the beginning of R_i (Fig. 4(a) and (c)) or not. If c is occupied, then r recognizes it as not being free and will not try to move there. If c is unoccupied at the beginning of R_i and it is not a free cell (Fig. 4(b)), there was a robot r' in c at the beginning of R_{i-1} and moved from it to a neighbor of c which is not visible by r. Then, r'' (the successor of r') will enter c in R_i to follow r'. At the beginning of the next round (R_{i+1}), c will be occupied by r'' (as in Fig. 4(c)) and r will not move there.

As a result two consecutive rounds, R_i and R_{i+1}, are sufficient for r to identify c as a non-free cell. In other case, c is identified as a free cell and r can move to c in R_{i+1}.   □

Lemma 4. *At most one Leader is present in the area.*

Proof. Recall the event *transfer of the leadership*. The Leader r switches to Stopped state during R_i. Its successor r' can only switch to Leader state once it has detected that r is in Stopped state, which will be in the first step of R_{i+1}.   □

Lemma 5. *The proposed method fills the area.*

Proof. For contradiction, assume a cell c is left unoccupied after the algorithm is terminated. There are two cases: (i) c is unoccupied but not free (some robots visited c already but c left it unoccupied), (ii) c is a free cell.

For (i), assume r is the last robot which left the cell c. The successor of r must have followed r and occupied c, which is a contradiction.

For (ii), consider a cell c' which is a closest occupied cell to c. There is a path between c and c' consisting only of unoccupied cells (or c' is a neighbor of c). By (i) all unoccupied cells are free cells, as well. Let r be the robot in c', which is in Stopped state as the algorithm has been already terminated. However, r has a free neighboring cell, therefore, r can move there. This prevents r from switching to Stopped state, which is a contradiction.

□

Theorem 1. *By using the presented algorithm a connected orthogonal area with a single door is filled in $O(n)$ rounds without collisions by robots with visibility range of 1 hop and $O(1)$ bits of persistent memory.*

Proof. According to the previous lemmas, the algorithm fills the area and prevents collisions. The robots move in their even rounds. Whenever a robot r is placed at the door, its first odd round is started with the first N-step, i.e. if r is placed in a E-, S-, or W-step, it waits for the next N-step. Then in the odd round it just observes and in the next even round it moves, allowing the placement of a new robot. Therefore, a new robot can be placed in every third round. Since the number of cells in the area is n, placing n robots requires $3n$ rounds, which takes $12n$ LCM cycles.

The persistent memory of the robots must store the state of the robots (2 bits), the parity of the current round (1 bit), and the step of the current round (2 bits). Leaders must additionally store information about the occupancy of the neighboring cells observed in the odd round (4 bits). Followers must store the direction they will move in their next even round (2 bits), which is the direction their predecessor is in their odd round. Additionally, Followers must store where the predecessor moves in their odd round (2 bits) and the information about the occupancy of the neighboring cells for the case when the Follower has to switch to Leader state.

□

4.3 Multiple Door Case

In the multiple door case, also called k-door filling, the robots are entering the area through $k \geq 1$ doors. The $k = 1$ case is the single door case. In the k-door filling we assume that each door has enough robot, thus, when a robot leaves a door, a new one can be placed there.

During the analysis of the multiple door case, we examine how multiple Leaders and the set of Followers following them interact with each other. We prove that robots entering from distinct doors cannot collide or block each other.

An invariant of the algorithm is that each Follower only follows its predecessor or becomes a Leader. First, we have to prove that robots in Leader state are not colliding with other robots. Then, we have to show that the paths of the Leaders do not cross each other (for the single door case, we have already shown that the leader does not cross its own path).

Lemma 6. *A Leader cannot collide with a Follower.*

Proof. Lemma 3 still holds: each robot in Leader state can determine which cells are free within two rounds. Therefore, they will not go to cells where a Follower would. □

Fig. 5. Leaders can not collide. The Leader r_1 occupies c in the N-step. In the E-step r_2 sees c as an occupied cell. An important aspect is that in this scenario the robot r_2 still has to move to a cell c' different from c, if it can. Then it remains a Leader. If c' is not free, then r_2 cannot move and in step s_4 and switches to Stopped state.

Lemma 7. *A Leader cannot collide with another Leader.*

Proof. Assume a scenario where two leaders r_1 and r_2 would move to the same cell c (see Fig. 5). Furthermore, assume both r_1 and r_2 are in their even round and recognized c as free cell. Let R_i be this round, consisting of steps $s_1 \dots s_4$, and as R_{i+1} is their even round, they both will try to move there. Since c is visible by r_1 and r_2 from different directions, therefore they try to move to c in different steps, which means one of the robots (w.l.o.g. r_1) will move and the other one (r_2) will be prohibited to move to c after it has been occupied by r_1. □

Lemma 8. *Paths of different Leaders cannot cross each other.*

Proof. Lemma 3 still holds for the multiple door case. It means that the Leader only moves to free cells, which implies that a Leader will not cross the path, where Followers are already moving. In Lemma 7 we have shown that two Leaders cannot collide (and will not cross the path of each other). Therefore, the path of the Leaders will not cross each other during the dispersion. □

Theorem 2. *By using the presented algorithm a connected orthogonal area with multiple doors is filled in $O(n)$ rounds without collisions by robots with visibility range of 1 hop and $O(1)$ bits of persistent memory.*

Proof. As in the single door case, the robots do not collide with each other. Furthermore, the paths of the Leaders cannot cross, they do not block robots entering from other doors. □

Lemma 9. *The algorithm is asymptotically worst-case optimal in the multiple door case.*

Proof. Consider a region A, as depicted in Fig. 6(b). Removing the cell c from A disconnects the region, such that all doors are separated from the major connected component A' of $A \setminus c$. Therefore, all robots must enter to A' through c. Since collision is not allowed, only one robot per round can enter to A' in any algorithm. If A' contains $\Omega(n)$ cells, then each algorithm requires $\Omega(n)$ rounds. Our algorithm fills A in $O(n)$ time.   □

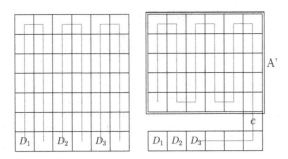

Fig. 6. Best-case (*left*) and worst-case (*right*) examples for the make-span of multiple door filling. The best-case is, when the robots entering at disjoint doors fill equal sized partitions of the area. A worst case input, when single door is used to fill almost the whole area. D_1, D_2, D_3 are the doors. Red represents the path of the first Leader entering through the given door. (Color figure online)

For any input, even in the best-case, at most k robots per round can be placed in the area. Therefore, $\Omega(n/k)$ is a lower bound on the filling time of any algorithm for each input. Consequently, for each input the $O(n)$ running time of our algorithm is at most k times the running time of the optimal algorithm.

Note: for certain inputs our algorithm also reaches k speedup (see Fig. 6(a)), When robots entering from different doors fill equal sized partitions of the region, the runtime is $O(n/k)$.

5   Summary

We have considered the filling problem in an unknown, connected, orthogonal region, where the robots enter the region through several doors and they have to disperse in order to reach full coverage. The robots are autonomous, anonymous, silent, they have a limited visibility range of 1 hop, and use $O(1)$ bits of persistent memory. They have a common notion of North, South, East, and West. Collision of the robots is not allowed. We have presented an algorithm solving the filling problem with multiply doors in $O(n)$ time steps in the synchronous model, where n is the number of cells in the area. This algorithm is optimal in terms of visibility range, and asymptotically optimal in size of persistent memory used by the robots. The running time is asymptotically optimal for the single door case. For the multiple door case, our algorithm is asymptotically worst-case optimal, and its running time is at most k times the running time of the optimal algorithm for any input, where k is he number of doors.

References

1. Barrameda, E.M., Das, S., Santoro, N.: Deployment of asynchronous robotic sensors in unknown orthogonal environments. In: Fekete, S.P. (ed.) ALGOSENSORS 2008. LNCS, vol. 5389, pp. 125–140. Springer, Heidelberg (2008). https://doi.org/10.1007/978-3-540-92862-1_11

2. Barrameda, E.M., Das, S., Santoro, N.: Uniform dispersal of asynchronous finite-state mobile robots in presence of holes. In: Flocchini, P., Gao, J., Kranakis, E., Meyer auf der Heide, F. (eds.) ALGOSENSORS 2013. LNCS, vol. 8243, pp. 228–243. Springer, Heidelberg (2014). https://doi.org/10.1007/978-3-642-45346-5_17

3. Das, S., Flocchini, P., Prencipe, G., Santoro, N., Yamashita, M.: Autonomous mobile robots with lights. Theor. Comput. Sci. **609**, 171–184 (2016)

4. Flocchini, P., Prencipe, G., Santoro, N.: Distributed Computing by Oblivious Mobile Robots: Synthesis Lectures on Distributed Computing Theory. Morgan & Claypool Publishers, San Rafael (2012)

5. Hsiang, T.-R., Arkin, E.M., Bender, M.A., Fekete, S.P., Mitchell, J.S.B.: Algorithms for rapidly dispersing robot swarms in unknown environments. In: Boissonnat, J.-D., Burdick, J., Goldberg, K., Hutchinson, S. (eds.) Algorithmic Foundations of Robotics V. STAR, vol. 7, pp. 77–93. Springer, Heidelberg (2004). https://doi.org/10.1007/978-3-540-45058-0_6

6. Lukovszki, T., Meyer auf der Heide, F.: Fast collisionless pattern formation by anonymous, position-aware robots. In: Aguilera, M.K., Querzoni, L., Shapiro, M. (eds.) OPODIS 2014. LNCS, vol. 8878, pp. 248–262. Springer, Cham (2014). https://doi.org/10.1007/978-3-319-14472-6_17

Gathering Anonymous, Oblivious Robots on a Grid

Matthias Fischer, Daniel Jung$^{(\boxtimes)}$, and Friedhelm Meyer auf der Heide

Computer Science Department, Heinz Nixdorf Institute, Paderborn University,
Fürstenallee 11, 33102 Paderborn, Germany
{mafi,daniel.jung,fmadh}@uni-paderborn.de

Abstract. We consider a swarm of n autonomous mobile robots, distributed on a 2-dimensional grid. A basic task for such a swarm is the gathering process: All robots have to gather at one (not predefined) place. A common local model for extremely simple robots is the following: The robots do not have a common compass, only have a constant viewing radius, are autonomous and indistinguishable, can move at most a constant distance in each step, cannot communicate, are oblivious and do not have flags or states. The only gathering algorithm under this robot model, with known runtime bounds, needs $\mathcal{O}(n^2)$ rounds and works in the Euclidean plane. The underlying time model for the algorithm is the fully synchronous $\mathcal{FSYNC}$ model. On the other side, in the case of the 2-dimensional grid, the only known gathering algorithms for the same time and a similar local model additionally require a constant memory, states and "flags" to communicate these states to neighbors in viewing range. They gather in time $\mathcal{O}(n)$.

In this paper we contribute the (to the best of our knowledge) first gathering algorithm on the grid that works under the same simple local model as the above mentioned Euclidean plane strategy, i.e., without memory (oblivious), "flags" and states. We prove its correctness and an $\mathcal{O}(n^2)$ time bound in the fully synchronous $\mathcal{FSYNC}$ time model. This time bound matches the time bound of the best known algorithm for the Euclidean plane mentioned above. We say gathering is done if all robots are located within a 2×2 square, because in $\mathcal{FSYNC}$ such configurations cannot be solved.

Keywords: Gathering problem · Autonomous robots
Distributed algorithms · Local algorithms · Mobile agents
Runtime bound · Swarm formation problems

1 Introduction

Swarm robotics considers large swarms of relatively simple mobile robots deployed to some two- or three-dimensional area. These robots have very limited

This work was partially supported by the German Research Foundation (DFG) within the Collaborative Research Centre "On-The-Fly Computing" (SFB 901).

© Springer International Publishing AG 2017
A. Fernández Anta et al. (Eds.): ALGOSENSORS 2017, LNCS 10718, pp. 168–181, 2017.
https://doi.org/10.1007/978-3-319-72751-6_13

sensor capabilities; typically they can only observe aspects of their local environment. The objective of swarm robotics is to understand which global behavior of a swarm is induced by local strategies, simultaneously executed by the individual robots. Typically, the decisions of the individual robots are based on local information only.

In order to formally argue about the impact of such local decisions of the robots on the overall behavior of the swarm, many simple models of robots, their local algorithms, the space they live in, and underlying time models are proposed. For a survey see the book [12] by Flocchini et al.

A basic desired global behavior of such a swarm is the gathering process: All robots have to gather at one (not predefined) place. Local algorithms for this process are defined and analysed for a variety of models [1,3,5–7,12].

A common local model for extremely simple robots is the following: There is no global coordinate system. The robots do not have a common compass, only have a constant viewing radius, are autonomous and indistinguishable, can move at most a constant distance in each step, cannot communicate, are fully oblivious and do not have flags or lights to communicate a state to others. In this very restricted robot model, a robot's decision about its next action can only be based on the current relative positions of the otherwise indistinguishable other robots in its constant sized viewing range, and independent on past decisions or information (oblivious).

The only gathering algorithm under this robot model, with known runtime bounds, needs $\mathcal{O}(n^2)$ rounds and works in the Euclidean plane. The underlying time model for the algorithm is the fully synchronous $\mathcal{FSYNC}$ model (see [2,12]). In $\mathcal{FSYNC}$, all robots are always active and do everything synchronously. Time is subdivided into equally sized rounds of constant lengths. In every round all robots simultaneously execute their operations in the common *look-compute-move model* [4] (Sect. 3).

In the discretization of the Euclidean plane, the two-dimensional grid, under the same time and robot model, no runtime bounds for gathering are known. The concept of the Euclidean algorithm [6] cannot be transferred to the grid, because it must be able to compute the center of the minimum enclosing circle of the robots in its viewing range (and then move to this position) and furthermore move arbitrary small distances. This clearly is impossible on the grid. Instead, completely different approaches are needed.

In the only known gathering algorithms on the grid under the same time and a similar robot model, the robots need states (so-called runs) and flags to communicate these states to neighbors, and have to be able to memorize a fixed number of steps [1,5]. There, a robot with an active run state can further move this state to a neighboring robot. This allows coordinated robot operations over several consecutive rounds. In [1,5], these operations are crucial for total running time proof ($\mathcal{O}(n)$ rounds).

In the current submission, we drop the additional robot capabilities memory and flags or lights to communicate a state to others. Then, analogously to the Euclidean strategy [6], explained above, a robot's decision about its next action can only be based on the current relative positions of the otherwise indistinguishable other robots in its constant sized viewing range, and independent on past decisions or information (oblivious). Especially coordinated robot operations over several consecutive rounds that are used in [1,5] cannot be performed under this more restricted model.

To the best of our knowledge we present the first strategy under this restricted model on the grid and prove a total running time of $\mathcal{O}(n^2)$ rounds which complies with the best known running time for the Euclidean strategies in this model [6]. More precisely, the running time of our strategy depends quadratically on the outer boundary length of the swarm. The outer boundary is the seamless sequence of neighboring robots that encloses all the others robots inside.

We conjecture that $\Omega(n^2)$ is a lower bound for the number of rounds needed for our algorithm and, more generally, even for any algorithm within our restricted model. At least for our algorithm, we conjecture that a worst case instance is a configuration with robots on the boundary of an axis-parallel square. Experiments support this conjecture. Full version of our paper: [11].

2   Related Work

There is vast literature on robot problems researching how specific coordination problems can be solved by a swarm of robots given a certain limited set of abilities. The robots are usually point-shaped (hence collisions are neglected) and positioned in the Euclidean plane. They can be equipped with a memory or are *oblivious*, i.e., the robots do not remember anything from the past and perform their actions only on their current views. If robots are anonymous, they do not carry any IDs and cannot be distinguished by their neighbors. Another type of constraint is the compass model: If all robots have the same coordinate system, some tasks are easier to solve than if all robots' coordinate systems are distorted. In [13,14] a classification of these two and also of dynamically changing compass models is considered, as well as their effects regarding the gathering problem in the Euclidean plane. The operation of a robot is considered in the *look-compute-move model* [4]. How the steps of several robots are aligned is given by the *time model*, which can range from an asynchronous $\mathcal{ASYNC}$ model (e.g., see [4]), where even the single steps of the robots' steps may be interleaved, to a fully synchronous $\mathcal{FSYNC}$ model (e.g., see [2]), where all steps are performed simultaneously. A collection of recent algorithmic results concerning distributed solving of basic problems like gathering and pattern formation, using robots with very limited capabilities, can be found in the book [12] by Flocchini et al.

One of the most natural problems is to gather a swarm of robots in a single point. Usually, the swarm consists of point-shaped, oblivious, and anonymous robots. The problem is widely studied in the Euclidean plane. Having point-shaped robots, collisions are understood as merges/fusions of robots and interpreted as gathering progress [5–7]. In [3] the first gathering algorithm for the $\mathcal{ASYNC}$ time model with multiplicity detection (i.e., when a robot can detect if other robots are also located at its own position) and global views is provided. Gathering in the local setting was studied in [2]. In [18] situations when no gathering is possible are studied. The question of gathering on graphs instead of gathering in the plane was considered in [8,15,17]. In [20] the authors assume global vision, the $\mathcal{ASYNC}$ time model and furthermore allow unbounded (finite) movements. They show optimal bounds concerning the number of robot movements for special graph topologies such as trees and rings.

Concerning the gathering on grids, in [10] it is shown that multiplicity detection is not needed and the authors further provide a characterization of solvable gathering configurations on finite grids. In [21], these results are extended to infinite grids, assuming global vision. The authors characterize *gatherable* grid configurations concerning exact gathering in a single point. Under their robot model and the $\mathcal{ASYNC}$ time model, the authors present an algorithm which gathers *gatherable* configurations optimally concerning the total number of movements.

Assuming only local capabilities of the robots, esp. only local vision and no compass, makes gathering challenging. For example, a given global vision, the robots could compute the center of the globally smallest enclosing square or circle and just move to this point. For gathering with presence of a global compass, the authors in [19] provide a simple gathering algorithm: The robots from the left and right swarm boundaries keep moving towards the swarm's inside. In some kind of degenerated cases, instead the robots on the top and bottom boundaries do this.

In the $\mathcal{FSYNC}$ time model, the total running time is a quality measure of an algorithm. In this time model there exist several results that prove runtime bounds [1,5,6,9,16]. For local robot models, the locality strongly restricts the robot capabilities: no global control, no unique IDs, no compass, only local vision (i.e., they can only see other robots up to a constant distance) and no (global) communication. But even under this strongly local model, the presence of remaining local capabilities such as allowing a constant number of states, constant memory or locally visible states (flags, lights), can drastically change running times by even more than the factor n [1,5,16]. The price for this improvement then is many more complicated strategies.

One example are strategies that maintain and shorten a communication chain between an explorer and a base camp. The *Hopper* and *Manhattan Hopper* strategies [16] solve this problem in time $\mathcal{O}(n)$, in the Euclidean plane and on the grid, respectively, using robots with a constant number of states, a constant memory and the capability to communicate states to local neighbors (flags, lights). Without these additional robot capabilities, the simple Euclidean *Go-To-The-Middle* strategy [9] needs notably more time $\mathcal{O}(n^2 \log(n))$ for solving the same problem. Concerning the gathering under this restricted model, the simple, Euclidean *Go-To-The-Center* strategy [6] needs time $\mathcal{O}(n^2)$. (A faster strategy for the Euclidean plane does not exist, yet, and it is still unknown if this bound is tight.) On the grid, two asymptotically optimal $\mathcal{O}(n)$ strategies exist that, solve the gathering of an arbitrary connected swarm [5] and the gathering of a closed chain of robots [1], respectively. Like the above communication chain strategies, they require more complex robots with a const. number of states, a const. memory and the capability to communicate states to local neighbors (flags, lights). Strategies without these additional capabilities do not exist, yet.

In the present paper, we deliver such an algorithm that uses the same strongly restricted model as the Euclidean *Go-To-The-Center* gathering strategy [6]. Our strategy gathers in time $\mathcal{O}(|\text{outer boundary}|^2) \subseteq \mathcal{O}(n^2)$, where |outer boundary| denotes the length of the swarm's outer boundary (cf. Fig. 4(*i*)) which naturally is $\in \mathcal{O}(n)$. This is comparable to the $\mathcal{O}(n^2)$ bound for the Euclidean gathering [6]. We conjecture that $\mathcal{O}(|\text{outer boundary}|^2)$ is also tight on the grid.

3   Our Local Model

Our mobile robots need very few and simple capabilities: A robot moves on a two-dimensional grid and can change its position to one of its eight horizontal, vertical or diagonal neighboring grid cells. It can see other robots only within a constant *viewing radius* of 7 (measured in L_1-distance). We call the range of visible robots the *viewing range*. Within this viewing range, a robot can only see the relative positions of the viewable robots. The robots have no compass, no global control, and no IDs. They cannot communicate, do not have any states (no flags, lights) and are oblivious.

Our algorithm uses the fully synchronous time model $\mathcal{FSYNC}$, in that all robots are always active and do everything synchronously. Time is subdivided into equally sized rounds of constant lengths. In every round all robots simultaneously execute their operations in the common *look-compute-move model* [4], which divides one operation into three steps. Every round contains only one cycle of these steps: In the *look* step, the robot gets a snapshot of the current scenario from its own perspective, restricted to its constant-sized viewing range. During the *compute* step, the robot computes its action, and eventually performs it in the *move* step. If a robot has moved to an occupied grid cell, the robots from then on behave like one robot. We say they *merge* and remove one of them.

We say gathering is done if all robots are located within a 2×2 square, because such configurations cannot be solved in our time model.

The swarm must be connected. In our model, two robots are connected if they are located in horizontal or vertical neighboring grid cells. The operations of our algorithm do not destroy this connectivity.

4   The Algorithm

A robot decides to hop on one of its 8 neighboring grid cells only dependent on the current robot positions within its viewing range. We distinguish diagonal (Sect. 4.1) and horizontal/vertical (Sect. 4.2) hops. The hops are intended to achieve the gathering progress by modifying the swarm's *outer boundary*. Figure 4(*i*) defines the swarm's boundaries: Black and hatched robots are *boundary* robots. The boundary on which the black robots are located borders the swarm and is called the swarm's *outer boundary*. In the figure, all other robots are colored grey. White cells are empty.

4.1   Diagonal Hops

If a robot r (marked black in Figs. 1 and 2) checks whether it can execute a diagonal hop, it compares the patterns of Figs. 1 and 2 to the robot positions in its viewing range: Robot r checks if one of the Diag-$\{A, B\}$ Hop patterns matches the current scenario from its own perspective. Patterns that are created by an arbitrary horizontal and vertical mirroring and an arbitrary 90° rotation of the three patterns of Fig. 1 are also valid and have to be checked.

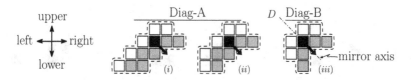

Fig. 1. Hop patterns: One of the Diag-A or Diag-B Hop patterns must match the relative robot positions within the black robot's viewing range. This is the hop criterion, necessary for allowing the black robot to perform the depicted diagonal hop. In this paper, the notation Diag-$\{A, B\}$ means "Diag-A or Diag-B".

Depending on the matching Hop pattern the robot does the following:

1. If a Diag-A pattern matches, then robot r checks if, using the same rotation and mirroring, any of the Inhibit patterns match the *upper-right area* of its viewing range. If at least one Inhibit pattern matches, then the Diag-A hop of robot r is not executed. Otherwise, if none of the Inhibit patterns match, the robot r hops according to the matching Diag-A pattern.
2. If the Diag-B pattern matches, then the robot r checks if, using the same rotation and mirroring, also any of the Inhibit patterns match the *upper-right area* and the *lower-left area* of its viewing range. However, in case of the *lower-left area*, the Inhibit pattern has to be mirrored at the diagonal mirroring axis D shown in Fig. 1(iii). If for both areas matching Inhibit patterns have been found, then the Diag-B hop of robot r is not executed. Otherwise the robot r hops according to the matching Diag-B pattern.

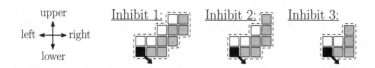

Fig. 2. Inhibit patterns: Patterns that, in case they match, inhibit the black robot's hop.

4.2   Horizontal and Vertical Hops (HV Hops)

Robots can also hop in vertical or horizontal direction (HV hop). We allow these
hops for length 1 and 2 (cf. Fig. 3). For length 2, horizontal or vertical hops,
respectively, are a joint operation of two neighbouring robots. If for a robot
a horizontal and a vertical HV hop apply at the same time (see $b^\star$ of Fig. 3),
then it instead performs a diagonal hop as shown in the figure. After a HV hop,
every target cell contains at least two robots. We let these robots *merge*: i.e., we
remove all but one of the robots at the according cell.

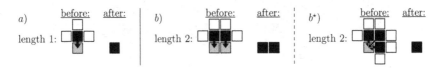

Fig. 3. The black robots simultaneously hop downwards. Afterwards, robots that are
located at the same position *merge*.

Diag-$\{A, B\}$ and HV hops are executed simultaneously in the same step of
the algorithm. For the pseudocode, see [11].

5   Measuring the Gathering Progress

The gathering progress measures that we will use for the analysis of our strategy
are heavily dependent on the length and shape of the swarm's outer boundary. In
order to analyze these measures, we need the terms boundary, outer boundary,
length, as well as convex and concave vertices.

Swarm's boundary. The swarm's boundary is the set of all robots that have
at least one empty adjacent cell in a horizontal, vertical, or diagonal direction.
Figure 4(i) shows an example: Black and hatched robots are boundary robots.
The empty cells contain no robot and are colored in white. When speaking about
a *subboundary*, we mean a connected sequence of robots of some boundary.

Swarm's outer boundary. The swarm's outer boundary is the boundary that
borders on the outside of the swarm. In Fig. 4(i) the black robots belong to
outer boundary. All other robots are not part of the boundaries, i.e., they have
an adjacent robot in all directions (horizontal, vertical, and diagonal). In Fig. 4(i)
they are colored grey.

Outer boundary's length. We measure the outer boundary's length as follows: We start at a cell of the outer boundary and perform a complete walk
along this boundary while we define the *length* as the total number of steps
that we performed during this walk. This means that if the swarm is hourglass-
or cross-shaped, for example, some robots are counted multiple (up to four)

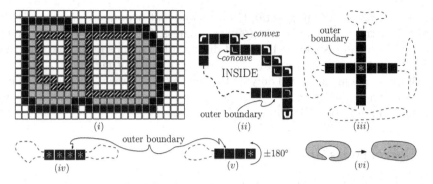

Fig. 4. (i): Definition of the (*outer*) *boundary*. Outer boundary: black robots; (ii): Definition of *convex* and *concave vertex*. Convex vertices: fat curves; (iii): Cross shape. The "∗" marked robot is counted four times; (iv): Hourglass shape: The "∗" marked robots are counted twice; (v): ±180° rotation. The "∗" marked robot is counted twice.; (vi): During the gathering, inner bubbles can be developed.

times (cf. Fig. 4(iii, iv)). Furthermore, robots on which a turn by ±180° is performed during the walk are counted twice (cf. Fig. 4(v)). We denote this length by |outer boundary|.

Convex and concave vertices. On the boundary we further distinguish *convex* and *concave vertices*. A vertex of the boundary is a robot that looks like a corner. In Fig. 4(ii) fat curves mark *convex vertices* of the swarm's outer boundary, while the thin curves mark the *concave vertices* of the swarm's outer boundary.

Outline of the running time proof—How we get gathering progress. We distinguish three kinds of progress measures that help us to prove the quadratic running time.

Boundary: Length of the swarm's outer boundary.
Convex: Difference between the number of convex vertices on the swarm's outer boundary and its maximum value.
Area: Included area.

We have designed the hops in such a way that the length of the outer boundary (*Boundary*) never increases. But it can remain unchanged over several rounds. Then, instead, we measure the progress by *Convex*. As we draw robots as squares, the total number of convex vertices on the swarms' outer boundary is naturally upper bounded by its maximum value 4|outer boundary|. *Convex* is the difference between this maximum value and the actual number of convex vertices on the outer boundary. We will show that also *Convex* never increases.

In rounds in which both *Boundary* and *Convex* do not achieve progress, we instead measure the gathering progress by *Area*. We measure *Area* as the number of robots on the swarm's outer boundary plus the number of inside cells (occupied as well as empty ones). In contrast to the other progress measures, *Area* does not decrease monotonically in general, but we show that it decreases monotonically in rounds without *Boundary* and *Convex* progress. We upper bound the size by that the *Area* can instead be increased during other rounds and show that this makes the total running time worse at most by a constant factor.

All three measures depend only on the length of the swarm's outer boundary. While *Boundary* and *Convex* are linear, *Area* is quadratic. This then leads us to a total running time $\mathcal{O}(|\text{outer boundary}|^2)$.

6   Correctness and Running Time

In this section, we formally prove the correctness of the progress measures and finally the total running time (Theorem 1).

6.1   Progress Measure *Boundary*

Lemma 1. *During the whole gathering,* Boundary *is monotonically decreasing.*

Proof. As the definition of HV hops requires that the robots hop onto occupied cells, such hops naturally cannot increase the number of robots on the outer boundary. So we consider Diag-$\{A, B\}$ hops in which robots hop towards the swarm's outside. In order to increase the number of robots on the outer boundary, the target cell of such hops must be empty. But then, the hopped robot has also been part of the outer boundary before the hop, so that the boundary length did not increase. □

6.2   Impact of Inhibit Patterns: *Collisions*

For the proofs of the progress measures *Convex* and *Area*, we need a deeper insight why certain robot hops are inhibited by Inhibit patterns (cf. Fig. 2). When proving that the *Convex* progress is monotonically decreasing (Lemma 2), we analyze the change of the total number of convex vertices that is induced by the robot hops. This is only possible if certain simultaneously hopping robots are not too close together. Inhibit patterns ensure this minimum distance.

Cf. Figs. 1 and 2. From a more global point of view, the Inhibit patterns ensure that the black robot only performs its Diag-A or Diag-B hop, respectively, if the next robot(s) at distance 2 along the boundary does (do) not perform a hop in the opposite direction. If the hop of the black robot is blocked by Inhibit patterns, we will say the robots *collide*. This will be used in the proofs of our progress measures. Figure 5 shows significant collision examples: (i): For both, r and r' Diag-A matches. Without inhibition patterns, r would hop to the lower right, while r' would hop in the opposite direction to the upper left. But the Inhibit 1 pattern inhibits the hop of r: r collides with r'. Analogously also the hop of r' can be inhibited. (ii): If Diag-B matches for r, then the hop is inhibited if concerning both r' and r'' an inhibition pattern matches. In this example, for r' a Diag-A pattern and for r'' a Diag-B pattern could else enable hops in the opposite direction than the hop of r, but the matching Inhibit 1 and 2 patterns inhibit the hop of r: r collides with r', r''. A more detailed analysis of collisions is provided in the proofs for Lemma 3.

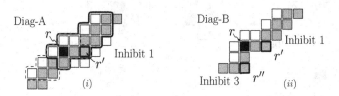

Fig. 5. *Collisions.* (i): For r, Diag-A and Inhibit 1 matches. r does not hop. (ii): For r, Diag-B and Inhibit 1 and 3 match. r does not hop.

6.3   Progress Measure *Convex*

Lemma 2. *During the whole gathering,* Convex *is monotonically decreasing.*

Proof. If we say "convex/concave vertices", we consider only the outer boundary. First, we analyze the HV hops. Here, a HV hop can reduce the number of convex vertices by at most 2. At the same time, the outer boundary becomes shorter by at least 2. Then, *Convex* either remains unchanged or decreases.

Concerning the Diag-{A, B} hops, we look at Fig. 6: The figure shows how the diagonal hops can (locally) change the number of convex vertices on the outer boundary. (In the figure, (ii) shows the hops from (i), but for switched INSIDE and OUTSIDE.) In all cases, the Inhibit patterns ensure that the robots a, a' do not move (cf. Sect. 6.2). In the figure, we distinguish Diag-A and Diag-B hops, while A, A* refer to Diag-A and B, B* to Diag-B. We distinguish the case that the robot b does not hop (A, B) and the other case, that it performs a hop (A*, B*). The result of the case distinction is that in column (i) the number of convex vertices never decreases. In column (ii), this is also the case if the white marked cells s, t, u are empty.

If instead not all of s, t, u are empty, the number of convex vertices might also become smaller. But even in this case, still *Convex* progress does not increase: We now show that then also |outer boundary| becomes smaller as well as the maximum value for the total number of convex vertices, so that by definition *Convex* progress is not increased, i.e., it still behaves monotonically: Fig. 7 shows the relevant cases. With reference to Fig. 6(ii), all hops are performed towards the swarms' outside. In column (i), only the cell s contains a robot. There we see, that in all cases the outer boundary becomes shorter by at least 2. In case of A*, this can be even more than 2 for the case that after the hop the cell below b (in Fig. 6(ii), this was the cell u) also contains a robot. Column (ii) shows the cases where cell t is (also) occupied. Because the swarm is always connected, the hatched robot must be connected to the rest of the swarm. This can be either via the subboundary α or β. The hop shortens the outer boundary by forming inner bubbles (Fig. 4(vi)).   □

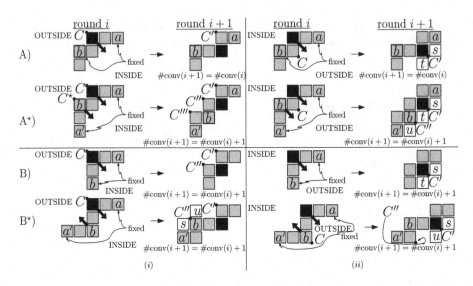

Fig. 6. Local effect of all kinds of hops on the number of convex vertices on the outer boundary. C, C', C'' denote the counted convex vertices. #conv(i) denotes the number of convex vertices in round i.

6.4   Progress Measure *Area*

The third progress measure *Area* does not behave monotonically. It can be increased during rounds where we get *Boundary* or *Convex* progress. But we use it for estimating the number of the remaining rounds (Lemma 3). And in the proof of Theorem 1 we show that the increased amount of the *Area* progress measure does not worsen the asymptotic running time. For the proof of Lemma 3, see [11].

Lemma 3. *If in a step of the gathering process neither* Boundary *nor* Convex *has progress, then instead* Area *has progress by at least* -8.

6.5   Total Running Time

Now we can combine all three progress measures *Boundary*, *Convex* and *Area* for the running time proof (Theorem 1).

Theorem 1. *A connected swarm of n robots on a grid can be gathered in* $\mathcal{O}(|\text{outer boundary}|^2) \subseteq \mathcal{O}(n^2)$ *many rounds.*

Proof. Let B be the initial length of the swarm's outer boundary. We know from Lemma 1 that *Boundary* decreases monotonously. Then, progress in *Boundary* happens at most B times. By Lemma 2, *Convex* also decreases monotonously. As every robot on the swarm's outer boundary can provide at most 4 convex vertices, *Convex* progress happens at most $4B$ times.

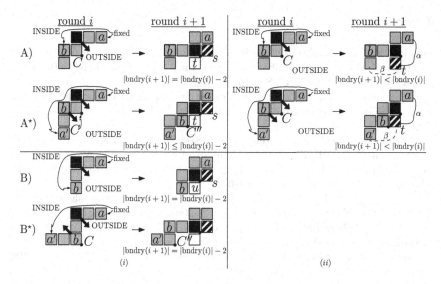

Fig. 7. Diagonal hops can change the outer boundary's length. C, C', C'' denote relevant convex vertices. $\#\mathrm{bndry}(i)$ denotes the outer boundary length in round i.

We estimate the rounds without *Boundary* and *Convex* progress via the size of the included area, i.e., the *Area* progress. By Lemma 3, we know that in every round without *Boundary* and *Convex* progress, the area becomes smaller by at least -8. But, *Area* is not a monotone progress measure, in rounds with *Boundary* or *Convex* progress, the included area can increase: While HV hops cannot increase the included area, Diag-$\{A, B\}$ hops can. First, we assume that the according Diag-$\{A, B\}$ hops do not change the outer boundary length. Then, every time *Convex* has progress, the area can become larger by at most B, because every robot hop on the outer boundary can increase the area by at most 1. As *Convex* happens at most $4B$ times, this in total is upper bounded by $4B^2$.

If the outer boundary length changes, i.e., becomes shorter, then the included area can increase (cf. proof of Lemma 2 and Fig. 4(vi)). Then, a *Boundary* progress by ℓ can also increase the included area by $\Delta A \leq \ell^2 \leq \ell B$. But as *Boundary* progress is monotonically decreasing, the sum of all these ΔA is upper bounded by B^2.

Summing it up, during the whole process of the gathering, the area can be increased by at most $(4+1)B^2 = 5B^2$. Together with the initial area of at most B^2, *Area* progress happens at most $6B^2$. Then, the gathering is done after at most $B + 4B + 6B^2$ rounds.   □

References

1. Abshoff, S., Cord-Landwehr, A., Fischer, M., Jung, D., Meyer auf der Heide, F.: Gathering a closed chain of robots on a grid. In: IPDPS 2016, pp. 689–699 (2016). https://doi.org/10.1109/IPDPS.2016.51
2. Ando, H., Suzuki, Y., Yamashita, M.: Formation and agreement problems for synchronous mobile robots with limited visibility. In: ISIC 1995, pp. 453–460, August 1995
3. Cieliebak, M., Flocchini, P., Prencipe, G., Santoro, N.: Solving the robots gathering problem. In: Baeten, J.C.M., Lenstra, J.K., Parrow, J., Woeginger, G.J. (eds.) ICALP 2003. LNCS, vol. 2719, pp. 1181–1196. Springer, Heidelberg (2003). https://doi.org/10.1007/3-540-45061-0_90
4. Cohen, R., Peleg, D.: Robot convergence via center-of-gravity algorithms. In: Královič, R., Sýkora, O. (eds.) SIROCCO 2004. LNCS, vol. 3104, pp. 79–88. Springer, Heidelberg (2004). https://doi.org/10.1007/978-3-540-27796-5_8
5. Cord-Landwehr, A., Fischer, M., Jung, D., Meyer auf der Heide, F.: Asymptotically optimal gathering on a grid. In: SPAA 2016, pp. 301–312 (2016). http://doi.acm.org/10.1145/2935764.2935789
6. Degener, B., Kempkes, B., Langner, T., Meyer auf der Heide, F., Pietrzyk, P., Wattenhofer, R.: A tight runtime bound for synchronous gathering of autonomous robots with limited visibility. In: SPAA 2011, pp. 139–148 (2011)
7. Degener, B., Kempkes, B., Meyer auf der Heide, F.: A local $O(n^2)$ gathering algorithm. In: SPAA 2010, pp. 217–223 (2010)
8. Dessmark, A., Fraigniaud, P., Kowalski, D.R., Pelc, A.: Deterministic rendezvous in graphs. Algorithmica 46(1), 69–96 (2006)
9. Dynia, M., Kutylowski, J., Lorek, P., Meyer auf der Heide, F.: Maintaining communication between an explorer and a base station. In: IFIP TC10, pp. 137–146, 1 January 2006
10. D'Angelo, G., Di Stefano, G., Klasing, R., Navarra, A.: Gathering of robots on anonymous grids without multiplicity detection. In: Even, G., Halldórsson, M.M. (eds.) SIROCCO 2012. LNCS, vol. 7355, pp. 327–338. Springer, Heidelberg (2012). https://doi.org/10.1007/978-3-642-31104-8_28
11. Fischer, M., Jung, D., Meyer auf der Heide, F.: Gathering anonymous, oblivious robots on a grid. CoRR abs/1702.03400 (2017). http://arxiv.org/abs/1702.03400
12. Flocchini, P., Prencipe, G., Santoro, N.: Distributed Computing by Oblivious Mobile Robots. Synthesis Lectures on Distributed Computing Theory. Morgan & Claypool, San Rafael (2012)
13. Izumi, T., Souissi, S., Katayama, Y., Inuzuka, N., Défago, X., Wada, K., Yamashita, M.: The gathering problem for two oblivious robots with unreliable compasses. SICOMP 41(1), 26–46 (2012)
14. Katayama, Y., Tomida, Y., Imazu, H., Inuzuka, N., Wada, K.: Dynamic compass models and gathering algorithms for autonomous mobile robots. In: Prencipe, G., Zaks, S. (eds.) SIROCCO 2007. LNCS, vol. 4474, pp. 274–288. Springer, Heidelberg (2007). https://doi.org/10.1007/978-3-540-72951-8_22
15. Klasing, R., Markou, E., Pelc, A.: Gathering asynchronous oblivious mobile robots in a ring. TCS 390(1), 27–39 (2008)
16. Kutylowski, J., Meyer auf der Heide, F.: Optimal strategies for maintaining a chain of relays between an explorer and a base camp. TCS 410(36), 3391–3405 (2009)
17. Martínez, S.: Practical multiagent rendezvous through modified circumcenter algorithms. Automatica 45(9), 2010–2017 (2009)

18. Prencipe, G.: Impossibility of gathering by a set of autonomous mobile robots. TCS **384**(2–3), 222–231 (2007)
19. Saadatmand, S., Moazzami, D., Moeini, A.: A cellular automaton based algorithm for mobile sensor gathering. JAC **47**(1), 93–99 (2016)
20. Di Stefano, G., Navarra, A.: Optimal gathering of oblivious robots in anonymous graphs. In: Moscibroda, T., Rescigno, A.A. (eds.) SIROCCO 2013. LNCS, vol. 8179, pp. 213–224. Springer, Cham (2013). https://doi.org/10.1007/978-3-319-03578-9_18
21. Di Stefano, G., Navarra, A.: Optimal gathering on infinite grids. In: Felber, P., Garg, V. (eds.) SSS 2014. LNCS, vol. 8756, pp. 211–225. Springer, Cham (2014). https://doi.org/10.1007/978-3-319-11764-5_15

A Continuous Strategy for Collisionless Gathering

Shouwei Li, Christine Markarian,
Friedhelm Meyer auf der Heide, and Pavel Podlipyan[⊠]

Department of Computer Science, Heinz Nixdorf Institute, Paderborn University,
Fürstenallee 11, 33102 Paderborn, Germany
{shouwei.li,christine.markarian,fmadh,pavel.podlipyan}@upb.de

Abstract. We consider continuous strategies for swarms of robots in the Euclidean plane. In such a strategy, each robot continuously observes its local neighborhood, and continuously adapts speed and direction following a local rule. We present two main results. The first defines a class of strategies, the contracting strategies, that perform gathering in time $O(nd)$, where d is a diameter of the initial configuration. Several well-known strategies belong to this class. Our second result is about collisions in such strategies. We present a contracting strategy which ensures that no collisions occur. This strategy needs the robots to have some additional capabilities.

1 Introduction

The study of gathering swarms of robots has been ongoing since decades. The goal here is to find provably fast strategies that put together the robots, initially distributed on the Euclidean plane, into a single predefined point.

Given a group of n autonomous, dimensionless, deterministic, and anonymous robots, with bounded viewing range, a *Gathering* algorithm is an algorithm that eventually gathers all n robots into one point. Gathering algorithms studied thus far have been analyzed in two different time models: *discrete* and *continuous*. While most of these consider the discrete time model in which robots act in rounds (see [12] for a comprehensive survey), only few of them are analyzed in the continuous time model. In the continuous model, robots continuously adjust their speed and direction to the current relative positions of the robots within their viewing range, while obeying a speed limit normalized to 1. The first Gathering algorithm for the continuous time model was presented by Gordon et al. in [14] and then analyzed by Kempkes et al. in [15].

In many existing approaches (e.g., in [1,6,9–11]), a *collision*, i.e., a merge or fusion during gathering, is often interpreted as a success for the Gathering algorithm since it represents progress in gathering. In this paper, motivated by applications in which collisions must be avoided, we provide Gathering algorithms that can gather robots without collisions as well.

This work was partially supported by the German Research Foundation (DFG) within the Collaborative Research Center "On-The-Fly Computing" (SFB 901) and the International Graduate School "Dynamic Intelligent Systems".

© Springer International Publishing AG 2017
A. Fernández Anta et al. (Eds.): ALGOSENSORS 2017, LNCS 10718, pp. 182–197, 2017.
https://doi.org/10.1007/978-3-319-72751-6_14

In the discrete time model, Lukovszki et al. in [18] proposed collisionless algorithms for a variant of the Gathering problem in which all robots know the gathering point. In fact, the need to avoid collisions becomes clearer when robots gain extent. Robots with an extent were first studied by Czyzowicz et al. in [7]. In this work, robots are represented by unit disks instead of points. The aim of gathering is to form a configuration for which the union of all disks representing the robots is connected. For the same model as in [7], a general solution for more than four robots is presented by Agathangelou et al. in [2]. Furthermore, Pagli et al. in [20] study a discrete (near-)Gathering problem with an asynchronous activation model. This problem is solved for the point like chiral robots that agree on common direction (i.e. agree on common coordinate system).

In the continuous time model, the first attempts towards collisionless gathering were given by Li et al. in [17] where we introduce and evaluate the continuous Go-To-The-Gabriel-Center algorithm that gathers robots in a one-dimensional Euclidean space without collisions in time $O(n)$.

Our Contribution. In this paper, we study the Gathering problem in the continuous time model and introduce a simple convergence criterion along with a corresponding class of algorithms that perform gathering in time $O(nd)$, where d is a diameter of initial configuration.

Moreover, we show that known algorithms belong to this class, and propose two new ones, namely Go-To-The-Relative-Center algorithm (GTRC) and Safe-Go-To-The-Relative-Center algorithm (S-GTRC). Both are contracting and thus gather in quadratic time. The second one is proven to perform no collision. It uses slightly more complex robots: They are non oblivious, chiral, and luminous (i.e. they have visible external memory, proposed by Das et al. in [8]).

Our techniques are inspired by ideas from a number of previously known approaches. GTRC extends the Go-To-The-Center algorithm (GTC) introduced in [3], studied in the discrete time model, to the continuous time model variant. We use the relative neighborhood graph proposed by Toussaint in [22] to modify the original algorithm. Our modified algorithm considers only the neighbors of a robots w.r.t. the relative neighborhood subgraph of the visibility graph. Due to space constraints, the proofs are omitted and all Figures are moved to Appendix A.3. The full version of the paper can be found in [16].

2   Problem Description

We consider the common robot model used in most previous approaches (e.g., [3,14,15,17]). We are given a set $R = \{r_1, \ldots, r_n\}$ of n autonomous mobile robots with viewing range of 1. We denote by $r_i(t) \in \mathbb{R}^2$ the position of robot r_i at time t. Robots agree on the unit distance. Each robot has its own local coordinate system. Robots are oblivious, meaning that they act depending only on the information about the current point of time. They are anonymous, meaning that they do not have IDs. They are silent, meaning that they do not communicate.

The Euclidean distance between two robots r_i and r_j at time t is represented by $|r_i(t), r_j(t)|$. Two robots are open unit disk graph (open UDG) or unit disk graph (UDG) neighbors at time t, if $|r_i(t), r_j(t)| < 1$. The set of robots that consists of the robot r_i itself and all its UDG neighbors at time t is called UDG neighborhood of r_i and denoted by $UDG_t(r_i)$. The UDG defined on all robots at some point in time t is denoted by $UDG_t(R) = (R, E_t)$, where $(r_i, r_j) \in E_t$ iff $|r_i(t), r_j(t)| < 1$. We skip t in the notation of UDG neighborhood and UDG unless it needs to be mentioned explicitly.

The disposition of robots at some point in time t on the plane is called a *configuration*. The disposition of robots at time $t = 0$ is called the *initial configuration*. Initial configurations are arbitrary except that the UDG over all robots at time $t = 0$ ($UDG_0(R)$) is connected and all robots have distinct positions. The goal is to gather all robots at one point, which is not predefined.

We consider the continuous time model, first introduced by Gordon et al. in [14] and later studied by Kempkes et al. in [15]. The velocity of the robot depends solely on the relative positions of neighboring robots at the current point of time. It may change in a non-continuous manner since robots measure the relative positions of their neighbors without delay and instantly adjust their own movement with respect to the measurements. The maximum speed of the robot is assumed to be 1.

3   Contracting Algorithms

In this section, for continuous time model we introduce a class of algorithms that perform gathering on the Euclidean plane in time $O(nd)$, where d is the diameter of the initial configuration.

Let us first define the progress measure. Let $H_t(R) \subset \mathbb{R}^2$ be the closed convex hull around the positions of all robots at time t. We are particularly interested in robots that are corners of a convex hull. Namely, we consider the set of robots $CH_t(R) = \{c_i \in H_t(R) : \alpha_i(t) \in [0, \pi), i \in [1, k], k \le n\}$, where n is the total number of robots and k is the number of robots that belong to the boundary of the convex hull and have internal angle $\alpha_i(t) \in [0, \pi)$. We refer to $CH_t(R)$ as the *corner set* of the convex hull $H_t(R)$. Unless explicitly needed, we skip t in the notation of the convex hull and corner set.

Definition 1 (Contracting algorithm). *In continuous time model a Gathering algorithm for n robots on the Euclidean plane is a contracting if for every time t such that cardinality of $CH_t(R)$ is strictly greater than 1, every robot from $CH_t(R)$ moves with speed 1 in the direction that points into $H_t(R)$.*

If an algorithm is contracting, then we can bound the speed with which the length of the convex hull is decreasing. Let $l(t)$ be the length of the convex hull boundary at time t. Using the corner set we express the length as follows:

$$l(t) = \sum_{i=1}^{k} \left| c_i, c_{(i \bmod k)+1} \right|, \tag{1}$$

where k is the number of robots in the corner set. Let $l'(t)$ be the speed with which the length $l(t)$ of the convex hull boundary changes. The length of the convex hull boundary will be the progress measure for contracting algorithms.

Lemma 1. *If a group of n robots executes a contracting Gathering algorithm and the robots are not yet gathered (i.e. cardinality of $CH_t(R)$ is strictly greater than 1), then the length $l(t)$ of the convex hull boundary at any point in time t decreases with speed $l'(t) \geq 8/n$.*

The proof of Lemma 1 is easily derived from the runtime analysis in [17]. It is well known that the length of the convex hull boundary is $O(d)$, where d is the diameter of the initial configuration. Hence, the theorem below follows directly.

Theorem 1. *Every contracting Gathering algorithm solves the Gathering problem in time $O(nd)$, where d is the diameter of an initial configuration.*

The main steps of a gathering algorithm are described as follows. Each robot r continuously computes the target point $T(r)$ and moves with speed 1 towards it. If it reaches the target point, it stays there (follows the target point) until the position of the target point changes discontinuously (e.g., target point may "jump" due to new neighbors).

There are many ways to calculate the target point on the Euclidean plane. Consider, for example, the Go-On-Bisector algorithm (GOB) proposed by Gordon et al. in [14] and later studied by Kempkes et al. in [15]. In the Go-On-Bisector algorithm, robots that are on the boundary of the local convex hull (around their UDG neighborhood) move inside the local convex hull along the bisector of the internal angle. It is not difficult to see that if robots are not yet gathered, then every robot $r \in CH(R)$ moves inside the convex hull $H(R)$ with speed 1. In other words, GOB is a contracting algorithm.

We can also calculate the target point by the average of the positions of all robots in the neighborhood. This algorithm is known as the Go-To-The-Gravity-Center algorithm (GTGrC), due to Cohen et al. in [5]. It is easy to see that GTGrC is a contracting algorithm, as well.

Another way to calculate the target point is by the center of the minimum enclosing circle around the UDG neighborhood of the robot. This algorithm is known as the Go-To-The-Center algorithm (GTC), due to Ando et al. in [3]. The continuous version of GTC is considered in [17] by Li et al. Using argumentation from [17], we can easily prove that GTC is a contracting algorithm. Moreover, we can calculate a minimum enclosing circle with respect to the unit Gabriel graph due to Gabriel et al. in [13]. Consequently, the algorithm Go-To-The-Gabriel-Center (GTGC) proposed in [17], that considers a unit Gabriel graph, is also a contracting algorithm.

Note that the diameter of a connected UDG with n vertices is at most $n - 1$. Therefore, Theorem 1 and the above discussion yield the following.

Corollary 1. *Given an initial configuration that is a connected UDG, a contracting Gathering algorithm solves the Gathering problem in time $O(n^2)$.*

In the next section, we are going to present one more contracting algorithm, namely GTRC. This is modification of GTC proposed by Ando et al. in [3]. In this algorithm, instead of the UDG neighborhood we calculate minimum enclosing circle with respect to the unit Relative neighborhood graph proposed by Toussaint in [22].

4 Go-To-The-Relative-Center Algorithm (GTRC)

We take another viewpoint on the well-known Go-To-The-Center algorithm due to Ando et al. in [3]. In the Go-To-The-Center algorithm, robots move towards the center $T_t(r)$ of the *minimum enclosing circle* $C_t(r)$, which is the smallest circle that contains all the robots in $UDG_t(r)$.

Let us consider a robot $r \in R$ at some point in time t and its UDG neighborhood $UDG_t(r)$. It is shown in [4] that either (I) there are two robots $m_1, m_2 \in UDG_t(r)$ on the circumference of $C_t(r)$ such that the line segment $(m_1 m_2)$ is the diameter of $C_t(r)$ or (II) there are three points $m_1, m_2, m_3 \in UDG_t(r)$ such that $C_t(r)$ circumscribes $\triangle m_1 m_2 m_3$ and the center of $C_t(r)$ is inside $\triangle m_1 m_2 m_3$. Accordingly, we refer to the set $MEC_t(r) = \{m_1, m_2\}$ or $MEC_t(r) = \{m_1, m_2, m_3\}$ as the *minimum enclosing set* of robot r at time t. We say that the robots of $MEC_t(r)$ *form* the minimum enclosing circle $C_t(r)$ of robot r at time t. In fact, robot r might belong to its own minimum enclosing set (e.g., $MEC_t(r) = \{m_1, r\}$).

Note that the minimum enclosing circle of a point set is unique [4] and can be found in linear time in the Euclidean space of any constant dimension [19]. In case there is more than one minimum enclosing set $MEC_t(r)$ that may form $C_t(r)$, then we assume that the robot selects one of them arbitrarily. Unless explicitly needed, we skip t in the notation of minimum enclosing set and circle.

Instead of calculating $C(r)$ with respect to $UDG(r)$, we use the Relative neighborhood graph proposed in [22]. The latter is defined in the two-dimensional Euclidean space as follows:

Definition 2 (Relative neighborhood graph criterion). *Any two robots u, v are connected iff there does not exist any robot $w \in R$ satisfying $|u, w| < |u, v|$ and $|v, w| < |u, v|$.*

We denote by $RNG_t(r)$ (at time t) the subgraph obtained from the UDG neighborhood $UDG_t(r)$ by applying the Relative neighborhood graph criterion. We call $RNG_t(r)$ the unit *Relative graph neighborhood* of robot r. The Relative neighborhood graph (RNG) defined on all robots at time t is denoted by $RNG_t(R)$.

The set of points on the Euclidean plane that corresponds to the intersection of open unit disks of all robots in $RNG(r)$ is denoted by $Q(r)$. The set of points $Q(r)$ is open and convex since it is the intersection of open unit disks that are convex [21]. The circle $C_Q(r)$ with center at r inscribed into $Q(r)$ is the *connectivity circle*. The radius of connectivity circle is denoted by ρ_Q. Unless needed explicitly, we skip t in the notation of the Relative neighborhood graph, neighborhoods, etc.

Now we are ready to present the GTRC. Its pseudo-code can be found as Algorithm 2 in Appendix A.2.

4.1 Correctness and Runtime Analysis of GTRC

We begin this section with a simple proposition and then move to the connectivity property of GTRC.

Proposition 1. *For every robot $r \in R$ at any fixed point in time, $T(r) \in Q(r)$.*

Proposition 1 holds since the radius of $C(r)$ is less than 1 and $C(r)$ encircles all the robots in $RNG(r)$. The argumentation of the connectivity property is similar to the one used in [17] for Go-To-The-Center algorithm (GTC) and Go-To-The-Gabriel-Center algorithm (GTGC). The proof of Lemma 2 is based on contradiction by assuming that connectivity does not hold.

Lemma 2. *Given a group of robots R on the Euclidean plane executing GTRC. If $\{u, w\}$ is an edge in the open Relative neighborhood graph $RNG(R)$ at time 0, then there is a path from u to w in $RNG(R), \forall t \geq 0$.*

Using the argumentation in [17], we can show that, at any point of time t before the robots gather, the robots of $CH(R)$ move only inside the convex hull, with speed 1. Therefore GTRC is a contracting algorithm. Moreover, the length of the convex hull boundary around the initial configuration is shown, in [17], to be not greater than $2(n - 1)$, where n is the number of robots. Hence, we conclude the following theorem.

Theorem 2. *A group of n robots executing GTRC gathers in time $O(n^2)$.*

4.2   Collisions During Gathering with GTRC

In this section, we show that GTRC performs gathering with collisions. Recall that if two or more robots have the same position at the same point in time, then there is collision. We refer to the set of robots in the minimum enclosing set $MEC(r)$ of a robot $r \in R$ as a *crash* point. We are able to show that collisions in GTRC take place only at crash points.

Definition 3 (Crash point). *For any robot r with minimum enclosing set $MEC(r)$, the midpoint between any two robots $m_1 m_2 \in MEC(r)$ is a crash point $p(r)$. We say that robots m_1 and m_2 define the crash point $p(r)$.*

Note that if a minimum enclosing set consists of three robots $MEC(r) = \{m_1, m_2, m_3\}$, then the midpoint of every edge in $\triangle m_1 m_2 m_3$ is a crash point.

The set of Relative neighborhood edges adjacent to robot r is denoted as $E_{RNG}(r)$. The set of all edges of the Relative neighborhood graph is denoted by $E_{RNG}(R)$. We define, in the same way, the set $E_{UDG}(r)$ of the unit disk graph edges adjacent to robot r and the set $E_{MEC}(r)$ of the Relative neighborhood edges between robot r and the members of the minimum enclosing set $MEC(r)$.

For the Relative neighborhood graph $RNG(R) = (R, E)$, we define the set-valued map $\mathfrak{R} : X \rightsquigarrow Y$, where $X \subset E$ and $Y \subset \mathbb{R}^2$ relate the RNG edges to the area occupied by the corresponding RNG lenses. For example, $\mathfrak{R}(\{r, u\}, \{r, w\})$ is the union of the RNG lenses that correspond to the RNG edges $\{r, u\}, \{r, w\}$.

Definition 4 (Minimum enclosing set cover). *The convex hull around robot r and the members of $MEC(r)$ form the minimum enclosing set cover $K(r)$ of robot $r \in R$.*

Let us now consider a robot $r \in R$ and its minimum enclosing sets. We show that, in any case, the minimum enclosing set cover is a subset of the union of the RNG lenses that correspond to the RNG edges between robot w and the robots of $MEC(w)$, namely $K(r) \subset \mathfrak{R}(E_{MEC}(r))$. Using basic geometric arguments, we conclude Lemma 3.

Lemma 3. *If $MEC(r) = \{m_1, m_2, m_3\}$ is the minimum enclosing set over $RNG(r)$, then $\triangle m_1 m_2 m_3$ is completely covered by the RNG lenses that correspond to the RNG edges between r and the robots of $MEC(r)$.*

We can show similar results as to Lemma 3 for the minimum enclosing sets with different structures.

Lemma 4. *If $MEC(r) = \{m_1, m_2\}$ is the minimum enclosing set over $RNG(r)$, then $\triangle r m_1 m_2$ (Fig. 1b) is completely covered by the RNG lenses between r and the robots of $MEC(r)$.*

Lemma 5. *If $MEC(r) = \{m_1, m_2, r\}$ is the minimum enclosing set over $RNG(r)$, then $\triangle r m_1 m_2$ is completely covered by the RNG lenses between r and the robots of $MEC(r)$.*

It remains to consider the case where $MEC(r) = \{m_1, r\}$ is the minimum enclosing circle over $RNG(r)$. We observe that the minimum enclosing circle $C(r)$ is a proper subset of the RNG lens between m_1 and r. This observation together with Lemmata 3, 4, and 5 yield the following corollary.

Corollary 2. *The union of the RNG lenses between robot r and the robots of $MEC(r)$ is a superset of the minimum enclosing set cover $K(r)$, i.e. $K(r) \subset \mathfrak{R}(E_{MEC}(r))$.*

Note that the RNG lenses between the unit disk graph neighbors that satisfy the RNG criterion do not contain any other robots. If there is a robot inside one of the RNG lenses, then the corresponding part of the RNG will change.

Let us now consider, in details, the collisions that occur between the robots executing GTRC. The set of robots $M \subset R$ that has a collision at time t_* is represented by a single robot u for any $t \geq t_*$. We call u the *representative* of M.

Observation 1 (Early collision). *The set of robots $M \subset R$ had a collision at time t_* if the minimum enclosing circle around the RNG neighborhood of the representative u has a diameter greater than zero.*

Note that the opposite of early collision a *final collision*. In other words, assume that the set of robots $M \subseteq R$ had a collision at time t_* and the minimum enclosing circle around the RNG neighborhood of the representative u has diameter zero. This implies that, after the final collision, there are no other robots in the unit disk graph neighborhood of the representative u. There are also no robots other than those in the unit disk graph neighborhood of robot u due to the connectivity property of S-GTRC (shown in Lemma 2). Thus, if final collision takes place, then $M = R$. Obviously, there can only be one final collision. We say that the gathering is *collisionless* if there are no early collisions.

In the next Lemma 6, we utilize Corollary 2 to show that a robot w can have early collision only at the crash point $p(w)$. The proof is a result of checking carefully all possible dispositions of w and $MEC(w)$.

Lemma 6. *Let us consider robot w during arbitrary time interval $[a, b]$, such that during this interval, diameter d_w of the minimum enclosing circle of robot w is strictly greater than zero. If robot w collides with some other robots $M \subset R$ during $[a, b]$, then:*

1. *collision takes place at the crash point $p(w)$ of robot w;*
2. *robot w does not belong to its own minimum enclosing set $MEC(w)$.*

5   Collisionless Gathering with Extended Robot Model

In this section, we show that it is possible to perform gathering without collision by allowing robots to have some additional capabilities.

We extend the robot model and design the Safe-Go-To-The-Relative-Center algorithm (S-GTRC) using the contracting conditions from Lemma 1 and the structural properties from Lemma 6. The goal of S-GTRC is to gather without *early collisions* all the robots at one not predefined point for any initial configuration. The initial configuration is arbitrary except that $UDG_0(R)$ is connected and all the robots have distinct positions.

The extended robot model is described as follows. The viewing range of a robot is 2. This will be needed to avoid collisions. Thus, a robot can see the UDG neighbors of its UDG neighbors. Note that S-GTRC preserves connectivity with respect to UDG. Open two-unit disk graph is defined analogously to open unit disk graph. For all robots in R, we define $2\text{-}UDG(R) = (R, 2\text{-}E_t)$, where $(r_i, r_j) \in 2\text{-}E_t$ iff for r_i and r_j, it holds that $|r_i(t), r_j(t)| < 2$.

As in the common model, robots are anonymous and each robot has a local coordinate system that is not aligned with the coordinate systems of other robots. Unlike in the common model, robots here are chiral, i.e., they all agree either on left- or right-hand orientation. Robots are equipped with synchronized clocks. Robots are luminous, i.e. they have one bit of visible external memory like in [8] by Das et al. The maximum speed of the robot is assumed to be $s \geq 1$.

Next, we describe S-GTRC and show that it performs gathering in the continuous time model for the extended robot model without early collisions, in time $O(n^2)$.

The main idea of S-GTRC is described as follows. Robots are separated into two groups/states: *regular* and *safe*. In the regular group, robots execute GTRC and do not take into account the robots in the safe group. If some robots in the regular group are about to collide, at least one of them switches to the safe state and independently from other robots, moves towards some specific, closely situated fixed point. This point is selected in such a way that collision is not possible. The state of the robot is automatically visible to its neighbors via the visible external memory.

From Lemma 6 we know that robot r collides only at the crash point of some other robot w. Therefore, shortly before collision, both robots r and w together with their target points $T(w)$ and $T(r)$ are inside the relatively small ball B with the center at the crash point $p(w)$ and radius $1/2m|m_1, m_2|$, where m is a positive parameter. Ball B is depicted in Fig. 3. In order to avoid collision, we let at least one of the robots move independently from the other robots. For a short period of time, the target point of the robot will be selected from the circle D with radius $1/k|w, r|$ centered at the position of r, where $k > 1$ is a positive parameter that will help us preserve connectivity. Circle D together with B are depicted in Fig. 3. Circle D is depicted in Fig. 2. We refer to the arc A of D as the *target arc*. In the safe state, robots move towards the midpoint of the target arc. We construct the target arc in such a way that robots in the safe state satisfy the contracting conditions of Lemma 1. This is shown in Lemma 8.

The target arc is constructed as follows. Let us first consider arc $B \subset D$ such that the central angle $\angle arc = 3\pi/4$ and the bisector of $\angle arc$ coincide with the bisector of $\angle m_1 r m_2$. Then, we draw the line rg perpendicular to $m_1 m_2$. With respect to this line, we either take left or right, depending on the chirality part of B, i.e. $B_L \subset B$. Finally we subtract from B_L, the arc that corresponds to the central angle $\angle grb = \pi/20$. What is left is *target arc* A, that we depict in Fig. 2.

In order to select the suitable moment for the independent motion of robot r, we check whether r is close to the crash point $p(w)$ of some robot w. Namely, robot r checks if $\exists w : w, T(w), MEC(r) \in B_{1/2m|m_1, m_2|}(p(w))$, where robots m_1, m_2 define $p(w)$ AND $|w, r| = \min_{u \in UDG(r) \setminus r}\{|r, u|\}$, where $UDG(r)$ consists of robots in both states (regular and safe) AND $|w, r| \le \rho_Q$ AND r is the leftmost (rightmost, depending on chirality) robot with respect to the direction towards $p(w)$.

We show in Lemma 9 that there exists a point in time, shortly before collision takes place, at which this condition is satisfied. We refer to the logical expression above as the *safety condition*, denoted by the function $\mathfrak{S}_r : X \to \mathbb{Z}_2$, where $X \subset \mathbb{R}^2$, i.e. $\mathfrak{S}_r(w) = true$ means that robot w at its current position violates the safety condition with respect to the crash point $p(w)$ of robot w.

At every point in time t, each robot can read the positions and states of its neighboring robots in 2-$UDG(R)$. Every robot has access to the synchronized clock. Besides that, each robot can read and write into the two variables. These will be used to store either the position on the Euclidean plane, denoted by $M(r)$, or the point in time $\Delta(r)$ together with an additional state $L(r)$. The variable of robot r that represents its state is called $S(r)$. This is set by default to *regular*. Our algorithm has three positive parameters, used by all of the robots: s, m and k. Parameter m defines how close to the crash point we check the safety condition. Parameter s tells us the ratio between the speed of the robots in the safe and regular state. Robots in the safe mode are assumed to be faster. Parameter k tells us what portion of the minimum distance to other robots does the robot cover during the motion in the safe state. The initial state at time 0 of every robot is regular. The initial value of the memory slots that correspond to $M(r)$, $\Delta(r)$ and $L(r)$ are *undefined*. The Safe-Go-To-The-Relative-Center algorithm is presented in pseudo-code as Algorithm 1 in Appendix A.1.

The main difference between S-GTRC and GTRC lies in the states: regular and safe. In regular state, a robot moves according to GTRC. Early collisions may take place only in the regular branch of the algorithm. The safe branch is designed to avoid early collisions. The collision that takes place in S-GTRC without safe branch is called *potential* early collision. Next, we show that the safe branch avoids potential early collisions.

First, we show that in the *safe* state, the robots preserve connectivity. Then, we consider the runtime and show that the robots in the safe state move only inside the convex hull, with speed s. In Lemma 7, we show that if parameter k is big enough (i.e. the radius of circle D is small enough), then at the end of the independent motion in the safe state, the robot will still have the same unit disk graph neighbors.

Lemma 7. *If robot w is in $UDG(r)$ at time 0, when robot r switches to safe state, then w is still in $UDG(r)$ at time $t > 0$, when robot r switches back to regular state.*

From Lemmas 2 and 7, we conclude that S-GTRC preserves connectivity.

Corollary 3. *Let us consider a group of robots R on the Euclidean plane executing S-GTRC. If $\{u, w\}$ is an edge in the open Relative neighborhood graph $RNG(R)$ at time 0, then $\{u, w\}$ is an edge in $RNG(R)$ at $\forall t \geq 0$ or there is a path from u to v in $RNG(R), \forall t \geq 0$.*

Next, we analyze the runtime. We know that the robots in the regular state (i.e. those executing GTRC) satisfy the contracting conditions of Lemma 1. It remains to show that in the safe state the robots also satisfy the contracting conditions. In Lemma 8, we show that the target arc is constructed in such way that its middle point is always inside the local and consequently also a global convex hull.

Lemma 8. *If robot r is in the safe state, then the target point $M(r)$ is inside the convex hull $H(R)$ over all robots, and it does not coincide with the position of the robot $r(t)$ at least until it switches into regular state again.*

Lemma 8 implies that in the safe state any robot r moves with speed $s \geq 1$ inside the convex hull. This means that S-GTRC is a contracting algorithm. Besides that, the length of the convex hull boundary around the initial configuration is not greater then $2(n-1)$, where n is a number of robots, as shown in [17].

Theorem 3. *The group of n robots executing S-GTRC gathers in time $O(n^2)$.*

Next, we investigate the collisions of robots executing S-GTRC. Robots that execute S-GTRC can be in one of the three states: regular, locked (regular) or safe, where a locked robot is a robot r that is positioned at its own crash point $p(r)$ and $\exists w : \mathfrak{S}_w(r) = true$. It actually is in the regular state but the presence of w in the proximity of its $p(r)$ prevents r from switching to the safe state.

Proposition 2. *If robot r is in the safe state, then it does not collide with any other robot.*

This proposition holds since according to S-GTRC, during the motion in the safe state, a robot covers at most a distance $1/k|w,r| = 1/k \min_{u \in UDG(r)\setminus r}\{|r,u|\}$. Due to the speed limit s, all other robots during the safe motion of r can cover at most the same distance. For $k \geq 3$, the position of r will never coincide with the position of some other robot. Similar argumentation works in the Proposition 3 for the locked regular state.

Proposition 3. *If robot r is in the locked state, then it does not collide with any other robot.*

In order to have a collision, according to Corollary 2, robot r needs to reach the crash point $p(w)$ of some other robot w. An example of disposition of robots r and w shortly before a collision at $p(w)$ is depicted in Fig. 4.

Robot r needs to pass through a shrinking gap between two lenses corresponding to the RNG edges between w and m_1, m_2. Next, in Lemma 9, we show that the safe move always triggers before the potential collision by carefully analyzing the safety condition. In Lemma 10, we show that at the end of the motion in the safe sate, a robot avoids potential collision by showing that during the run, a robot cannot reach the crash point of any other robot.

Lemma 9. *If a set of robots $M \subset R$ has a potential collision with robot w at time t_* in $p(w)$, then there exists robot $r \in M$ and point in time $t < t_*$ such that $\mathfrak{S}_r(w) = true$.*

Lemma 10. *If a set of robots $M \subset R$ has a potential collision with robot w at time t_* in $p(w)$, then there exists $r \in M$ such that the safety condition is triggered by $p(w)$ at time $t_1 < t_*$ and for $k = 4$, $m \geq 500$, $s \geq 10$ at the end of the safe motion at time $t_2 \in (t_1, t_*)$ the crash point $p(w)$ does not exist.*

Using Propositions 2 and 3 together with Lemmas 6, 9, 10 and Theorem 3, we prove one of our main results, Theorem 4.

Theorem 4. *For $s \geq 10$, $m \geq 500$, and $k = 4$, the Safe-Go-To-The-Relative-Center algorithm performs gathering in $O(n^2)$ without collisions.*

6   Future Work

In this paper, we propose algorithms that can gather robots *with* and *without* collisions. In fact, collisions may be the outcome of certain start configurations of swarms and so cannot be avoided for some instances. Within this context, we conjecture the following: "Start configurations of swarms yielding unavoidable collisions are singular in a sense that a random perturbation of robot positions within arbitrarily small circles around their original positions results in a collision-free configuration with probability one".

A   Appendix

A.1   Safe-Go-To-The-Relative-Center algorithm

Algorithm 1. S-GTRC

Require: Initial configuration, parameters m and k, velocity s.

1: Robot r observes the positions of all its neighbors (*regular* and *safe*) in 2-$UDG(r)$:

2: **if** $\max_{a,b \in 2\text{-}UDG(r)} |a,b| \geq 1$ **then**

3:    Robot r observes the positions of its *regular* neighbors in $UDG(r)$ and calculates $RNG(r)$.

4:    Robot r computes the minimum circle $C(r)$ enclosing $RNG(r)$. The center $T(r)$ of $C(r)$ is the *target point* of r.

5:    Robot r observes the positions of all *regular* robots in 2-$UDG(r)$. and calculates crash points for every robot in $UDG(r)$.

6:    Robot r checks:

7:    **if** Robot r is at its own crash point, i.e. $r = p(r)$ AND $\exists w : \mathfrak{S}_w(r) = true$ AND $S(w) = safe$ **then**

8:       Set states: $L(r) := locked$, $S(r) := regular$,

9:       $\Delta(r) := \frac{|r, M(r)|}{s} + \#clock$.

10:    **else**

11:       **if** $\exists w : \mathfrak{S}_r(w) = true$ **then**

12:          **if** $M(r) = undefined$ **then**

13:             $S(r) := safe$,

14:             $M(r) :=$ mid point of *target arc* $A \subset D$, where D is the circle with center at r and radius radius $\frac{1}{k}|w, r|$.

15:          **else**

16:             $S(r) := regular$

17:       **if** $L(r) = locked$ AND $\#clock = \Delta(r)$ **then**

18:          Set states: $S(r) := undefined$, $\Delta(r) := undefined$.

19:    Robot r moves:

20:    **if** $S(r) = regular$ **then**

21:       **if** r is already at $T(r)$ **then**

22:          Robot r remains at $T(r)$ and moves in the same way as the target point does.

23:       **else**

24:          Robot r moves with maximum speed 1 towards $T(r)$.

25:    **else**

26:       **if** r is already at $M(r)$ **then**

27:          Set states: $S(r) := regular$, $M(r) := undefined$.

28:       **else**

29:          Robot r moves with maximum speed s towards $M(r)$.

A.2   Go-To-The-Relative-Center algorithm

Algorithm 2. GTRC

Require: Initial configuration
1: Robot r observes the positions of all neighbors in $UDG(r)$ and checks whether:
2: Robot r observes the positions of its unit disk graph neighbors and calculates $RNG(r)$.
3: Robot r computes the minimum circle $C(r)$ enclosing $RNG(r)$. The center $T(r)$ of $C(r)$ is the *target point* of r.
4: Robot r moves:
5: **if** r is already at $T(r)$ **then**
6:     Robot r remains at $T(r)$ and moves in the same way as the target point does.
7: **else**
8:     Robot r moves with speed 1 towards $T(r)$.

A.3   Figures

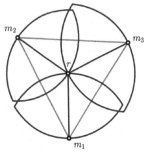
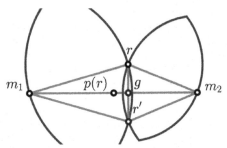

(a) Robot r together with its minimum enclosing set $MEC(r)$ and RNG lenses, that correspond to the RNG edges between r and members of $MEC(r)$.

(b) Robot r together with a crash point $p(r)$ and the robots m_1, m_2 that define this crash point. Besides that here we depict RNG lenses that correspond to the RNG edges of $\{m_1, r\}$ and $\{m_2, r\}$.

Fig. 1. Lenses of the Relative neighborhood graph

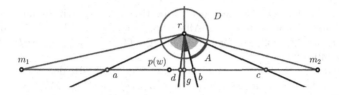

Fig. 2. The construction of *target arc* $A \subset D$.

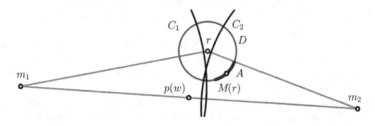

Fig. 3. Robot r in the safe state moves towards target point $M(r)$ the mid point of the target arc $A \subset D$. Black arcs C_1, C_2 are the parts of according RNG lenses.

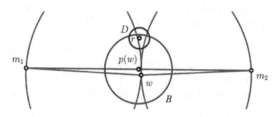

Fig. 4. Robots w with RNG lenses that correspond to RNG edges $\{w, m_1\}$ and $\{w, m_2\}$. Both robots are inside of the circle B with the center at the crash point $p(w)$ and radius $^1/_2m|m_1, m_2|$. If robot r performs safe move, then the target point during the safe move belongs to the circle D.

References

1. Abshoff, S., Cord-Landwehr, A., Fischer, M., Jung, D., Meyer auf der Heide, F.: Gathering a closed chain of robots on a grid. In: Proceedings of the 30th International Parallel and Distributed Processing Symposium (IPDPS), pp. 689–699. IEEE, May 2016
2. Agathangelou, C., Georgiou, C., Mavronicolas, M.: A distributed algorithm for gathering many fat mobile robots in the plane. In: Proceedings of the 2013 ACM Symposium on Principles of Distributed Computing, PODC 2013. ACM, New York, pp. 250–259 (2013)
3. Ando, H., Suzuki, I., Yamashita, M.: Formation and agreement problems for synchronous mobile robots with limited visibility. In: Proceedings of the 1995 IEEE International Symposium on Intelligent Control, 1995, pp. 453–460 (1995)
4. Chrystal, G.: On the problem to construct the minimum circle enclosing n given points in a plane. In: Proceedings of the Edinburgh Mathematical Society, Third Meeting, pp. 30–35 (1885)

5. Cohen, R., Peleg, D.: Convergence properties of the gravitational algorithm in asynchronous robot systems. In: Albers, S., Radzik, T. (eds.) ESA 2004. LNCS, vol. 3221, pp. 228–239. Springer, Heidelberg (2004). https://doi.org/10.1007/978-3-540-30140-0_22

6. Cord-Landwehr, A., Fischer, M., Jung, D., Meyer auf der Heide, F.: Asymptotically optimal gathering on a grid. In: Proceedings of the 28th ACM Symposium on Parallelism in Algorithms and Architectures (SPAA), pp. 301–312. ACM, July 2016

7. Czyzowicz, J., Gąsieniec, L., Pelc, A.: Gathering few fat mobile robots in the plane. In: Shvartsman, M.M.A.A. (ed.) OPODIS 2006. LNCS, vol. 4305, pp. 350–364. Springer, Heidelberg (2006). https://doi.org/10.1007/11945529_25

8. Das, S., Flocchini, P., Prencipe, G., Santoro, N., Yamashita, M.: The power of lights: synchronizing asynchronous robots using visible bits. In: 2012 IEEE 32nd International Conference on Distributed Computing Systems, Macau, China, June 18–21, pp. 506–515 (2012)

9. Degener, B., Kempkes, B., Meyer auf der Heide, F.: A local O(n2) gathering algorithm. In: Proceedings of the 22nd ACM Symposium on Parallelism in Algorithms and Architectures, SPAA 2010, pp. 217–223. ACM, New York (2010)

10. Degener, B., Kempkes, B., Langner, T., Meyer auf der Heide, F., Pietrzyk, P., Wattenhofer, R.: A tight runtime bound for synchronous gathering of autonomous robots with limited visibility. In: Proceedings of the 23rd ACM Symposium on Parallelism in Algorithms and Architectures, SPAA 2011. ACM, New York, pp. 139–148 (2011)

11. Fischer, M., Jung, D., Meyer auf der Heide, F.: Gathering anonymous, oblivious robots on a grid. CoRR, abs/1702.03400 (2017)

12. Flocchini, P.: Distributed Computing by Oblivious Mobile Robots. Synthesis Lectures on Distributed Computing Theory, no. 10. Morgan and Claypool, San Rafael (2012)

13. Gabriel, R.K., Sokal, R.R.: A new statistical approach to geographic variation analysis. Syst. Biol. **18**(3), 259–278 (1969)

14. Gordon, N., Wagner, I.A., Bruckstein, A.M.: Gathering multiple robotic a(ge)nts with limited sensing capabilities. In: Dorigo, M., Birattari, M., Blum, C., Gambardella, L.M., Mondada, F., Stützle, T. (eds.) ANTS 2004. LNCS, vol. 3172, pp. 142–153. Springer, Heidelberg (2004). https://doi.org/10.1007/978-3-540-28646-2_13

15. Kempkes, B., Kling, P., Meyer auf der Heide, F.: Optimal and competitive runtime bounds for continuous, local gathering of mobile robots. In: Proceedinbgs of the 24th ACM Symposium on Parallelism in Algorithms and Architectures, SPAA 2012. ACM, New York, pp. 18–26 (2012)

16. Li, S., Markarian, C., Meyer auf der Heide, F., Podlipyan, P.: A continuous strategy for collisionless gathering. Full version, March 2017. https://www.hni.uni-paderborn.de/pub/9531

17. Li, S., Meyer auf der Heide, F., Podlipyan, P.: The impact of the gabriel subgraph of the visibility graph on the gathering of mobile autonomous robots. In: Chrobak, M., Fernández Anta, A., Gąsieniec, L., Klasing, R. (eds.) ALGOSENSORS 2016. LNCS, vol. 10050, pp. 62–79. Springer, Cham (2017). https://doi.org/10.1007/978-3-319-53058-1_5

18. Lukovszki, T., Meyer auf der Heide, F.: Fast collisionless pattern formation by anonymous, position-aware robots. In: Aguilera, M.K., Querzoni, L., Shapiro, M. (eds.) OPODIS 2014. LNCS, vol. 8878, pp. 248–262. Springer, Cham (2014). https://doi.org/10.1007/978-3-319-14472-6_17

19. Megiddo, N.: Linear-time algorithms for linear programming in $\mathbb{R}^3$ and related problems. SIAM J. Comput. **12**(4), 759–776 (1983)
20. Pagli, L., Prencipe, G., Viglietta, G.: Getting close without touching. In: Even, G., Halldórsson, M.M. (eds.) SIROCCO 2012. LNCS, vol. 7355, pp. 315–326. Springer, Heidelberg (2012). https://doi.org/10.1007/978-3-642-31104-8_27
21. Singer, I.: Abstract Convex Analysis. Wiley, Hoboken (1997)
22. Toussaint, G.T.: The relative neighbourhood graph of a finite planar set. Pattern Recogn. **12**, 261–268 (1980)

Maximizing Barrier Coverage Lifetime
with Static Sensors

Menachem Poss and Dror Rawitz[✉]

Faculty of Engineering, Bar-Ilan University, 52900 Ramat Gan, Israel
menachemposs@gmail.com, dror.rawitz@biu.ac.il

Abstract. We study variants of the STRIP COVER problem (SC) in which sensors with limited battery power are deployed on a line barrier, and the goal is to cover the barrier as long as possible. The energy consumption of a sensor depends on its sensing radius: energy is drained in inverse proportion to the sensor radius raised to a constant exponent $\alpha \geq 1$. In the SET ONCE STRIP COVER (ONCESC) the radius of each sensor can be set once, and the sensor can be activated at any time. SC_k and $ONCESC_k$ are variants of SC and ONCESC, resp., in which each sensor is associated with a set of at most k predetermined radii.

It was previously known that ONCESC is NP-hard when $\alpha = 1$, and the complexity of the case where $\alpha > 1$ remained open. We extend the above mentioned NP-hardness result in two ways: we show that ONCESC is NP-hard for every $\alpha > 1$ and that ONCESC is *strongly* NP-hard for $\alpha = 1$. In addition, we show that $ONCESC_k$, for $k \geq 2$, is NP-hard, for any $\alpha \geq 1$, even for uniform radii sets. On the positive side, we present (i) a $5\gamma^\alpha$-approximation algorithm for $ONCESC_k$, for $k \geq 1$, where γ is the maximum ratio between two radii associated with the same sensor; (ii) a 5-approximation algorithm for SC_k, for every $k \geq 1$; and (iii) a $5 \cdot 2^\alpha$-approximation algorithm for STRIP COVER. Finally, we present an $O(n \log n)$-time algorithm for a variant of $ONCESC_k$ in which all sensors must be activated at the same time.

1 Introduction

Sensor networks are used in a broad range of applications related to surveillance, military, health care, and environmental monitoring. One of the main challenges in designing and operating such networks is coping with the limited resources of the sensors. In this paper we focus on one of these limited resources – energy. More specifically, we study a network of static sensors with limited battery power that is supposed to cover a line barrier, where the goal is to maximize network lifetime, namely to provide coverage as long as possible.

More formally, we study the setting where there are n static sensors located on a barrier represented by the interval $[0, N]$, where $N \in \mathbb{N}$. The input consist of battery charges $b \in \mathbb{N}^n$ and locations $x \in ([0, N] \cap \mathbb{N})^n$. We consider the *set once* model [2] in which we can set the *sensing radius* $\rho_i \in \mathbb{N}$ and the *activation*

D. Rawitz—Supported by the Israel Science Foundation (grant no. 497/14).

A. Fernández Anta et al. (Eds.): ALGOSENSORS 2017, LNCS 10718, pp. 198–210, 2017.
https://doi.org/10.1007/978-3-319-72751-6_15

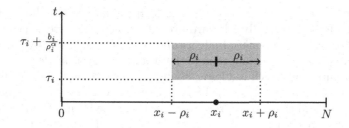

Fig. 1. Sensor i covers the interval $[x_i - \rho_i, x_i + \rho_i]$ during $[\tau_i, \tau_i + b_i/\rho_i^\alpha]$.

time $\tau_i \in \mathbb{R}_+$ (where $\mathbb{R}_+ = \mathbb{R} \cap [0, \infty)$) of each sensor i only once, namely the radius of a sensor remains the same once activated. The *energy consumption rate* of sensor i with a covering radius of ρ_i is proportional to ρ_i^α, where $\alpha \geq 1$ is a *path-loss exponent* [9]. Thus sensor i covers the region $[x_i - \rho_i, x_i + \rho_i]$ during the time interval $[\tau_i, \tau_i + b_i/\rho_i^\alpha]$. It is common to assume a super-linear dependence of the consumption rate on the radius, i.e., to assume that $\alpha > 1$. However, the case of $\alpha = 1$ is still interesting, since there are special settings where a linear dependence may be appropriate [10]. The goal is to find a solution (ρ, τ) for any input (x, b), such that the barrier is completely covered for as long as possible. An example of the coverage of one sensor is given in Fig. 1.

We consider two problems that differ in the types of radial assignments. In the *variable* radii case the radius of a sensor i can be any integer, while in the *fixed* radii case each sensor i is associated with a set R_i of possible values given in the input. In the variable radii case the problem is called SET ONCE STRIP COVER (abbreviated ONCESC), and in the fixed case, if all radii sets are of size at most k the problem is denoted by ONCESC$_k$. If all radii sets are the same, i.e., $R_i = R_j$, for every $i \neq j$, we refer to the instance as having *uniform radii sets*. Such an instance can be obtained if we use uniform sensors (say, from the same manufacturer) that previously took part in other tasks, and therefore may have non-uniform battery powers. We also consider the STRIP COVER problem (SC) in which the radius of each sensor can be set any finite number of times. In this case, a solution is a vector of functions where the ith function determines the radius of sensor i at any given moment. SC$_k$ is the fixed radii variant of SC.

Related work. Buchsbaum et al. [3] considered the RESTRICTED STRIP COVER problem (RSC) where the input consists of a set of rectangles that can be moved vertically but not horizontally, and the goal is to maximize the height at which an interval region is fully covered. This problem is equivalent to ONCESC$_1$. Notice that since there is only one radius per sensor in ONCESC$_1$, the value of α is irrelevant. Buchsbaum et al. [3] gave an $O(\log \log \log n)$-approximation algorithm for RSC and a $(2 + \varepsilon)$-approximation algorithm for the uniform radii case. They also showed that it is strongly NP-hard using a reduction from the DYNAMIC STORAGE ALLOCATION problem. Stockmeyer (see [6, Problem SR2]) showed that DYNAMIC STORAGE ALLOCATION is strongly NP-hard using a reduction from 3-PARTITION. (The reduction is given in [4].) In fact, in the above reduction implies that RSC is strongly NP-hard even if all rectangles are of height either 1 or 2. Later, a 5-approximation algorithm for RSC was given by Gibson and Varadarajan [7]. It is important to note that their analysis is made in comparison to the minimum load which is a lower bound on the optimal height (or lifetime).

Bar-Noy and Baumer [1] studied STRIP COVER. Assuming unit batteries and $\alpha = 1$, they showed that the approximation ratio of the Round Robin algorithm, which simply forces each sensor to successively cover the barrier for as long as possible by itself, is between $\frac{3}{2}$ and 1.825. Bar-Noy et al. [2] defined the *set once* model. They proved that ONCESC is NP-hard when $\alpha = 1$. They analyzed Round Robin for the case where $\alpha = 1$ and showed that its approximation ratio is $\frac{3}{2}$, even for STRIP COVER, thus closing the gap left by [1]. In addition, they gave an algorithm for a simpler variant, called SET RADIUS STRIP COVER, where all sensors must be activated at the same time.

Lev-Tov and Peleg [8] explored the DISK COVER problem in which the input consists of a set of *points* and a set of sensor locations on the plane, and the goal is to cover the points so as to minimize the sum of radii (i.e., $\alpha = 1$). For the one-dimensional case with variable radii they provided a polynomial time algorithm and a linear-time 4-approximation algorithm. The former algorithm works for any $\alpha \in \mathbb{N}$. They also presented two PTASs for DISK COVER, one for variable radii, and the other for fixed radii, assuming that the set of points is a subset of the set of sensor locations. Li et al. [5] studied barrier coverage with static sensors, where the objective is to minimize the total energy consumption rate, namely the goal is to find a radii assignment ρ that minimizes $\sum_i \rho_i^\alpha$. They presented a polynomial time algorithm for the fixed multiple radii case, a 2-approximation algorithm for the variable radii case. They also presented an FPTAS and an NP-hardness result for variable radii and $\alpha = 1$.

Our results. We provide NP-hardness results in Sect. 3. We show that ONCESC_k is NP-hard, for any $k \geq 2$ and $\alpha \geq 1$, even for the case of uniform radii sets. This result is obtained using a reduction from PARTITION. We use the same construction, but with a more elaborate analysis, to show that ONCESC is NP-hard, for any $\alpha \geq 1$. We use a reduction from 3-PARTITION to show that ONCESC is strongly NP-hard in the case where $\alpha = 1$. The last two results improve the NP-hardness result for ONCESC with $\alpha = 1$ from [2].

Several algorithms are given in Sect. 4. We present a $5\gamma^\alpha$-approximation algorithm for ONCESC_k, for $k \geq 1$, where γ is the maximum ratio between two radii that belong to the same sensor, i.e., $\gamma \triangleq \max_i \{r_{i,k_i}/r_{i,1}\}$, where $k_i = |R_i|$. Observe that if $\gamma = O(1)$, then the approximation ratio is also $O(1)$. We present a 5-approximation algorithm for SC_k, for every $k \geq 1$. In addition, we give a $5 \cdot 2^\alpha$-approximation algorithm for STRIP COVER. The algorithms are based on the 5-approximation algorithm RSC (ONCESC_1) by Gibson and Varadarajan [7].

Finally, we consider the fixed radii version of SET RADIUS STRIP COVER, for which we give an $O(n \log n)$-time algorithm for every $k \geq 1$. (Recall that in this problem all sensors must be activated at the same time.) This section was removed for lack of space.

2   Preliminaries

The SET ONCE STRIP COVER problem (abbreviated ONCESC) is defined as follows. Let $U = \{0, \ldots, N\}$, where $N \in \mathbb{N}$, be the discrete barrier that we aim to cover. A ONCESC instance consists of a vector $x \in U^n$ of n sensor locations on the barrier and a corresponding vector $b \in \mathbb{N}^n$ of battery charges. We assume that $x_i \leq x_{i+1}$ for every $i \in \{1, \ldots, n-1\}$. A solution is an assignment of radii and activation times to sensors. More specifically, a solution is a pair $(\rho, \tau) \in \mathbb{N}^n \times \mathbb{R}_+^n$, where ρ_i is the *sensing radius* of sensor i and τ_i is the *activation time* of sensor i. We assume that the radius of each sensor cannot be reset, therefore sensor i becomes active at time τ_i, covers the *interval* $[x_i - \rho_i, x_i + \rho_i]$ for b_i/ρ_i^α time units, and then becomes inactive since it has exhausted its battery.

Any solution can be visualized by a space-time diagram in which each coverage assignment can be represented by a rectangle. It is customary in such diagrams to view the sensor locations as forming the horizontal axis, with time extending upwards vertically. In this case, the coverage of a sensor located at x_i and assigned the radius ρ_i beginning at time τ_i is depicted by a rectangle with lower-left corner $(x_i - \rho_i, \tau_i)$ and upper-right corner $(x_i + \rho_i, \tau_i + b_i/\rho_i^\alpha)$. Let the set of all points contained in this rectangle be denoted as $\mathrm{Rect}(\rho_i, \tau_i)$. A point (u, t) in space-time is *covered* by a solution (ρ, τ) if $(u, t) \in \bigcup_i \mathrm{Rect}(\rho_i, \tau_i)$. The *lifetime* of the network in a solution (ρ, τ) is the maximum value T such that every point $(u, t) \in [0, N] \times [0, T]$ is covered. We denote this value by $T(\rho, \tau)$. In ONCESC the goal is to find a solution (ρ, τ) that maximizes the lifetime. Given an instance (x, b), the optimal lifetime is denoted by $\mathrm{OPT}(x, b)$ (or by OPT).

The ONCESC$_k$ problem is a variant of ONCESC in which the instance also contains a set $R_i = \{r_{i,1}, \ldots, r_{i,k_i}\}$, for every sensor i, containing $k_i \leq k$ radii that are possible for sensor i. That is, it is required that $\rho_i \in R_i \cup \{0\}$, for every i. We assume that $r_{i,j} < r_{i,j+1}$, for every j. Observe that ONCESC$_1$ is equivalent to RESTRICTED STRIP COVER. RADSC$_k$ is variant of ONCESC$_k$ in which all sensors become active at together, namely where $\tau = 0$. Hence, a solution is simply a radial assignment ρ.

In STRIP COVER (SC) the radius of a sensor may change any finite number of times, hence a solution is a vector of functions $[\rho$, where $\rho_i : \mathbb{R}^+ \to [0, N] \cap \mathbb{N}$ is a piecewise constant function that determines the radius of sensor i at any given moment. In this case the lifetime of a solution ρ is denoted by $T(\rho)$. SC$_k$ is the fixed radii variant of SC. (Notice that SC$_1$ is not the same as ONCESC$_1$).

3   Hardness Results

We provide several hardness results. We show that ONCESC$_2$ is NP-hard for any $\alpha \geq 1$, even for uniform radii sets, using a reduction from PARTITION. The same reduction, but with a more careful analysis, is used to show that ONCESC is NP-hard. A reduction from 3-PARTITION is used to show that ONCESC is strongly NP-hard in the case where $\alpha = 1$. We start the section by examining the structure of an optimal solution.

3.1   Structure of an Optimal Solution

We show that that there exists an optimal solution that has a certain structure. More specifically, given a radii assignment, we show that if there are sensors that cover the whole barrier by themselves, then we can change the activation times such that these sensors would first cover the barrier one by one, and then the rest of the sensors would cover the barrier. Moreover, this can be done without decreasing the lifetime of the network.

Given a radii assignment ρ, let $W(\rho)$ be the set of sensors that cover the whole barrier, namely define $W(\rho) \triangleq \{i : [0,N] \subseteq [x_i - \rho_i, x_i + \rho_i]\}$. Also define $t_W \triangleq \sum_{i \in W(\rho)} b_i / \rho_i^\alpha$. We first show that a sensor in $W(\rho)$ can be moved to the beginning of the schedule.

Lemma 1. *Let (b,x) be a* ONCESC *(or a* ONCESC$_k$*) instance, and let (ρ, τ) be a solution. Also, i be a sensor that covers the whole barrier, i.e., $[0,N] \subseteq [x_i - \rho_i, x_i + \rho_i]$. Then, there exists a vector τ', such that (i) (ρ, τ') is a feasible solution and $T(\rho, \tau') \geq T(\rho, \tau)$; and (ii) $\tau_i' = 0$ and $\tau_j' \geq \frac{b_i}{\rho_i^\alpha}$, for every $j \neq i$.*

Proof. The idea is to activate sensor i at time 0. This creates a coverage gap, and this gap is closed by activating sensors that were originally activated before i after i's battery is depleted. Formally, given (ρ, τ) we define τ' as follows:

$$\tau_j' = \begin{cases} 0 & i = j, \\ \max\{\tau_j, b_i/\rho_i^\alpha\} & \tau_j \geq \tau_i, \\ \tau_j + b_i/\rho_i^\alpha & \tau_j < \tau_i. \end{cases}$$

First observe that $\tau_i' = 0$ and $\tau_j' \geq \frac{b_i}{\rho_i^\alpha}$, for every $j \neq i$, by construction.

To prove that $T(\rho, \tau') \geq T(\rho, \tau)$, we show that any point $(u,t) \in [0,N] \times [0, T(\rho, \tau)]$ is covered in the solution (ρ, τ'). We consider three cases.

- First, if $t < b_i/\rho_i^\alpha$, then (u,t) is covered by sensor i.
- The second case is when $b_i/\rho_i^\alpha \leq t < \tau_i + b_i/\rho_i^\alpha$. Since $t - b_i/\rho_i^\alpha \geq 0$, the point $(u, t - b_i/\rho_i^\alpha)$ must be covered in the solution (ρ, τ) by some sensor j. Moreover, since $t - b_i/\rho_i^\alpha \leq \tau_i$, it must be that $j \neq i$ and $\tau_j < \tau_i$. It follows that $\tau_j' = \tau_j + b_i/\rho_i^\alpha$, and thus (u,t) is covered in (ρ, τ').
- If $t \geq \tau_i + b_i/\rho_i^\alpha$, then the point (u,t) is covered in (ρ, τ) by a sensor $j \neq i$. Hence $t \in [\tau_j, \tau_j + b_j/\rho_j^\alpha]$. We claim that (u,t) is covered in (ρ, τ') by the same sensor j. Since $\tau_j' \geq \tau_j$, for any $j \neq i$, it is enough to show that $t \geq \tau_j'$. First, if $\tau_j < \tau_i$ we have that $\tau_j' = \tau_j + b_i/\rho_i^\alpha < \tau_i + b_i/\rho_i^\alpha$, and it follows that $t > \tau_j'$. The other option is that $\tau_j \geq \tau_i$, and in this case we have that $\tau_j' = \max\{\tau_j, b_i/\rho_i^\alpha\}$. Since $t \geq \tau_j$ and $t > \tau_i + b_i/\rho_i^\alpha$ we are done.

The lemma follows.   □

The following result is obtained using induction and Lemma 1.

Lemma 2. *Let (b,x) be a* ONCESC *(or a* ONCESC$_k$*) instance, and let (ρ, τ) be a solution. Then, there is a vector τ' such that (i) (ρ, τ') is a feasible solution, (ii) $T(\rho, \tau') \geq T(\rho, \tau)$, and (iii) if $i \in W(\rho)$, then $\tau_i' + \frac{b_i}{\rho_i^\alpha} \leq t_W$, and otherwise, $\tau_i' \geq t_W$.*

3.2 Multiple Fixed Radii

In this section we show that ONCESC_2 is NP-hard even in the case of uniform radii sets. This is done using a reduction from PARTITION.

Theorem 1. ONCESC_2 *is NP-hard, even for uniform radii sets.*

Proof. Consider a PARTITION instance $A = \{a_1, \ldots, a_n\}$. Define $B \triangleq \frac{1}{2} \sum_i a_i$. We construct a ONCESC_2 instance as follows. First assume that N is a multiple of 6. The instance contains $n + 2$ sensors, where $b_1 = b_{n+2} = B$ and $b_{i+1} = a_i$, for every i. The locations are: $x_1 = \frac{N}{6}$, $x_i = \frac{N}{2}$, for every $i \in \{2, \ldots, n+1\}$, and $x_{n+2} = \frac{5N}{6}$. The radii subsets are: $R_i = \{\frac{N}{6}, \frac{N}{2}\}$, for every i.

We claim that $A \in$ PARTITION if and only if $\text{OPT}(b, x) = B \cdot \frac{6^\alpha + 2^\alpha}{N^\alpha}$.

Consider an optimal solution (ρ, τ) to such an instance. Observe that $W(\rho) = \{i : \rho_i = N/2 \text{ and } i \neq 1, n+2\}$. By Lemma 2 we may assume that any sensor $i \in W(\rho)$ works alone when covering the barrier. The remaining sensors are activated not earlier than t_W.

Observe that the remaining sensors that are located at $\frac{N}{2}$ cannot cover 0 with a radius of $\frac{N}{6}$. Thus if $\rho_1 = \frac{N}{2}$, sensor 1 cannot be helped by sensors that are located at $\frac{N}{2}$. Moreover, it must get help from sensor $n + 2$, and in this case the network lifetime is: $T(\rho, \tau) = \sum_{i \in W(\rho)} \frac{a_i}{(N/2)^\alpha} + \frac{B}{(N/2)^\alpha}$. This value is maximized when $w(\rho) = \{2, \ldots, n+2\}$, and in this case $T(\rho, \tau) = B \cdot \frac{3 \cdot 2^\alpha}{N^\alpha}$. A similar argument can be used if $\rho_{n+2} = \frac{N}{2}$.

Now consider the case where $\rho_1 = \frac{N}{6}$ and $\rho_{n+2} = \frac{N}{6}$. Sensors 1 and $n+2$ cannot cover the barrier by themselves, since they do not cover the interval $[\frac{N}{3}, \frac{2N}{3}]$. However, sensors in $\{2, \ldots, n+1\} \setminus W(\rho)$ can do that, and the best way to do that is with no overlaps. It follows that

$$T(\rho, \tau) = \sum_{i \in W(\rho)} \frac{a_i}{(N/2)^\alpha} + \min\left\{\frac{B}{(N/6)^\alpha}, \frac{C}{(N/6)^\alpha}\right\} = \frac{2B - C}{(N/2)^\alpha} + \frac{\min\{B, C\}}{(N/6)^\alpha},$$

where $C = \sum_{i \in \{2, \ldots, n+1\} \setminus W(\rho)} a_i$. If $C \leq B$, we have that $T(\rho, \tau) = 2B \cdot \frac{2^\alpha}{N^\alpha} + C \cdot (\frac{6^\alpha}{N^\alpha} - \frac{2^\alpha}{N^\alpha})$, which is an increasing function of C. If $C \geq B$, then $T(\rho, \tau) = B \cdot (2\frac{2^\alpha}{N^\alpha} + \frac{6^\alpha}{N^\alpha}) - C \cdot \frac{2^\alpha}{N^\alpha}$, which is decreasing function of C. It follows that the maximum value is obtain when $C = B$ and in this case $T(\rho, \tau) = B \cdot (\frac{2^\alpha}{N^\alpha} + \frac{6^\alpha}{N^\alpha}) = B \cdot \frac{2^\alpha + 6^\alpha}{N^\alpha}$.

Since $3 \cdot 2^\alpha < 2^\alpha + 6^\alpha$, for $\alpha \geq 1$, it follows that the best we could hope for is $B \cdot \frac{2^\alpha + 6^\alpha}{N^\alpha}$.

If $A \in$ PARTITION, there exists I such that $\sum_{i \in I} a_i = B$. We can obtain a lifetime of $B \cdot \frac{2^\alpha + 6^\alpha}{N^\alpha}$ by assigning $\rho_{i+1} = \frac{N}{2}$ if $i \in I$, and $\rho_i = \frac{N}{6}$ otherwise. On the other hand, if a lifetime of $B \cdot \frac{2^\alpha + 6^\alpha}{N^\alpha}$ is attainable, then the set $W(\rho)$ induces a partition. More specifically, $\sum_{i \in I} a_i = B$ for $I = \{i : i + 1 \in W(\rho)\}$. $\square$

This result can be extended to any $k > 2$.

Theorem 2. ONCESC_k *is NP-hard, for any $k > 2$, even for uniform radii sets.*

Proof. We use a similar reduction to the one from the proof of Theorem 1, where the only difference is that $R_i = \{\frac{N}{2}, \frac{N}{6}, k - 2, \ldots, 1\}$ (assuming that $k < \frac{N}{6}$). Since the smaller radii cannot be used to cover the barrier, they are redundant. The theorem follows.   □

3.3   Variable Radii

We show that ONCESC is NP-hard for any $\alpha \geq 1$. This is done using the reduction from PARTITION that was defined in the proof of Theorem 1.

Theorem 3. ONCESC *is NP-hard, for any $\alpha \geq 1$.*

Proof. We use the same construction (not including the R_is) that was used in the proof of Theorem 1, and again we claim that $A \in$ PARTITION if and only if $\mathrm{OPT}(b, x) = B \cdot \frac{2^\alpha + 6^\alpha}{N^\alpha}$.

Consider an optimal solution (ρ, τ) to such an instance. First notice that we may assume without loss of generality that $\rho_1, \rho_{n+2} \leq \frac{5}{N}$ and that $\rho_i \leq \frac{N}{2}$, for every $i \in \{2, \ldots, n+1\}$. Observe that $W(\rho) = \{i : \rho_i \geq \frac{N}{2}\}$. By Lemma 2 we may assume that any sensor $i \in W(\rho)$ works alone when covering the barrier.

We examine the sensors 1 and $n + 2$. There are four cases corresponding to whether or not they belong to $W(\rho)$.

1. First consider the case where $1, n + 2 \in W(\rho)$. By Lemma 2 we may assume that both sensors cover the whole barrier by themselves, each during a time interval of length $\frac{B}{(5N/6)^\alpha} = \frac{6^\alpha B}{5^\alpha N^\alpha}$. Observe that if they work together, each with radius $\frac{N}{3}$ they cover the barrier for $\frac{B}{(N/3)^\alpha} = \frac{3^\alpha B}{N^\alpha}$ time. This increases the lifetime, since $\frac{2 \cdot 6^\alpha}{5^\alpha} < 3^\alpha$, for $\alpha \geq 1$.
2. Next we consider the case where $1 \notin W(\rho)$ and $n+2 \in W(\rho)$. Since $1 \notin W(\rho)$, it follows that $\rho_1 < \frac{5N}{6}$ which means that 1 does not cover the right endpoint of the barrier, namely the point N. Since $n + 2 \in W(\rho)$, sensors $n + 2$ never collaborates with 1. Hence, to cover the barrier, sensor 1 needs help from another sensor $i \notin W(\rho)$ located at $\frac{N}{2}$. Assuming that this happens, i needs to cover the right endpoint, and we have that $\rho_i \geq \frac{N}{2}$, which means that $i \in W(\rho)$ and we get a contradiction.
3. Similarly we may eliminated the option, where $1 \in W(\rho)$ and $n + 2 \notin W(\rho)$.

It follows that we may assume that $1, n + 2 \notin W(\rho)$.

Now consider the sensors that are not in $W(\rho)$. We know that $\rho_i < \frac{N}{2}$, for every $i \in \{2, \ldots, n+1\} \setminus W(\rho)$. Hence, any such sensor i cannot reach both barrier end points, and therefore it must rely on the help of both 1 and $n + 2$. We also know that $\rho_1, \rho_{n+2} < \frac{5N}{6}$, and this means that the two sensors must work together to obtain coverage. It follows that after t_W the barrier is covered only if both 1 and $n + 2$ are activated. Also, we may assume that $\rho_1, \rho_{n+2} \geq \frac{N}{6}$, since otherwise the endpoints remain uncovered.

If $\rho_1 + \rho_{n+2} \geq \frac{2N}{3}$, both sensors cover the barrier by themselves, and they achieve maximum lifetime when $\rho_1 = \rho_{n+2} = N3$. In this case we get

$$T(\rho,\tau) = \sum_{i \in W(\rho)} \frac{a_i}{(N/2)^\alpha} + \frac{B}{(N/3)^\alpha} = \frac{2B-C}{(N/2)^\alpha} + \frac{B}{(N/3)^\alpha} = \frac{(2 \cdot 2^\alpha + 3^\alpha)B - 2^\alpha C}{N^\alpha}$$

where $C = \sum_{i \in \{2,\ldots,n+1\} \setminus W(\rho)} a_i$. This function decreasing in C and therefore the maximum is obtained when $C = 0$. In this case

$$T(\rho,\tau) = B \cdot \frac{2 \cdot 2^\alpha + 3^\alpha}{N^\alpha}. \tag{1}$$

Next we consider the case where $\rho_1 + \rho_{n+2} < \frac{2N}{3}$. In this case sensors 1 and $n+2$ need the help of sensors located at $\frac{N}{2}$. Let S be the set of sensors that collaborate with 1 and $n+2$. Since $\rho_1, \rho_{n+2} \geq \frac{N}{6}$, we may assume that $\rho_i \leq \frac{N}{6}$, for every $i \in S$. Let $\ell = \min_{i \in S} \rho_i$. We may assume that $\rho_1, \rho_{n+2} = \frac{N}{3} - \ell \geq \frac{N}{6} \geq \ell$. Hence the lifetime that is obtained is:

$$T(\rho,\tau) = \sum_{i \in W(\rho)} \frac{a_i}{(N/2)^\alpha} + \min\left\{\frac{C}{\ell^\alpha}, \frac{B}{(N/3-\ell)^\alpha}\right\}$$

$$= (2B-C) \cdot \frac{2^\alpha}{N^\alpha} + \min\left\{\frac{C}{\ell^\alpha}, \frac{B}{(N/3-\ell)^\alpha}\right\}$$

If $C \geq B$, then $\frac{C}{\ell^\alpha} \geq \frac{B}{(N/3-\ell)^\alpha}$, since $\ell \leq \frac{N}{3} - \ell$. Hence, $T(\rho,\tau) = (2B - C) \cdot \frac{2^\alpha}{N^\alpha} + \frac{B}{(N/3-\ell)^\alpha}$. This expression is maximized when $\ell = \frac{N}{6}$, and we get that $T(\rho,\tau) = B \cdot \frac{2^\alpha + 6^\alpha}{N^\alpha} - C \cdot \frac{2^\alpha}{N^\alpha}$. This function is decreasing in C, and it is maximized when $C = B$. The lifetime in this case is $T(\rho,\tau) = B \cdot \frac{2^\alpha + 6^\alpha}{N^\alpha}$.

On the other hand, if $C \leq B$, then ρ_1 and ρ_{n+2} will increase to extend the lifetime. Assume that $\ell \in \mathbb{R}$ is allowed. In this case the maximum is obtained when $\frac{C}{\ell^\alpha} = \frac{B}{(N/3-\ell)^\alpha}$. It follows that $(\frac{N}{3\ell} - 1)^\alpha = \frac{B}{C}$ or $\ell = \frac{N}{3(\sqrt[\alpha]{B/\alpha}+1)}$. Hence,

$$T(\rho,\tau) = (2B-C) \cdot \frac{2^\alpha}{N^\alpha} + \frac{C}{\ell^\alpha} = 2B \cdot \frac{2^\alpha}{N^\alpha} + C \cdot \left(\frac{3^\alpha(\sqrt[\alpha]{B/C}+1)^\alpha - 2^\alpha}{N^\alpha}\right).$$

The derivative is strictly positive, since

$$\frac{\partial T(\rho,\tau)}{\partial C} = \frac{1}{N^\alpha}\left(\left[3^\alpha(\sqrt[\alpha]{B/C}+1)^\alpha - 2^\alpha\right] - C \cdot 3^\alpha \alpha(\sqrt[\alpha]{B/C}+1)^{\alpha-1}\frac{\sqrt[\alpha]{B/C}}{\alpha C}\right)$$

$$= \left([3^\alpha(\sqrt[\alpha]{B/C}+1)^\alpha - 2^\alpha] - 3^\alpha(\sqrt[\alpha]{B/C}+1)^{\alpha-1} \cdot \sqrt[\alpha]{B/C}\right)/N^\alpha$$

$$= \left(3^\alpha(\sqrt[\alpha]{B/C}+1)^{\alpha-1}\left[(\sqrt[\alpha]{B/C}+1) - \sqrt[\alpha]{B/C}\right] - 2^\alpha\right)/N^\alpha$$

$$= \left(3^\alpha(\sqrt[\alpha]{B/C}+1)^{\alpha-1} - 2^\alpha\right)/N^\alpha$$

$$\geq (3^\alpha 2^{\alpha-1} - 2^\alpha)/N^\alpha,$$

where the inequality is implied by $C \leq B$ and $\alpha \geq 1$. It follows that the function is increasing and reaches its maximum when $C = B$. As mentioned above, the lifetime in this case is $T(\rho, \tau) = B \cdot \frac{2^\alpha + 6^\alpha}{N^\alpha}$.

It remains to compare the above mentioned value to (1). Since $\alpha \geq 1$, we have that $B \cdot \frac{2 \cdot 2^\alpha + 3^\alpha}{N^\alpha} < B \cdot \frac{2^\alpha + 6^\alpha}{N^\alpha}$. Hence, $B \cdot \frac{2^\alpha + 6^\alpha}{N^\alpha}$ is the best we can hope for.

To finalize the proof, if $A \in$ PARTITION, there exists I such that $\sum_{i \in I} a_i = B$. A lifetime of $B \cdot \frac{2^\alpha + 6^\alpha}{N^\alpha}$ can be achieved by assigning $\rho_{i+1} = \frac{N}{2}$ if $i \in I$, and $\rho_i = \frac{N}{6}$ otherwise. On the other hand, if a lifetime of $B \cdot \frac{2^\alpha + 6^\alpha}{N^\alpha}$ is attainable, then the set $W(\rho)$ induces a partition. $\qquad \square$

3.4 Strong NP-Hardness When $\alpha = 1$

We show that ONCESC with $\alpha = 1$ is strongly NP-hard using a reduction from 3-PARTITION. To do that we need the following observation that was given in [2].

Observation 1 [2]. *The lifetime of a ONCESC instance (x, b) is at most $2 \sum_i b_i / N$. Moreover, if OPT $= 2 \sum_i b_i / N$, then there are no coverage overlaps or coverage outside the barrier.*

Theorem 4. ONCESC *with $\alpha = 1$ is strongly NP-hard.*

Proof. Consider a 3-PARTITION instance $A = \{a_1, \ldots, a_n\}$, where $n = 3m$ and $\frac{1}{m} \sum_i a_i = Q$. (Recall that 3-PARTITION is NP-hard even if $a_i \in (\frac{Q}{4}, \frac{Q}{2})$, for every i, and Q is polynomial in n.) Define an instance of ONCESC as follows. First, let $N = 8m$. The instance contains $n + 2m = 5m$ sensors, where

$$x_i = \begin{cases} m + i - 1 & i \in [1, m], \\ 4m & i \in [m + 1, 4m], \\ 2m + i & i \in [4m + 1, 5m], \end{cases} \qquad b_i = \begin{cases} Q \cdot \frac{m + i - 1}{2m - 2i + 2} & i \in [1, m], \\ a_{i-m} & i \in [m + 1, 4m], \\ Q \cdot \frac{6m - i}{2i - 8m} & i \in [4m + 1, 5m]. \end{cases}$$

We claim that $A \in$ 3-PARTITION if and only if OPT $= 2 \sum_i b_i / N = \frac{Q}{2} \cdot H_m$, where H_m is the ith Harmonic number.

Supposed $A \in$ 3-PARTITION, i.e., there exists a partition $A_1, \ldots, A_m$ of A into m triples such that $\sum_{a \in A_j} a = Q$. We construct a solution (ρ_i, τ_i) as follows:

$$\rho_i = \begin{cases} m + i - 1 & i \in [1, m], \\ 2j & i \in [m + 1, 4m] \text{and } a_{i-m} \in A_j, \\ 6m - i & i \in [4m + 1, 5m], \end{cases}$$

and

$$\tau_i = \begin{cases} \frac{Q}{2} \cdot H_{m-i} & i \in [1, m], \\ \frac{Q}{2} \cdot H_{j-1} + \frac{\sum_{q=0}^{d-1} b_i}{\rho_i} & i \in [m + 1, 4m] \text{and } i - m = i_{j,d}, \\ \frac{Q}{2} \cdot H_{i - 4m - 1} & i \in [4m + 1, 5m], \end{cases}$$

where $A_j = \{a_{i_{j,0}}, a_{i_{j,1}}, a_{i_{j,2}}\}$. Observe that there are five sensors that are activated in the time interval $[Q/2 \cdot H_{\ell-1}, Q/2 \cdot H_\ell)$, for $\ell \leq m$. First, sensor $m - \ell + 1$ and sensor $4m + \ell$ are activated exactly at $Q/2 \cdot H_{\ell-1}$. Also, the three sensors that correspond to A_ℓ are also activated in this interval:

- Sensor $m + i_{\ell,0}$ is activated at $Q/2 \cdot H_{\ell-1}$.
- Sensor $m + i_{\ell,1}$ is activated at $Q/2 \cdot H_{\ell-1} + b_{i_{\ell,0}}/\rho_{i_{\ell,0}}$.
- Sensor $m + i_{\ell,2}$ is activated at $Q/2 \cdot H_{i-1} + b_{i_{\ell,0}}/\rho_{i_{\ell,0}} + b_{i_{\ell,1}}/\rho_{i_{\ell,1}}$.

Observe that $\rho_{m-\ell+1} = \rho_{4m+\ell} = 2m - \ell$ and $\rho_{i_{\ell,d}} = 2\ell$, for $d \in \{0,1,2\}$. In addition, notice that sensor $m + i_{\ell,1}$ is activated exactly when the battery of sensor $m + i_{\ell,0}$ is depleted, and sensor $m + i_{\ell,2}$ is activated exactly when the battery of sensor $m + i_{\ell,1}$ is depleted. Moreover, since $x_{m-\ell+1} = 2m - \ell$, $x_{4m+\ell} = 6m + \ell$, and $x_{i_{j,d}} = 4m$, for $d \in \{0,1,2\}$, the barrier is covered with no overlaps by the five sensors during the time interval $[Q/2 \cdot H_{\ell-1}, Q/2 \cdot H_\ell)$. If follows that there are no overlaps, and hence $\text{OPT} = 2 \sum_i b_i/N = \frac{Q}{2} \cdot H_m$.

Now, assume that there exists an optimal solution (ρ', τ') without excess coverage whose lifetime is $2 \sum_i b_i/N = \frac{Q}{2} \cdot H_m$. Consider a sensors i, where $i \leq m$. If $\rho'_i > \rho_i$, then energy is wasted for coverage beyond the left end-point of the barrier. If $\rho'_i < \rho_i$, then there are several options. If sensor i is active after the solution lifetime, then energy is wasted. Otherwise, since sensor i fails to cover the left end-point of the barrier, the endpoint must be covered by another sensor i' and hence their coverage intervals must intersect, which again means a waste of energy. It follows that $\rho'_i = \rho_i$, for every $i \leq m$. By a symmetric argument we have that $\rho'_i = \rho_i$, for every $i \geq 4m + 1$.

Since the rest of the sensors are located at $4m$, the sensors in the two sets $\{1, \ldots, m\}$ and $\{4m + 1, \ldots, 5m\}$ must work in the same pairs as in (ρ, τ), since otherwise energy is wasted due to overlaps. More specifically, if at some time a sensor $i \leq m$, does not work with sensor $5m - i + 1$, then if it covers the barrier only with a sensor located at $4m$, then i is redundant and there is a coverage overlap. If it covers the barrier with a sensor $i' \geq 4m+1$, such that $i' \neq 5m-i+1$, then they need help from a sensor located at $4m$, and its covering interval will overlap with the interval of i or with the interval of i'.

It follows that the solution can be partitioned into m temporal strips each corresponding to a sensors pair i and $5m - i + 1$. In each such strip, sensors located at $4m$ fill the gap in coverage $[2m + 2i - 2, 6m - 2i + 2]$ that lasts $\frac{b_i}{\rho_i} = \frac{Q}{2m-2i+2}$ time units left by the ith sensor pair. Since each such gap must be filled exactly by sensors located at $4m$, there is a set of sensors filling the gap with radii $2m - 2i + 2$. Since they need to cover the gap for $\frac{Q}{2m-2i+2}$ time, they combined batteries should add up to Q. Hence, the strips induce a partition of the sensors into m subsets, such that the total power of a subset is Q. The partition induces a partition of A, and hence $A \in 3\text{-PARTITION}$.   □

4   Approximation Algorithms

We present several algorithmic results. We provide a $5\gamma^\alpha$-approximation algorithm for ONCESC_k, where γ is the maximum ratio between two radii associated with the same sensor. In addition, we give a 5-approximation algorithm for SC_k, for every $k \geq 1$, and a $5 \cdot 2^\alpha$-approximation algorithm for STRIP COVER. The algorithms are based on the 5-approximation algorithm RSC from [7]. Most proofs of this section were omitted for lack of space.

4.1 Algorithm for Set Once Strip Cover with Fixed Radii

We start by presenting a $5\gamma^\alpha$-approximation algorithm for ONCESC_k, for $k \geq 1$, where γ is the maximum ratio between two radii that belong to the same sensor. That is $\gamma \triangleq \max_i \{r_{i,k_i}/r_{i,1}\}$, where $k_i = |R_i|$.

The next lemma bounds the performance of an assignment that uses only the largest radii. The idea is to shrink the lifetime such that the batteries would be able to support the larger radii.

Lemma 3. *Given an instance of* ONCESC_k*, there exists a* γ^α*-approximate solution* (ρ, τ)*, where* $\rho_i \in \{0, r_{i,k_i}\}$*, for every* i*.*

It follows that we can use an algorithm for RSC with the largest radii to obtain an approximate solution for ONCESC_k.

Theorem 5. *There is a polynomial time* $5\gamma^\alpha$*-approximation algorithm for* ONCESC_k*, for* $k \geq 2$*.*

4.2 Strip Cover with Multiple Fixed Radii

Next we provide a 5-approximation algorithm for SC_k, for every $k \geq 1$, that is based on the 5-approximation algorithm RSC by Gibson and Varadarajan [7].

Given an RSC (i.e., ONCESC_1) instance, define $I_i \triangleq [x_i - r_i, x_i + r_i]$ and define the load of a sensor i by $\ell_i \triangleq b_i/r_i$. Also, define $\text{load}(p) \triangleq \sum_{i:p\in I_i} \ell_i$, for every $p \in U$.

Observation 2 [7]*. Given an RSC instance,* $\text{OPT} \leq \min_{p\in U} \text{load}(p)$*.*

Given an SC_k instance, define $\ell_{i,j} \triangleq b_i/r_{i,j}^\alpha$ and $I_{i,j} \triangleq [x_i - r_{i,j}, x_i + r_{i,j}]$. Also, let $P = \bigcup_{i,j} \{x_i - r_{i,j}, x_i + r_{i,j}\}$, and let $p_1, \ldots, p_{|P|}$ be a non decreasing ordering of the point in P. Define

$$\mathcal{I} \triangleq \left\{[0, p_1], [p_{|P|}, N]\right\} \cup \bigcup_{j=1}^{|P|-1} \left\{[p_j, p_{j+1}]\right\}.$$

Observe that $|P|, |\mathcal{I}| = O(kn)$. We are now ready to present an integer program for the problem of finding the maximum load:

$$
\begin{aligned}
\max \quad & L \\
\text{s.t.} \quad & \sum_{i,j:I\subseteq I_{i,j}} \ell_{i,j} z_{i,j} \geq L \quad && \forall I \in \mathcal{I} \\
& \sum_j z_{i,j} \leq 1 \quad && \forall i \\
& z_{i,j} \in \{0,1\} \quad && \forall i,j
\end{aligned}
\tag{2}
$$

where $z_{i,j}$ stands for whether i is used with radius $r_{i,j}$. The first set of constraints ensure that the barrier is covered, while the second set makes sure that sensors are not over used. An LP-relaxation is obtained by replacing the intergrality constraints by $z_{i,j} \geq 0$, for every i and j.

Observation 3. *Given an* SC_k *instance, we have that* $L^* \geq$ OPT, *where* (z^*, L^*) *is an optimal fraction solution of* (2).

Proof. An optimal solution induces a solution (z, L) such that OPT $\leq L$.   □

We use an optimal solution to (2) to obtain an approximate solution for SC_k.

Theorem 6. *There exists a polynomial time 5-approximation algorithm for* SC_k, *for every* $k \geq 1$. *Moreover, each sensor uses each of its radii at most once in the computed solution.*

4.3   Strip Cover

Observe that an STRIP COVER instance can be view as a SC_k instance with $k = N$ and $R_i = \{1, \ldots, N\}$, for every i. Hence, Theorem 6 implies a 5-approximation algorithm if N is polynomial in the input size.

Theorem 7. *There exists a 5-approximation algorithm for* STRIP COVER *whose running time is polynomial in the input size and in* N. *Moreover, each sensor uses each of its radii at most once in the computed solution.*

The next step is to remove the dependency on N.

Lemma 4. *Let ρ be a solution for* SC_k *where $k = N$ and $R_i = \{1, \ldots, N\}$, for every i. Then, there exists a solution ρ' such that $\rho'(t) \in \{2^j : j = 1, \ldots, \log N\}$ and $T(\rho') \geq 2^{\alpha} T(\rho)$.*

Using Lemma 4 one may reduce the running time at the expense of the approximation ratio.

Theorem 8. *There exists a $5 \cdot 2^{\alpha}$-approximation algorithm for* STRIP COVER *whose running time is polynomial in the input size and in* $\log N$. *Moreover, only radii from the set $\{2^j : j = 1, \ldots, \log N\}$ are used, and each sensor uses each of these radii is used at most once.*

5   Discussion and Open Questions

Bar-Noy et al. [2] showed that ONCESC with $\alpha = 1$ is NP-hard and that it has a $\frac{3}{2}$-approximation algorithm. We show that, when $\alpha = 1$, ONCESC is strongly NP-hard. We also prove that ONCESC is NP-hard if $\alpha > 1$. Presenting an approximation algorithm for the case where $\alpha > 1$ remains an open question.

We showed that ONCESC$_k$, for any $k \geq 2$ and $\alpha \geq 1$, is NP-hard, even for the case of uniform radii sets. On the other hand, we presented a $5\gamma^{\alpha}$-approximation algorithm for ONCESC$_k$, for $k \geq 1$. If $\gamma = O(1)$, the approximation ratio is $O(1)$. It would be interesting to come up with an approximation algorithm whose ratio is independent of γ and/or of α.

As for STRIP COVER, there is a $\frac{3}{2}$-approximation algorithm for the case where $\alpha = 1$ [2], and we presented a $5 \cdot 2^{\alpha}$-approximation algorithm for $\alpha > 1$. In addition, we gave a 5-approximation algorithm for SC_k, for every $k \geq 1$. However, it is unclear whether either one of these problems is NP-hard.

References

1. Bar-Noy, A., Baumer, B.: Average case network lifetime on an interval with adjustable sensing ranges. Algorithmica **72**(1), 148–166 (2015)
2. Bar-Noy, A., Baumer, B., Rawitz, D.: Brief announcement: set it and forget it - approximating the set once strip cover problem. In: 25th ACM Symposium on Parallelism in Algorithms and Architectures, pp. 105–107 (2013). To appear in Algorithmica
3. Buchsbaum, A.L., Efrat, A., Jain, S., Venkatasubramanian, S., Yi, K.: Restricted strip covering and the sensor cover problem. In: 18th Annual ACM-SIAM Symposium on Discrete Algorithms, pp. 1056–1063 (2007)
4. Buchsbaum, A.L., Efrat, A., Jain, S., Venkatasubramanian, S., Yi, K.: Restricted strip covering and the sensor cover problem. Technical report arXiv:cs/0605102v1, CoRR (2008)
5. Fan, H., Li, M., Sun, X., Wan, P., Zhao, Y.: Barrier coverage by sensors with adjustable ranges. ACM Trans. Sens. Netw. **11**(1), 14:1–14:20 (2014)
6. Garey, M.R., Johnson, D.S.: Computers and Intractability: A Guide to the Theory of NP-Completeness. W.H. Freeman and Company, New York (1979)
7. Gibson, M., Varadarajan, K.: Decomposing coverings and the planar sensor cover problem. In: 50th Annual IEEE Symposium on Foundations of Computer Science, pp. 159–168 (2009)
8. Lev-Tov, N., Peleg, D.: Polynomial time approximation schemes for base station coverage with minimum total radii. Comput. Netw. **47**(4), 489–501 (2005)
9. Moscibroda, T., Wattenhofer, R., Zollinger, A.: Topology control meets SINR: the scheduling complexity of arbitrary topologies. In: 7th ACM Interational Symposium on Mobile Ad Hoc Networking and Computing, pp. 310–321 (2006)
10. Wan, P.-J., Chen, D., Dai, G., Wang, Z., Yao, F.F.: Maximizing capacity with power control under physical interference model in duplex mode. In: 31st Annual IEEE International Conference on Computer Communications, pp. 415–423 (2012)

Independent Sets in Restricted Line of Sight Networks

Pavan Sangha[✉], Prudence W. H. Wong, and Michele Zito

Department of Computer Science, University of Liverpool, Liverpool, UK
{p.sangha2,pwong,michele}@liverpool.ac.uk

Abstract. Line of Sight (LoS) networks were designed to model wireless networks in settings which may contain obstacles restricting visibility of sensors. A graph $G = (V, E)$ is a 2-dimensional LoS network if it can be embedded in an $n \times k$ rectangular point set such that a pair of vertices in V are adjacent if and only if the embedded vertices are placed on the same row or column and are at a distance less than ω. We study the Maximum Independent Set (MIS) problem in restricted LoS networks where k is a constant. It has been shown in the unrestricted case when $n = k$ and $n \to \infty$ that the MIS problem is NP-hard when $\omega > 2$ is fixed or when $\omega = O(n^{1-\epsilon})$ grows as a function of n for fixed $0 < \epsilon < 1$. In this paper we develop a dynamic programming (DP) algorithm which shows that in the restricted case the MIS problem is solvable in polynomial time for all ω. We then generalise the DP algorithm to solve three additional problems which involve two versions of the Maximum Weighted Independent Set (MWIS) problem and a scheduling problem which exhibits LoS properties in one dimension. We use the initial DP algorithm to develop an efficient polynomial time approximation scheme (EPTAS) for the MIS problem in restricted LoS networks. This has important applications, as it provides a semi-online solution to a particular instance of the scheduling problem. Finally we extend the EPTAS result to the MWIS problem.

1 Introduction

LoS network. A wireless network typically consists of wireless devices that use data connections to communicate wirelessly. Geometric graphs often provide a good model for such networks with vertices representing wireless devices, and edges representing communication between pairs of devices. Various types of geometric graphs have been proposed to model wireless sensor networks. The disk intersection model [5] is a commonly used one. Vertices are placed in some physical space with the communication range for a vertex represented by a circle of some prescribed radius. Edges are formed between pairs of vertices whose circles overlap. A benefit of this model is its ability to capture the constraint of communication range restriction. This restriction implies that vertices should be close in distance in order to communicate. Another constraint exhibited by real world wireless networks are line of sight restrictions, often due to the presence of a large number of obstacles, like those often found in urban settings

© Springer International Publishing AG 2017
A. Fernández Anta et al. (Eds.): ALGOSENSORS 2017, LNCS 10718, pp. 211–222, 2017.
https://doi.org/10.1007/978-3-319-72751-6_16

for example. With this restriction vertices can only communicate if they are both close in distance and also share a direct line of sight. While the presence of obstacles can be difficult to model, it is clear that a good model of wireless network should ideally incorporate both communication range restrictions and line of sight restrictions. Frieze et al. [9] introduced the notion of random Line of Sight networks to provide a model that incorporates both. Their work focused on structural properties of the model focusing on connectivity. Since then connectivity in higher dimensions, percolation and communication problems within the same model have been studied [3,6,8].

For positive integers k and n let $\mathbb{Z}_{n,k} = \{(x,y) : x \in \{1,\ldots,n\}, y \in \{1,\ldots,k\}\}$ denote the underlying point set. For two points $p_1 = (x_1,y_1), p_2 = (x_2,y_2) \in \mathbb{Z}_{n,k}$, let $d(p_1,p_2) = |x_1 - x_2| + |y_1 - y_2|$ denote the Manhattan distance between p_1 and p_2. The distance metric we use differs from the one used in [9] which relies on the underlying point set being toroidal to ease calculations. We say points p_1 and p_2 share a line of sight if $x_1 = x_2$ or $y_1 = y_2$. A graph $G = (V,E)$ is said to be a Line of Sight (LoS) network with parameters n, k and ω if there exists an embedding $f_G : V \to \mathbb{Z}_{n,k}$ such that $\{u,v\} \in E$ if and only if $f_G(u)$ and $f_G(v)$ share a line of sight and $d(f_G(u), f_G(v)) < \omega$. We refer to ω as the range parameter which is used to model the communication range restriction. The range parameter can be a constant or grow monotonically as a function of the parameters n or k.

MIS. We focus on the Maximum Independent Set (MIS) problem [7] in LoS networks. The MIS in a graph can be seen as a measure of network dispersion and independent sets share links with other important graph properties like vertex covers [7]. For $n \times k$ line of sight networks, large independent sets could help in covering scenarios like the following: "New York has many more streets than avenues. On parade days the police may want to dominate all junctions by also maximising presence". In general graphs finding the largest independent set is NP-hard [10] and even finding good approximation solutions are difficult [11]. For LoS networks in particular, if $\omega = 2$ or n, it is not difficult to see that the problem can be solved optimally. Sangha and Zito [12] showed on the other hand for $\omega = O(n^{1-\epsilon})$ where $0 < \epsilon < 1$ is fixed the problem is NP-hard. They also provide approximation results, showing the problem admits a 2-approximation for any ω and an efficient polynomial time approximation scheme (EPTAS) [4] for constant ω.

Additional applications. The abstract problem we study also finds application [2] in the following scheduling problems. Suppose an advertisement company manages advertisements from some number of clients (cf. k) over a long period of discrete time (cf. n). At any time advertisements of some subset of clients are available to be aired but the company can only select a certain number (cf. l) of them to advertise due to resource limitation. In addition some "advertisement diversity" policy requires that advertisements from the same client cannot be aired more than once in a given period of time (cf. ω). The goal of the company is to schedule the airing of these advertisements satisfying the constraints and maximise the number of advertisements aired (cf. MIS). This problem has one slight difference from the LoS problem in the sense that the restriction with ω only applies on one dimension (the time dimension) but not the other (the client dimension). Nevertheless, as to be showed later, the solution we develop can be adapted to solve this problem.

Our contribution. In this paper we study the MIS problem on restricted LoS networks. We focus on the restricted case that parameter k is a constant and $k < \omega$, in which case we show in Sect. 3 that the problem becomes polynomial time solvable via a dynamic programming (DP) algorithm. We also show in Sect. 4 that the DP algorithm can be extended to solve (i) the maximum weighted independent set problem (MWIS) on restricted LoS networks, (ii) MWIS on LoS networks with $k > \omega$ but both k and ω being constants, and (iii) the advertisement scheduling problem mentioned above. We then in Sect. 5 apply the DP algorithm to develop an EPTAS which improves the EPTAS in [12] by showing that when k is restricted to being constant the algorithm no longer requires ω to be constant. In addition we show that the EPTAS leads to a semi-online algorithm [1] with performance ratio $(1 + \epsilon)$, which requires a look-ahead distance dependent on ϵ. This gives a semi-online algorithm for the scheduling problem in the case where $k < \omega$ and $l = 1$.

2   Problem Definitions and Preliminaries

The dynamic programming algorithm uses arrays consisting of $0, 1$ elements. An array of size $x \times y$ consists of x rows and y columns. We start by introducing some notations.

- For any array A, a sub-array $A[i..j]$ contains columns $i, i + 1, \cdots, j$.
- For any two arrays A_1, A_2 of size $x \times y$, we say that A_1 *agrees with* A_2, denoted by $A_1 \leq_a A_2$, if $A_1[i,j] \leq A_2[i,j]$ for all $1 \leq i \leq x$ and $1 \leq j \leq y$.
- For any array A, we denote by $h(A)$ the "head" subarray of A containing all but the last column of A; and $t(A)$ the "tail" subarray of A containing all but the first column of A. That is, $h(A)$ and $t(A)$ have $y - 1$ columns if A has y columns.
- A_1 is said to be *tail-aligned* with A_2 if $t(A_1)$ is the same as $h(A_2)$, i.e., $t(A_1) \equiv h(A_2)$. In this case, we say A_2 is *head-aligned* with A_1.
- Let the *column sum* of an array A at column y be $cs(A, y) = \sum_x A[x, y]$.

Given a LoS network G in $\mathbb{Z}_{n,k}$ let array(G) satisfies array$(G)[i,j] = 1$ if and only if location $(j, i) \in \mathbb{Z}_{n,k}$ in the LoS embedding of G contains a vertex, otherwise array$(G)[i,j] = 0$ Fig. 1 provides an example. Given array(G) of size $k \times p$ where $\omega \leq p$, an independent array I of array(G) is any array of size $k \times p$ satisfying (i) $I \leq_a$ array(G) and (ii) I contains at most one 1 in each column and for any row i and distinct columns j_1, j_2, if $I[i, j_1] = 1$ and $I[i, j_2] = 1$ it must be the case that $|j_1 - j_2| \geq \omega$. We refer an independent array W specifically of size $k \times \omega$ as a "feasible array". Since any feasible array has exactly ω columns it contains at most one 1 per column but also at most one 1 per row. We denote by $\mathcal{F}$ the set of all feasible of arrays of size $k \times \omega$.

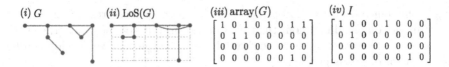

Fig. 1. Figure (i) is a graph G and Figure (ii) is its LoS embedding in $\mathbb{Z}_{8,4}$ with $\omega = 4$. Figure (iii) represents the array layout of G and Figure(iv) is an independent array of largest array sum, corresponding to the largest independent set in the graph G.

There is a clear connection between an independent array I of array(G) and an independent set $\mathcal{I}$ of the LoS network G. More precisely given a set $\mathcal{I} \subseteq V$ of a LoS network $G = (V, E)$ embedded in $\mathbb{Z}_{n,k}$ consider the array I satisfying $I[i,j] = 1$ if and only if location $(j, i) \in \mathbb{Z}_{n,k}$ contains a vertex in $\mathcal{I}$. We observe the following:

1. $\mathcal{I}$ is an independent set of G if and only if I is an independent array of the array G.
2. $|\mathcal{I}| = \sum_{i=1}^{k} \sum_{j=1}^{n} I[i,j]$.

We refer to the quantity $\sum_{i=1}^{k} \sum_{j=1}^{n} I[i,j]$ as the array sum of I which we denote by $as(I)$. Using the two observations above it follows that finding the maximum independent set in a LoS network G embedded in $\mathbb{Z}_{n,k}$ is equivalent to finding the independent array of the array G with the largest array sum, we refer to such an array as a largest independent array. In Sect. 3 we show how our DP algorithm finds the largest independent array of array(G) by scanning the feasible arrays of array(G).

3   Dynamic Programming

3.1   Algorithm

In this section we present a dynamic programming (DP) algorithm for finding a largest independent array. For simplicity we refer to array(G) as G. It will be clear from the context when referring to the LoS network G instead of the array G. Given the array G of size $k \times n$, for the purposes of the DP algorithm we prepend G with ω columns consisting entirely of 0's. We index these columns from $-(\omega - 1), \ldots, 0$. Thus the input array G is of size $k \times (\omega + n)$. The DP algorithm works by sequentially computing the array sums of $G[-(\omega - 1)..j]$ in G. The process keeps track of various sums $\text{MIS}(j, W)$ depending on the independent set choices in the right-most column of $G[-(\omega - 1)..j]$. For $j = 0, \ldots, n$ let $\mathcal{F}_{G,j} \subseteq \mathcal{F}$ be the set of feasible arrays W satisfying $W \leq_a G[j - \omega + 1..j]$. Note that in particular any independent array I of $G[-(\omega - 1)..j]$ for $1 \leq j \leq n$ satisfies $I[j - \omega + 1..j] \equiv W$ for some $W \in \mathcal{F}_{G,j}$.

Algorithm 1 describes how to compute the array sum on an independent array of $G[-(\omega - 1)..j]$ from information about $G[-(\omega - 1)..(j - 1)]$. More specifically

---

Algorithm 1. Computing the size of the largest independent array in G

1: **Initialise:** $\text{MIS}(0, \vec{0}) = 0$, where $\vec{0}$ is the $k \times \omega$ array of all 0's.
2: **for** $j = 1, \ldots, n$ **do**
3:     **for** $W \in \mathcal{F}$ **do**
4:         **if** $W \in \mathcal{F}_{G,j}$ **then**
5:             $\text{MIS}(j, W) = \max_{W' \in \mathcal{F}_{G,j-1}: t(W') \equiv h(W)} \text{MIS}(j-1, W') + \text{cs}(W, \omega)$
6:         **else**
7:             $\text{MIS}(j, W) = 0$
8:         **end if**
9:     **end for**
10:     $\text{MIS}(j) = \max_{W \in \mathcal{F}_{G,j}} \text{MIS}(j, W)$
11: **end for**

---

if $W \in \mathcal{F}_{G,j}$ we try to extend the independent arrays in $G[-(\omega - 1)..(j-1)]$ to independent arrays in $G[-(\omega - 1)..j]$. Let I' be an independent array in $G[-(\omega - 1)..(j-1)]$ such that $I[(j - \omega)..(j-1)] \equiv W'$ for some $W' \in \mathcal{F}_{G,j-1}$ and $t(W') \equiv h(W)$. By considering the next column of G, we extend I' to an independent array I of $G[-(\omega-1)..j]$ which satisfies $I[j - \omega + 1..j] \equiv W$. There are two cases for the sum $\text{cs}(W, \omega)$:

1. $\text{cs}(W, \omega)$ is 0 meaning that there is no entry in the last column of W. Then $\text{as}(I) = \text{as}(I')$ and hence the array sum for our independent set does not increase;
2. $\text{cs}(W, \omega)$ is 1 meaning that we can increase the array sum of our independent array by 1; note that since W is feasible, $\text{cs}(W, \omega)$ is at most 1.

The new independent array can be obtained by extending the one for $G[-(\omega - 1)..(j-1)]$ by appending the ω-th column of W. Figure 2 shows an example.

$$
(i) \quad G[1, \ldots, 8] = \begin{bmatrix} 1 & 1 & 1 & 0 & 0 & 1 & 1 & 0 \\ 0 & 0 & 1 & 1 & 1 & 1 & 1 & 0 \\ 1 & 1 & 0 & 0 & 1 & 0 & 1 & 1 \end{bmatrix} \quad I' = \begin{bmatrix} 1 & 0 & 0 & 0 & 0 & | & 1 & 0 & 0 \\ 0 & 0 & 1 & 0 & 0 & | & 0 & 1 & 0 \\ 0 & 1 & 0 & 0 & 1 & | & 0 & 0 & 1 \end{bmatrix} \quad W' = \begin{bmatrix} 1 & 0 & 0 \\ 0 & 1 & 0 \\ 0 & 0 & 1 \end{bmatrix}
$$

$$
(ii) \quad G[1, \ldots, 9] = \begin{bmatrix} 1 & 1 & 1 & 0 & 0 & 1 & 1 & 0 & 1 \\ 0 & 0 & 1 & 1 & 1 & 1 & 1 & 0 & 1 \\ 1 & 1 & 0 & 0 & 1 & 0 & 1 & 1 & 0 \end{bmatrix} \quad I = \begin{bmatrix} 1 & 0 & 0 & 0 & 0 & 1 & | & 0 & 0 & 1 \\ 0 & 0 & 1 & 0 & 0 & 0 & | & 1 & 0 & 0 \\ 0 & 1 & 0 & 0 & 1 & 0 & | & 0 & 1 & 0 \end{bmatrix} \quad W = \begin{bmatrix} 0 & 0 & 1 \\ 1 & 0 & 0 \\ 0 & 1 & 0 \end{bmatrix}
$$

Fig. 2. Figure (i) shows the first 8 columns of an array G and the independent array I' of $G[1..8]$ has the largest array sum satisfying $I'[6..8] \equiv W'$. In Figure (ii) the independent array I is the independent array of $G[1..9]$ which has the largest array sum satisfying $I[7..9] \equiv W$. Note $t(W') \equiv h(W)$ and that I can be obtained from I' by appending the last column of W to I'.

3.2   Correctness

In this section we prove the correctness of Algorithm 1. We first prove in Lemma 1 that it is sufficient to consider $\mathcal{F}_{G,j}$ and then the main result in Theorem 2.

Lemma 1. *For $j = 0, \ldots, n - 1$ for each $W_1' \in \mathcal{F}_{G,j}$ there exists $W_1 \in \mathcal{F}_{G,j+1}$ such that $t(W_1') \equiv h(W_1)$ and for each $W_2 \in \mathcal{F}_{G,j+1}$ there exists $W_2' \in \mathcal{F}_{G,j}$ such that $t(W_2') \equiv h(W_2)$.*

Proof. Take $W_1' \in \mathcal{F}_{G,j}$ and consider the simple feasible array W_1 satisfying (i) $t(W_1') \equiv h(W_1)$ and (ii) the last column of W_1 consists entirely of 0's. Combining (i) and (ii) with the fact that $W_1' \leq_a G[j - \omega + 1..j]$ we conclude $W_1 \leq_a G[j - \omega + 2..j+1]$ and thus $W_1 \in \mathcal{F}_{G,j+1}$. Similarly given $W_2 \in \mathcal{F}_{G,j+1}$ consider the feasible array W_2' satisfying (i) $t(W_2') \equiv h(W_2)$ and (ii) the first column consists entirely of 0's. Using similar reasoning to the first case we can conclude $W_2' \in \mathcal{F}_{G,j}$.   □

Theorem 2. *For $1 \leq j \leq n$, MIS(j) computed by Algorithm 1 gives the size of the maximum independent set in the graph $G[j]$ induced by the vertices of the first j columns in $\mathbb{Z}_{n,k}$.*

Proof. Recall that $\text{MIS}(j) = \max_{W \in \mathcal{F}_{G,j}} \text{MIS}(j, W)$ for all $1 \leq j \leq n$ and let OPT(j) denote the size of the maximum independent set in the graph $G[j]$ induced by the vertices of the first j columns of $\mathbb{Z}_{n,k}$ for all $1 \leq j \leq n$. We prove the theorem by induction on j. Firstly Lemma 1 proves that for each $W \in \mathcal{F}_{G,j}$ there exists W' such that $\max_{W' \in \mathcal{F}_{G,j-1}:t(W') \equiv h(W)} \text{MIS}(j - 1, W')$ is well defined. Next consider the case $j = 0$, since the first ω columns (indexed from $-(\omega-1), \ldots, 0$) consist entirely of 0's it implies that $\text{MIS}(0) = 0$. In addition let $\vec{0} \in \mathcal{F}$ denote the feasible array consisting entirely 0's then $\mathcal{F}_{G,0} = \{\vec{0}\}$ and consequently $\text{MIS}(0) = \text{MIS}(0, \vec{0}) = 0$.

Base case: If $\text{cs}(G[-(\omega - 1)..1], 1) = 0$ then OPT(1) = 0. $\mathcal{F}_{G,1} = \{\vec{0}\}$ and clearly $\text{MIS}(1, \vec{0}) = 0$. Using the facts that (i) $\mathcal{F}_{G,0} = \{\vec{0}\}$, (ii) $\text{MIS}(0, \vec{0}) = 0$, (iii) $\text{cs}(\vec{0}, \omega) = 0$ and (iv) $t(\vec{0}) = h(\vec{0})$ it follows that $\text{MIS}(1, \vec{0}) = \text{MIS}(0, \vec{0}) + cs(\vec{0}, \omega)$. For all $W \notin \mathcal{F}_{G,1}$, $\text{MIS}(1, W) = 0$ as it must be the case that $W \leq_a G[-(\omega - 2)..1]$. This means that $\text{MIS}(1) = 0$, which equals to OPT(1).

Otherwise $\text{cs}(G[-(\omega-1)..1], 1) > 0$ meaning OPT(1) = 1. In this case $\mathcal{F}_{G,1} \neq \{\vec{0}\}$. Thus there exists $W \in \mathcal{F}_{G,1}$ satisfying $t(\vec{0}) = h(W)$ and $W \neq \vec{0}$, and thus $\text{cs}(W, \omega) = 1$. Since W it head-aligned with $\vec{0}$ and contains a 1 in its final column we conclude $\text{MIS}(1, W) = 1$. Finally since $\mathcal{F}_{G,0} = \{\vec{0}\}$ and $\text{MIS}(0, \vec{0}) = 0$ it follows that $\text{MIS}(1, W) = \text{MIS}(0, \vec{0}) + cs(W, \omega)$. Again for all $W \notin \mathcal{F}_{G,1}$, $\text{MIS}(1, W) = 0$. Then $\text{MIS}(1)$ equals to 1, which equals OPT(1).

Inductive step: Suppose that the result holds for all columns of G indexed from $1, \ldots, j - 1$, i.e., $\text{MIS}(i)$ equals to OPT(i) for all $1 \leq i < j$. We show that this implies $\text{MIS}(j)$ equals to OPT(j). Assume on the contrary that there exists an independent array I in $G[-(\omega-1)..j]$ satisfying $\text{as}(I) > \text{MIS}(j)$ and suppose $W^* \in \mathcal{F}_{G,j}$ satisfies $W^* \equiv I[j - \omega + 1..j]$. Then $\text{as}(I) > \text{MIS}(j) \geq \text{MIS}(j, W^*)$ and thus

$$\text{as}(I) > \max_{W' \in \mathcal{F}_{G,j-1}:t(W') \equiv h(W^*)} \text{MIS}(j - 1, W') + \text{cs}(W^*, \omega). \tag{1}$$

Consider the independent array I' in $G[-(\omega-1)..j-1]$ obtained by removing the last column of I. Then it follows that $\text{as}(I') = \text{as}(I) - \text{cs}(I, j)$. Furthermore consider the simple feasible array $W'' \in \mathcal{F}_{G,j-1}$ satisfying $I'[j - \omega..j - 1] \equiv W''$, then $t(W'') \equiv h(W^*)$. In addition $cs(I, j) = cs(W^*, \omega)$ since the last column of I is the same as the ωth column of W^*. Thus $as(I') = as(I) - cs(W^*, \omega)$, substituting this into Inequality (1) we obtain

$$as(I') > \max_{W' \in \mathcal{F}_{G,j-1} : t(W') \equiv h(W^*)} MIS(j - 1, W').$$

Since $I'[j - 2 - \omega..j - 1] \equiv W''$ and $W'' \in \mathcal{F}_{G,j-1}$ satisfying $t(W'') \equiv h(W^*)$ this is a contradiction to the optimality of $\max MIS(j - 1, W')$ for $W' \in \mathcal{F}_{G,j-1}$ satisfying $t(W') \equiv h(W^*)$. □

3.3   Time Complexity

In this section we provide an upper bound on the worst case time complexity of the DP algorithm. We describe how the use of n separate bipartite graphs allow us to compute $MIS(j)$ for $1 \leq j \leq n$. Each bipartite graph consists of two sets of feasible arrays namely those which agree with $G[-(\omega - 1)..(j - 1)]$ and $G[-(\omega - 1)..j]$ respectively. Edges between classes represent pairs of arrays which are tail-head aligned.

For an array G and $0 \leq j \leq n - 1$ $B_j = (\mathcal{F}_{G,j}, \mathcal{F}_{G,j+1}, E)$ is a directed bipartite graph to model the tail-head alignment of arrays with classes $\mathcal{F}_{G,j}$ and $\mathcal{F}_{G,j+1}$. For a pair of simple feasible arrays $W \in \mathcal{F}_{G,j}$ and $W' \in \mathcal{F}_{G,j+1}$, $(W, W') \in E$ if and only if $t(W) \equiv h(W')$.

For a directed graph $G = (V, E)$ and vertex $v \in V$ let $N^-(v) = \{u \in V : (u, v) \in E\}$ and $d^-(u) = |N^-(u)|$. Similarly let $N^+(v) = \{u \in V : (v, u) \in E\}$ and $d^+(u) = |N^+(u)|$. Let $\Delta^+(G) = \max_{v \in V} d^+(v)$ denote the maximum out-degree of G. The bipartite graph B_j is used to compute $MIS(j + 1, W)$ for each $W \in \mathcal{F}_{G,j+1}$ by considering $N^-(W)$ and selecting the array $W' \in N^-(W)$ for which $MIS(j, W')$ is maximised.

An important piece of information required for the time complexity is to obtain an upper bound on $|\mathcal{F}_{G,j}|$ for all $0 \leq j \leq n$. We do this by obtaining an upper bound on $|\mathcal{F}|$. We show that $|\mathcal{F}| = \Theta(\omega^k)$ through the following two observations. Firstly $|\mathcal{F}| \leq (\omega + 1)^k$ since each simple feasible array contains at most one 1 per row, or not contain a 1 at all. Secondly $\frac{(\omega)!}{(\omega-k)!} \leq |\mathcal{F}|$ since there are precisely $\omega(\omega - 1)(\omega - 2) \cdots (\omega - (k - 1))$ simple feasible arrays with exactly one 1 in each row. Thus $|\mathcal{F}_{G,j}| = O(\omega^k)$. We make use of the following lemma in the calculation of the worst case running time of the algorithm.

Lemma 3. *For each B_j, the maximum out degree $\Delta^+(B_j) \leq k + 1$*

Proof. For each array in $W \in \mathcal{F}_{G,j}$ there are at most k arrays $W' \in \mathcal{F}_{G,j+1}$ with a single 1 in the final column satisfying $t(W) \equiv h(W')$ (one in each of the k possible locations). Combining this with the array consisting of entirely 0's in its final column gives us a $k + 1$ possible feasible arrays. □

Theorem 4. *The worst case running time of the DP algorithm is $O(nk\omega^k)$.*

Proof. For each $W \in \mathcal{F}_{G,j+1}$ using Algorithm 1 we obtain $\mathrm{MIS}(j+1, W)$ by comparing $\mathrm{MIS}(j, W')$ for each $W' \in N^-(W)$. Using Lemma 3 and the fact that $|\mathcal{F}_{G,j}| = O(\omega^k)$ it can be seen that there at most $O(k\omega^k)$ computations per bipartite graph B_j. Finally given that there are n bipartite graphs B_j we obtain a worst case running time of $O(nk\omega^k)$. $\qquad\square$

4   Extensions

In this section we extend our DP algorithm in Sect. 3 taking weight into account and show how it provides solutions to the following three problems: (i) The maximum weighted independent set problem for $k < \omega$ for constant k, (ii) The maximum weighted independent set problem for $k > \omega$ for constant k and ω, and (iii) A weighted version of the scheduling problem with parameter $1 \le l \le k$ for constant k.

Framework. W.l.o.g., we normalise the weight such that the minimum non-zero weight is 1, in other words, $G[i,j] = 0$ or $G[i,j] \ge 1$ for all $[i,j]$. Let $\mathcal{W}$ be the set of all $k \times \omega$ arrays with $0, 1$ entries. A basis set $\mathcal{F} \subseteq \mathcal{W}$ satisfies the followings (i) $\{\vec{0}\} \in \mathcal{F}$, (ii) If $W \in \mathcal{F}$ then $\exists W' \in \mathcal{F}$ where $t(W) \equiv h(W')$ and the last column of W' is all 0's and (iii) if $W \in \mathcal{F}$ then $\exists W'' \in \mathcal{F}$ where $t(W'') \equiv h(W)$ and the first column of W'' is all 0's. We extend some notations of Sect. 2 to account for the array generalisation. Given arrays G and I of the same size $I \le_a G$ if (i) $G[i,j] = 0 \Rightarrow I[i,j] = 0$ and (ii) $1 \le I[i,j] \le G[i,j]$ for all $G[i,j] \ne 0$. Since I may contain values greater than 1 and W contains only $0, 1$ we introduce an additional notion of equivalence, denoted by $\equiv_a$. Given an array I of size $k \times \omega$, we say that $I \equiv_a W$ provided $I[i,j] = 0$ if and only if $W[i,j] = 0$.

We extend G to an array of size $k \times (\omega + n)$ with columns indexed from $-(\omega-1), \dots, n$ and the first ω columns consisting of entirely 0's. Given a basis set $\mathcal{F}$ let $\mathcal{F}_{G,j}$ denote the set of feasible arrays $W \in \mathcal{F}$ satisfying $W \le_a G[j-\omega+1..j]$. The goal is to find an array I of maximum array sum satisfying $I \le_a G$ and $I[j-\omega+1..j] \equiv_a W$ for some $W \in \mathcal{F}_{G,j}$ for all $1 \le j \le n$. It is important to note that $\vec{0} \in \mathcal{F}_{G,j}$ for all $1 \le j \le n$ and so the array of size $k \times (\omega + n)$ consisting entirely of 0's satisfies the required conditions proving such an array always exists. Let $\mathrm{OPT}(j)$ denote the array sum of the largest array satisfying the conditions for $G[-(\omega - 1)..j]$ for all $1 \le j \le n$. We are required to compute $\mathrm{OPT}(n)$.

Algorithm 1 can be extended by modifying the main recurrence as follows: $F(0, \vec{0}) = 0$ and for each $W \in \mathcal{F}$ and $1 \le j \le n$ let

$$F(j, W) = \begin{cases} \displaystyle\max_{W' \in \mathcal{F}_{j-1}:t(W')\equiv h(W)} F(j-1, W') + W[\omega]^T \cdot G[j] & \text{if } W \in \mathcal{F}_{G,j}, \\ 0 & \text{otherwise.} \end{cases}$$

Note $W[\omega]^T \cdot G[j]$ denotes the dot product of the ωth column of W and the jth column of G. Let $F(j) = \max_{W \in \mathcal{F}_{G,j}} F(j, W)$, we use the following theorem to calculate $\mathrm{OPT}(n)$.

In the full paper we prove that the recurrence and the associated dynamic programming algorithm gives an optimal value.

Theorem 5. $F(j)$ *is equal to* OPT(j) *for all* $1 \leq j \leq n$.

We analyse the worst case running time of the algorithm in a similar way to the case for the maximum independent set problem let $B_j = (\mathcal{F}_{G,j}, \mathcal{F}_{G,j+1}, E)$ for $0 \leq j \leq n - 1$. Let $\Delta^+(B_j)$ denote the largest out-degree of an array in B_j for $0 \leq j \leq n - 1$ and let $\Delta^+(\mathcal{F}) = \max_j(\Delta^+(B_j))$. We then prove the following lemma regarding the running time.

Lemma 6. *The worst case running time of the generalised DP is* $O(n\Delta^+(\mathcal{F})|\mathcal{F}|)$.

Proof. Given B_j the number of computations required to compute $F(j + 1, W)$ for $W \in \mathcal{F}_{j+1}$ is proportional to the in-degree $d^-(W)$. Thus the total number of computations required for B_j is proportional to the number of edges which is at most $\Delta^+(\mathcal{F})|\mathcal{F}|$, since there are n bipartite graphs, the result follows. □

Applications of the extension. We now show how we solve the three problems introduced at the start of the section by choosing the basis set $\mathcal{F}$ corresponding to the problem.

Weighted independent set. We consider the maximum weighted independent set problem in LoS networks with $k < \omega$, where each weight assigned to a vertex has a value least 1. $G[i, j]$ is the weight assigned to the vertex in location $[i, j]$ of the LoS embedding of G. Since this is just the weighted version our initial problem we keep the same basis $\mathcal{F}$ which consists of all arrays with at most one 1 in each column and row.

Weighted independent set for larger k. We consider the maximum weighted independent set problem in LoS networks with parameter $k > \omega$ where $k \in \mathbb{N}$ is a constant. In this case an independent set $\mathcal{I}$ can have more than one 1 in a column. In particular $\mathcal{I}$ satisfies that (i) for distinct columns j_1, j_2 if $I[i, j_1] = 1$ and $I[i, j_2] = 1$ then $|j_1 - j_2| \geq \omega$ and (ii) for distinct rows i_1, i_2 if $I[i_1, j] = 1$ and $I[i_2, j] = 1$ then $|i_1 - i_2| \geq \omega$. Thus the basis set $\mathcal{F}$ represents the set of $k \times \omega$ arrays having at most one 1 per row and for each column the distance between any pair of 1's needs to be at least ω.

Weighted scheduling problem. We consider the scheduling problem where the parameter $1 \leq l \leq k$ with the addition that prices charged have different weights that are at least the value 1. Thus $G[i, j]$ contains the price charged to client i on day j. In this problem the basis set $\mathcal{F}$ is the set of all $k \times \omega$ arrays which consist of at most one 1 in each row and at most l, 1's in each column.

We use Lemma 6 to calculate the worst-case running times are $O(nk\omega^k)$ for the first problem, $O(nt^t(\omega)^{(t+1)\omega})$ for the second where $t = \lceil \frac{k}{\omega} \rceil$, and $O(nk^l\omega^k)$ for the third. Full details are provided in the full paper.

5   EPTAS

The DP algorithms in Sects. 3 and 4 give optimal solutions to an offline version of the MIS problem in LoS networks and scheduling problem where the entire input is known in advance. This is unrealistic for example the duration in the scheduling problem is large possibly spanning a year or more then it is likely the input evolves over time. It is desirable to take a more online approach with a good approximation performance. We improve the running time of the EPTAS in [12] based on the DP algorithms, providing a solution which is semi-online; in particular we assume we are allowed to observe the input up to a certain "look-ahead" distance. We show how the look-ahead distance influences the approximation ratio achieved.

Given a LoS network G, let G_j and $\overline{G_j} = G \backslash G_j$ denote the induced subgraph of G consisting of vertices which are embedded in the region with x-coordinates from 1 to $j\omega$ and from $j\omega+1$ to n, respectively. Let I_r be a maximum independent set in G_r. We determine a value r^* which is the "stopping point" of the overhead distance necessary to achieve $(1 + \epsilon)$-approximation. Precisely, we let r^* be the smallest integer such that $|I_{r^*+1}| < (1+\epsilon)|I_{r^*}|$. This means that for any $1 < r \le r^*$, $|I_r| \ge (1+\epsilon)|I_{r-1}|$ (note that $|I_r| \le kr$ due to the structural properties of a LoS network embedding). We first show an upper bound on r^* (proof is deferred to the full paper).

Lemma 7. $r^* \le \left(\frac{1+\sqrt{1+4\ln(1+\epsilon)\ln(k)}}{2\ln(1+\epsilon)} \right)^2$

To obtain a $(1+\epsilon)$-approximation to the maximum independent set once r^* is obtained we remove G_{r^*+1} from the graph G and apply the procedure iteratively to the graph $\overline{G_{r^*+1}}$. If I' is the independent set obtained from applying the procedure to $\overline{G_{r^*+1}}$ then we show that $I_{r^*} \cup I'$ is a $(1 + \epsilon)$-approximation to the maximum independent set in G.

Lemma 8. *Suppose that I' is a $(1 + \epsilon)$-approximate independent set in $\overline{G_{r^*+1}}$, then $I \equiv I_{r^*} \cup I'$ is $(1 + \epsilon)$-approximate for G.*

Proof. Recall that I_{r^*+1} is the largest independent set in G_{r^*+1}, since $|I_{r^*+1}| < (1+\epsilon)|I_{r^*}|$ it follows that I_{r^*} is a $(1+\epsilon)$-approximate independent set on G_{r^*+1}. Using the properties of LoS networks, for any vertex $v \in I_{r^*}$ the distance between v and a neighbour $u \in N(v)$ is at most ω. Thus the neighbourhood $\cup_{v \in I_{r^*}} N(v)$ belongs to G_{r^*+1} and we can conclude that $I_{r^*} \cup I'$ is an independent set in G. We denote by α the independence number. Finally, $\alpha(G) \le \alpha(G_{r^*+1}) + \alpha(\overline{G_{r^*+1}})) \le (1 + \epsilon)|I_{r^*}| + (1 + \epsilon)|I'| = (1 + \epsilon)|I_{r^*} \cup I'|$, giving us the required result.   □

Suppose we define one iteration of the algorithm as the process of reaching the first stopping point and second iteration as the next process of reaching the second stopping point and so on. Given $r^* = O(\omega)$ using Theorem 4 we can deduce computing I_r using the DP algorithm has a worst case running time of $O(\omega k \omega^k)$. Since we repeat this calculation r^* times in each iteration, the worst case running time of an iteration is $O(k\omega^{k+2})$. Finally since there are at most n iterations the EPTAS has a worst case running time of $O(nk\omega^{k+2})$.

We now turn our attention to the uses of the EPTAS for the scheduling application in the case where k is constant, $k < \omega$ and $l = 1$, note under these restrictions the goal of the scheduling problem is equivalent to the MIS problem. Suppose the advertisement company does not have a full schedule but would like to start processing a schedule given some partial information. The EPTAS shows a look-ahead distance of $c_1\omega$ where $c_1 = \left(\frac{1+\sqrt{1+4\ln(1+\epsilon)\ln(k)}}{2\ln(1+\epsilon)}\right)^2$ is sufficient. Once the we have computed the first stopping point r^* we can process the schedule up to $r^*(\omega+1)$ with a $(1+\epsilon)$-approximation guarantee. We then repeat this process when the next stopping point is computed and so on. A portion of the schedule of length $c_1\omega$ is sufficient to guarantee a stopping point is found. Hence we say that the EPTAS uses a $c_1\omega$ look-ahead distance.

Theorem 9. *For any $\epsilon > 0$, we have an EPTAS of time complexity $O(nk\omega^{k+2})$ provided we have a look-ahead of $c_1\omega$, where $c_1 = \left(\frac{1+\sqrt{1+4\ln(1+\epsilon)\ln(k)}}{2\ln(1+\epsilon)}\right)^2$.*

Extensions. Similar to Sect. 4, we show that with some small modifications our EPTAS can be used for the maximum weighted independent set problem. We assume however that it is known a priori that there exists some global parameter $w_{max} > 1$ which is constant such that for each vertex v in our graph $W(v) \le w_{max}$.

I_r is defined as the largest weighted independent set in G_r. We define the weight of I_r as $W(I_r) = \sum_{v \in I_r} W(v)$. Then r^* is defined as the smallest integer such that $W(I_{r^*+1}) < (1+\epsilon)W(I_{r^*})$. I.e., for any $1 < r \le r^*$, $W(I_r) \ge (1+\epsilon)W(I_{r-1})$ and $W(I_r) \le krw_{max}$. The proof of the following lemma follows from Lemma 7 by setting $k' = kw_{max}$.

Lemma 10. $r^* \le \left(\frac{1+\sqrt{1+4\ln(1+\epsilon)\ln(k')}}{2\ln(1+\epsilon)}\right)^2$

The proof of the following theorem is left for the full paper.

Theorem 11. *Suppose that I' is a $(1+\epsilon)$-approximate weighted independent set in $\overline{G_{r^*+1}}$, then $I \equiv I_{r^*} \cup I'$ is $(1+\epsilon)$-approximate for G.*

6   Conclusions

In this paper we study the WMIS problem on restricted LoS networks where parameter k is a constant, and propose a polynomial time dynamic programming algorithm for the problem. We also use the DP algorithm to develop an EPTAS that applies to a semi-online algorithm with performance ratio $(1+\epsilon)$ and a look-ahead distance dependent on ϵ, for any $\epsilon > 0$.

For future work there are various avenues to explore. One immediate direction is to study the LoS network with different ranges of various parameters. It is interesting to determine the complexity of the problem (polynomial time solvable or NP-hard) given different values of the parameters. We can also extend the problem definition such that the distance restriction ω may take two different values ω_1 and ω_2 for each of the two dimensions. Another direction is to study other optmisation problems on LoS networks. Furthermore, we can explore other scheduling problems with constraints that can modeled in a similar way as a LoS network and adapt solutions to these scheduling problems.

References

1. Albers, S.: Online algorithms: a survey. Math. Program. **97**(1–2), 3–26 (2003)
2. Bellman, R., Esogbue, A.O., Nabeshima, I.: Mathematical Aspects of Scheduling and Applications. Elsevier, Amsterdam (2014)
3. Bollobás, B., Janson, S., Riordan, O.: Line-of-sight percolation. Comb. Probab. Comput. **18**(1–2), 83–106 (2009)
4. Cesati, M., Trevisan, L.: On the efficiency of polynomial time approximation schemes. Inf. Process. Lett. **64**(4), 165–171 (1997)
5. Chiu, S.N., Stoyan, D., Kendall, W.S., Mecke, J.: Stochastic Geometry and Its Applications. Wiley, Hoboken (2013)
6. Czumaj, A., Wang, X.: Communication problems in random line-of-sight ad-hoc radio networks. In: Hromkovič, J., Královič, R., Nunkesser, M., Widmayer, P. (eds.) SAGA 2007. LNCS, vol. 4665, pp. 70–81. Springer, Heidelberg (2007). https://doi.org/10.1007/978-3-540-74871-7_7
7. Diestel, R.: Graph Theory. Springer, New York (2000)
8. Farczadi, L., Devroye, L.: Connectivity for line-of-sight networks in higher dimensions. Discret. Math. Theor. Comput. Sci. **15** (2013)
9. Frieze, A., Kleinberg, J., Ravi, R., Debany, W.: Line-of-sight networks. Comb. Probab. Comput. **18**(1–2), 145–163 (2009)
10. Garey, M.R., Johnson, D.S.: Computers and Intractability: A Guide to the Theory of NP-Completeness (1979)
11. Håstad, J.: Clique is hard to approximate within $n^{1-\varepsilon}$. Acta Math. **182**, 105–142 (1999)
12. Sangha, P., Zito, M.: Finding large independent sets in line of sight networks. In: Gaur, D., Narayanaswamy, N.S. (eds.) CALDAM 2017. LNCS, vol. 10156, pp. 332–343. Springer, Cham (2017). https://doi.org/10.1007/978-3-319-53007-9_29

Braid Chain Radio Communication

Jacek Cichoń, Mirosław Kutyłowski$^{(\boxtimes)}$, and Kamil Wolny

Faculty of Fundamental Problems of Technology,
Wrocław University of Science and Technology, Wrocław, Poland
{jacek.cichon,miroslaw.kutylowski,kamil.wolny}@pwr.edu.pl

Abstract. We consider data transmission in a wireless multi-hop network, where node failures may occur and it is risky to send over a single
path. As the stations may transmit at the same time, we apply the Cai-
Lu-Wang collision avoidance scheme. We assume that the nodes know
only their neighbors and there is no global coordination. The routing
strategy is to transmits to all nodes that are in some sense closer to the
destination node.

We analyze propagation strategy for the case that the nodes know only
the nodes in their propagation range and there is no global coordination
of the network. Instead, a node transmits to all nodes that appear in
some sense closer to the destination node. We investigate data propagation speed in such networks, namely the delay due to conflict resolution.

In order to understand the phenomena arising there we focus on a fundamental case called a *braid chain*. We prove that for a braid chain of n
layers the normalized delay due to anti-collision mechanism is $\approx 0.28n \cdot \Delta$
(while the expected time for a chain of single stations is $0.5n \cdot \Delta$), where
Δ stands for the time interval length of Cai-Lu-Wang scheme. We also
show that the behavior of the braid chain rapidly converges to its stationary distribution with the length of the chain. For these results we
develop analytical methods that can be applied for other networks, e.g.
for the case when the forwarding delays are chosen with exponential
distribution.

1 Introduction

Transmitting data over a multi-hop ad hoc radio network creates a number of
problems concerning reliability. In such an environment (e.g. in a sensor field)
the nodes might be likely to fail or become controlled by an adversary. So sending
a message via a single path could be risky – a failure of every single node disrupts
the communication channel.

On the other hand, if the data are sent over multiple nodes within the same
transmission range we are risking that the nodes may transmit data at the same
time, create radio channel collisions and interrupt the communication as well.
We may attempt to orchestrate the transmission times in a careful way, however

Supported by National Science Centre, project HARMONIA, decision DEC-
2013/08/M/ST6/00928.

A. Fernández Anta et al. (Eds.): ALGOSENSORS 2017, LNCS 10718, pp. 223–235, 2017.
https://doi.org/10.1007/978-3-319-72751-6_17

any detailed configuration by a central coordinator is problematic in case of ad hoc networks. In principle, the nodes should work autonomously without a detailed pre-configuration. This is particularly helpful if the network state may change due to the fact that the nodes join and leave the network at unpredictable moments.

Communication Model. We assume that each node knows its geographical coordinates (e.g. from its GPS receiver) as well as its neighbors and their coordinates.

The nodes communicate within their transmission range, however we assume that if nodes A and B are considered as *neighbors*, then there is a bidirectional connection between them (similarly as in the Unit Disk Graph model).

We assume also that the network assigns a separate input channel to each node (defined via a separate frequency band or time multiplexing). Since we are talking about a multi-hop network, each channel can be reused. We have to assure only that the nodes in the same transmission range are assigned to different channels. We assume that the input channels are assigned to the nodes in a distributed way – e.g. via distributed graph coloring algorithms. The crucial problem to address here are *collisions*: if more than one neighbor of a node A sends a message to A at the same time, then a collision occurs. In this case A might be unable to receive any of these message.

Cai-Lu-Wang Anti-Collision Mechanism designed and analyzed in [2], is based on carrier sensing and random delays. A node that intends to send a message, chooses t, the starting moment of a transmission, uniformly at random from the interval $[T, T + \Delta]$, where T is the current time and Δ is the protocol parameter. At time t, just before starting a transmission, the node checks whether the shared communication channel is already busy. If the carrier signal is not detected, it starts its own transmission. Otherwise, it tries to transmit later. Clearly, the Cai-Lu-Wang mechanism may fail, as there is a slight time delay δ between the time when a node observes that the channel is free and the time when it starts the transmission. So if another station decides to monitor the channel within the time interval $[T + t - \delta, T + t + \delta]$, then a collision will occur. Therefore the choice of Δ has to take into account acceptable level of collision probability on one side and average transmission delay $\frac{1}{2}\Delta$ on the other side.

In order to simplify the notation from now on we assume that $\Delta = 1$. In other words, we assume that the time is expressed in the number of Δ intervals.

Communication and Routing Strategy. We consider point-to-point communication. If a node A aims to deliver a message to a node B, then we assume that A knows the coordinates of B and can attach them to the header of the message. The simplest routing strategy would be to forward the message to a neighbor that is the closest one to the target destination. If the network is relatively dense and without holes, then this leads to fairly good results. However, a single node can cut the connection – due to either a fault or misbehavior. Another strategy would be to forward the message to all neighbors that are closer to the destination. This improves resilience to faults, but may flood the network with a large number of transmissions. So we need a kind of compromise.

An intermediate solution could be to define a kind of virtual route between the source A and destination B. Namely, as intermediate nodes we could take all nodes that are at distance at most D from the line connecting A and B. In this way we prevent flooding, while on the other hand we should enable multiple paths.

Note that this overall strategy communication strategy is quite simplistic and lacks any global coordination. This is one of the main arguments why it might be interesting for ad hoc networks – simplicity is one of the main quality criteria.

Braid Chain Communication. In order to understand the behavior of the proposed propagation strategy we investigate a special case of such an architecture that we call a *braid chain*. A braid chain of length n consists of $2n$ nodes which are arranged in n *layers*, with two nodes per layer, say U_i and D_i on the layer i. Each node from layer i can send a message to both nodes of layer $i+1$ (see Fig. 1). For receiving, each node uses its private input channel. As two nodes from layer i send messages to a node of layer $i+1$, the Cai-Wang anti-collision mechanism is used.

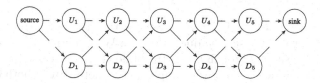

Fig. 1. Braid channel architecture

Now let us describe the communication strategy in a more formal way. First, the source node delivers the message to U_1 and D_1 at the same time. Let S_i denote a station at layer i - it can be U_i or D_i. It executes the following algorithm based on the Cai-Lu-Wang anti-collision mechanism ($\text{random}(0,1)$ stands for choosing $x \in [0,1]$ uniformly at random):

Braid Line Algorithm

1: start clock
2: **loop**
3:     **if** station S_i receives a message M at time T **then**
4:         $T_U \leftarrow T + \text{random}(0,1)$
5:         $T_D \leftarrow T + \text{random}(0,1)$
6:         **while** time not later than $\max\{T_U, T_D\}$ **do**
7:             **if** the current time is T_U **then**
8:                 **if** channel for U_{i+1} is free **then**
9:                     transmit message M to U_{i+1}
10:            **if** the current time is T_D **then**
11:                 **if** channel for D_{i+1} is free **then**
12:                     transmit message M to D_{i+1}

Let ζ denote the probability of collision during the transmission to a single node and let p stand for the probability of a node to fail. In general, the probability that the communication in a braid channel is not disconnected turn out to be higher than $(1 - (p^2 + \zeta^2 - (p\zeta)^2))^n$. If we disregard the probabilities of a failure due to collision on communication lines, then a braid chain communication channel is not disconnected with the probability $(1 - p^2)^n$. On the other hand, if we have two independent transmission lines, then the communication does not fail with the probability

$$1 - (1 - (1 - p)^n)^2 = (1 - p)^n(2 - (1 - p)^n).$$

Note that for $p = .1$ and $n = 100$, the braid chain communication channel is not disconnected with probability $\approx \frac{1}{e}$, while in the second case it happens with probability $\approx \frac{1}{e^{10}}$. Even if we increase the number of independent lines to 4, it will not help much as the probability of successful delivery would be $\approx \frac{4}{e^{10}}$ as it equals

$$1 - (1 - (1 - p)^n)^4 = (1 - p)^n(2 - (1 - p)^n)(-2(1 - p)^n + (1 - p)^{2n} + 2).$$

The difference is still striking, if we assume that the failure probability in a braid chain increases to, say, $2p$ due to Cai-Lu-Wang anti-collision mechanism failure. Then the success probability of the braid chain decreases to $(1 - (2p)^2)^n$ which is approximately $\frac{1}{e^4}$ for our example. Let us note that in both cases there might be collisions which are due to other transmissions. If we increase p to take them into account, then the advantage of the braid channel is even higher.

We may also consider the same architecture as for the braid chain, but for a single radio channel. Then instead of probability p of a failure of receiving by a node due to transmission collision, we have to consider the failure probability $p/2$ as we may double the length of the interval for the Cai-Lu-Wang method. Then the probability of a successful delivery now becomes $\approx (1 - p^2)^n \cdot (1 - \frac{p}{2})^n$ and it is significantly lower than $(1-(2p)^2)^n$. E.g. for $n = 100$, $p = .1$ and a single radio channel we get $(1 - p^2)^n \cdot (1 - \frac{p}{2})^n \approx \frac{1}{e} \cdot \frac{1}{e^5} = \frac{1}{e^6}$ while $(1 - (2p)^2)^n \approx \frac{1}{e^4}$.

Propagation Delay Problem. Since on each layer the nodes have to apply a random delay, the Cai-Lu-Wang anti-collision mechanism influences the data propagation speed over the braid chain. Our main goal is to investigate this speed, i.e. to learn what is the extra time necessary for a message sent from the source node to arrive at the sink node when transferred via a braid chain of the length n. By extra time we mean the time resulting from application of Cai-Lu-Wang anti-collision mechanism. Certainly, at each level this delay is between 0 and Δ (recall that we normalize time scale to get $\Delta = 1$), so the total delay during each successful execution is between 0 and n.

Unfortunately, despite simplicity of the protocol the underlying stochastic process turns out to be fairly difficult to analyze.

Related Work. Routing strategies and propagation of messages in multi-hop radio networks is a challenging topic. There are many papers dealing with routing strategies that have to be both reliable and efficient. Several message complexity problems for many different models connected with radio communication were presented and analyzed in [4]. Related problems of extreme propagation were also raised in [1]. Paper [9] presents a randomized gossiping protocol for information spreading. In [5] authors analyze the fault tolerant data propagation scheme where retransmitions are being performed when acknowledgment is not received. A somewhat similar channel architecture to the braid chain concept, but from a perspective of percolation problem, was investigated in [3]. In this paper several results for the quality of the transmission in such braid chain model were obtained with the presence of random failures.

Although many of these papers contain deep analytic results on algorithmic features of the considered protocols - most conclusions and details are related to common scenarios or "single step" cases. This is due to the fact that a time analysis for a process running on a multi-hop network usually requires deep understanding and effective mathematical tools for analyzing complex stochastic processes. Such tools are frequently unavailable.

Results Overview. The key problem to understand the propagation delay of braid channel communication is to understand the distribution of the time difference between the times when the nodes of a given layer learn the transmitted message – so called *layer delay*. In Sect. 2 we derive moments of this distribution and show that these distributions converge rapidly to a stationary distribution.

In Sect. 3 we analyze difference between times when a message arrives at layers i and $i+1$ based on the assumption that the layer delay at layer i is distributed according to the stationary distribution. Thereby we obtain the expected time for message delivery over a braid chain with a given number of layers.

Our main theoretical contribution is the way to extract constants by iterative integration (Sect. 3) and application of rapid mixing (Sect. 2). Apart from the analytic results on the braid chain communication, the proof techniques, especially from Sect. 3, might be useful elsewhere due to their relative simplicity and effectiveness (see, e.g. Sect. 4). They apply immediately for different network architectures, the flooding algorithms, etc.

2   Layer Delays

Let us consider a transmission of a message M over the braid channel.

Definition 1. *Let r_{U_i} and r_{D_i} denote the times when U_i, and respectively D_i receive a message M for the first time. Then by the delay at layer i we mean $d_i = |r_{U_i} - r_{D_i}|$.*

Note that $d_i \leq 1$. Indeed, let us consider the layer $i - 1$ and assume that $V \in \{U_{i-1}, D_{i-1}\}$ is the first node at layer $i - 1$ that receives M. Then M cannot be delivered to the first node of layer i before time $r(V)$, while on the other hand V transmits M to both U_i and D_i before time $r(V) + 1$.

Let us observe that the value of d_i influences the time needed to transmit M from the layer i to the layer $i+1$. Indeed, initially for time d_i only one node of layer i is ready for forwarding the message M to the layer $i+1$. After this initial time period d_i, the second node at layer i becomes ready for sending and the probability of starting a transmission to a node of layer $i+1$ increases. So we see that small values of d_i increase the propagation speed.

If $d_i = 1$, then the messages arriving at layer $i+1$ originate from the same node. The time to deliver the message from layer i to layer $i+1$ is then described by the minimum of two random variables distributed uniformly at random over the interval $[0,1]$, with the expected time $\frac{1}{3}$. As this case is the most pessimistic one, we get a rough estimation of the propagation speed. Below we will get a precise result.

2.1 Modelling the Layer Delays

In this section we present some theoretical points, strictly connected with the further analysis of the layer delays. We consider the following model. Let x, y, u, v be independent random variables with the uniform distribution on the interval $[0,1]$. Let $d \in [0,1]$. For a given choice of number d we consider random variables X_d, Y_d and Z_d defined as follows:

$$X_d = \min(x, d+y), \quad Y_d = \min(u, d+v), \quad Z_d = |X_d - Y_d|.$$

Lemma 1. X_d and Y_d are independent random variables with the same distribution with density function f_d given by the following formula:

$$f_d(z) = \begin{cases} 1 & \text{if } z \in [0, d], \\ 2 + d - 2z & \text{if } z \in [d, 1]. \end{cases}$$

Proof. Notice that the condition $\min(x, d+y) \geqslant z$ is equivalent to $(x \geqslant z) \wedge (y \geqslant z - d)$. Since x and y are independent, $\Pr(X > z) = (1 - z)(1 - \max(0, z - d))$. So:

$$\Pr(X < z) = \begin{cases} z & \text{if } z \in [0, d], \\ (2 + d)z - d - z^2 & \text{if } z \in [d, 1]. \end{cases}$$

By differentiating this cumulative distribution function we get the formula for f_d. $\square$

Let us consider the random variable $U_d = X_d - Y_d$. The density of U_d can be computed as the convolution of f_d and f_d. That is,

$$f_{U,d}(x) = \int f_d(u) f_d(u - x)\, du.$$

As X_d, Y_d are independent and have the same probability distribution, $f_{U,d}$ is symmetric. Therefore, the density function k_d of the random variable Z_d equals

$$k_d(x) = \begin{cases} 2 f_{U,d}(x) & \text{if } x \in [0, 1], \\ 0 & \text{otherwise.} \end{cases}$$

For $x \geq 0$ after some tedious calculations we get the following values for $k_d(x)$:

$$
\begin{array}{ll}
\frac{2}{3}(4 - 3d + 3d^2 - d^3 - 3x + 3dx - 3d^2x - 3x^2 + 2x^3) & \text{for } d \in (0, \frac{1}{2}] \wedge x \in [0, d] \\
\frac{2}{3}(-d^3 - 3d^2x + 3d^2 + 2x^3 - 6x + 4) & \text{for } d \in (0, \frac{1}{2}) \wedge x \in [d, 1-d] \\
2 + 2d - 4x - 2dx + 2x^2 & \text{for } d \in (0, \frac{1}{2}) \wedge x \in [1-d, 1] \\
\frac{2}{3}(2 - 3x + x^3) & \text{for } d = 0 \\
\frac{2}{3}(4 - 3d + 3d^2 - d^3 - 3x + 3dx - 3d^2x - 3x^2 + 2x^3) & \text{for } d (\in \frac{1}{2}, 1) \wedge x \in [0, 1-d] \\
2(1 - x) & \text{for } d \in (\frac{1}{2}, 1) \wedge x \in [1-d, d] \\
2 + 2d - 4x - 2dx + 2x^2 & \text{for } d \in (\frac{1}{2}, 1) \wedge x \in [d, 1] \\
0 & \text{otherwise}
\end{array}
$$

Let Z be a random variable with density function k_d. We will need exact expressions for the moments of Z. The following lemma provides us a compact answer.

Lemma 2. *Let Z be a random variable with the density function $k_d(z)$. Then the expected value for Z^n for $n \in \mathbb{N}$ equals:*

$$
\mathbf{E}[Z^n] = \frac{2}{(n+1)(n+2)}\left(d + d(1-d)^{3+n} + d^{2+n}(d-1) + \right.
$$
$$
\left. \frac{2 + 2(1-d)^{3+n}(1-2d) - 2d^{3+n}}{n+3} - \frac{4(1-d)^{4+n}}{(n+3)(n+4)}\right)
$$

Proof (Sketch). In order to prove Lemma 2 we have to use the tools of analytical combinatorics. A reader may refer to [7] for details.

Let $M_Z(t)$ denote the moment-generating function of the distribution with the density function $k_d(z)$, that is,

$$
M_Z(t) = \int_{-\infty}^{+\infty} e^{zt} k_d(z)\, dz.
$$

Then

$$
M_Z(t) = \sum_{n=0}^{\infty} \frac{t^n}{n!} \mathbf{E}[Z^n].
$$

Using the properties of the moment-generating function, we only have to retrieve the coefficient $[t^n]M_Z(t)$. Then $\mathbf{E}[Z^n] = n![t^n]M_Z(t)$. After tedious computations, using the formula for k_d, we get the formula stated in Lemma 2.[1]   □

By Lemma 2 and linearity of expected value we get in particular:

$$
\mathbf{E}[Z] = \frac{1}{15}(d^5 - 5d^3 + 5d^2 + 4)
$$

$$
\mathbf{E}[Z^5 - 5Z^3 + 5Z^2] = \frac{1}{378}(10d^9 - 63d^8 + 153d^7 - 252d^6 +
$$
$$
378d^5 - 315d^4 - 21d^3 + 144d^2 + 110)
$$

---

[1] A reader interested in deriving them is advised to use a symbolic computations program.

2.2   Stabilization of Probability Distribution of Layer Delays

In the previous subsection we have analyzed layer delay d_{i+1} given a specific value d for d_i. Our goal now is to check how the probability distribution of d_i evolves with i. Note that $d_1 = 0$, since the source node transmits the message to U_1 and D_1 at the same time. Indeed, there is no need for the Cai-Lu-Wang anti-collision mechanism, as there is no other node in the starting layer sending to U_1 and D_1. For d_i with $i > 1$ the situation gets complicated.

We consider a stochastic process $(d_i)_{i=0,1,...}$. As probability distribution of d_{i+1} depends only on the value of d_i, it is a Markov chain. The function $k(d, x) = k_d(x)$ computed in Sect. 2.1 is the stochastic density kernel of this chain, that is:

$$\Pr[d_{i+1} \in A | d_i = d] = \int_A k_d(x)\, dx.$$

Stationary Distribution. First observe that $k_d(x) > 0$ for each $d \in [0, 1]$ and $x \in [0, 1]$. Since every strictly positive kernel on a bounded closed interval is ergodic (see [6]) we deduce that the Markov Chain $(d_i)_{i=0,1,...}$ is ergodic. Therefore we conclude that.

Lemma 3. *There exists a probability distribution μ on the interval $[0, 1]$ such that for any probability distribution π, the sequence of iterates (π_n) defined by $\pi_1 = \pi$ and $\pi_{n+1}(x) = \int_0^1 \pi_n(t)k(t, x)\, dt$ converges uniformly to μ.*

Thereby the distribution μ from Lemma 3 is the stationary probability distribution for the Markov chain $(d_i)_{i=0,1,...}$.

Theorem 1. $\mathbf{E}\,[\mu] = 0.286067 + \epsilon$, *where* $0 \leqslant \epsilon \leqslant 0.006$,
    $\mathbf{Var}\,[\mu] = 0.126981 + \delta - (0.286067 + \epsilon)^2$, *where* $0 \leqslant \delta \leqslant 0.0005$.

Proof. For the stationary distribution μ, after making a single transition we get again the stationary distribution. That is, $\mu(x) = \int_0^1 \mu(t)k(t, x)\, dt$. Note that $k(t, x)$ is a piecewise polynomial function. Therefore we are able to make the following computations:

$$
\begin{aligned}
\mathbf{E}[\mu] &= \int_0^1 x\mu(x)\, dx = \int_0^1 x\left(\int_0^1 \mu(t)k(t, x)\, dt\right) dx \\
&= \int_0^1 \mu(t)\left(\int_0^1 xk(t, x)\, dx\right) dt \\
&= \tfrac{1}{15}\int_0^1 (t^5 - 5t^3 + 5t^2 + 4)\mu(t)\, dt \\
&= \tfrac{4}{15} + \tfrac{1}{15}\int_0^1 (t^5 - 5t^3 + 5t^2)\mu(t)\, dt &(1) \\
&= \tfrac{4}{15} + \tfrac{1}{15}\int_0^1 (t^5 - 5t^3 + 5t^2)\left(\int_0^1 \mu(s)k(s, t)\, ds\right) dt &(2) \\
&= \tfrac{4}{15} + \tfrac{1}{15}\int_0^1 \mu(s)\left(\int_0^1 (t^5 - 5t^3 + 5t^2)k(s, t)\, dt\right) ds \\
&= \tfrac{4}{15} + \tfrac{1}{15}\tfrac{110}{378} + \tfrac{1}{15}\tfrac{1}{378}\int_0^1 \mu(s)w(s)\, ds &(3)
\end{aligned}
$$

where $w(s) = 10s^9 - 63s^8 + 153s^7 - 252s^6 + 378s^5 - 315s^4 - 21s^3 + 144s^2$. It is easy to check that $0 \leqslant w(s) \leqslant 34$ for $s \in [0, 1]$. Therefore

$$\mathbf{E}[\mu] = \frac{4}{15} + \frac{1}{15}\frac{110}{378} + \epsilon, \tag{4}$$

where $0 \leqslant \epsilon \leqslant 0.00599647$. So finally, $\mathbf{E}[\mu] \approx 0.28$.

Remark 1. The main trick here was transition from (1) to (2) based on the fact that μ is the stationary distribution. As we can compute the integrals of the form $\int_0^1 t^i k(s,t)\, dt$ for a fixed i, in line (3) we get back an integral of the form $\int_0^1 \mu(s) w(s)\, ds$, where $w(s)$ is a polynomial, with some constant in front of the integral. Note that if we need a better precision, we can repeat the same trick to (3). So by repeating the same trick again and again, we can get an estimation of $\mathbf{E}[\mu]$ with any required precision via a simple numeric computation.

One can notice that method above may also be used for the computations of variance. For this purpose we use $\mathbf{Var}[\mu] = \mathbf{E}[\mu^2] - (\mathbf{E}[\mu])^2$, the above result on $\mathbf{E}[\mu]$ and perform analogous calculations for $\mathbf{E}[\mu^2]$. Instead of using an explicit formula for $\int_0^1 x k(t,x)\, dx$ we make calculations for $\int_0^1 x^2 k(t,x)\, dx$.

Rapid Mixing. Now we aim to show that the Markov chain $(d_i)_{i=0,1,\dots}$ rapidly approaches the stationary distribution μ. We show the following result:

Theorem 2. *For $t > \frac{(-\log \varepsilon + 1)}{2 - \log 3}$ the variation distance between the distributions μ and d_t is at most ε. That is $\frac{1}{2}\int_0^1 |\mu(x) - d_t(x)| dx < \varepsilon$.*

Proof. Let us recall the Coupling Lemma - a powerful and easy tool for showing convergence of Markov chains [8]. In order to estimate the convergence of a given process $(d_i)_{i=0,1,\dots}$ to its stationary distribution we create two *coupled* processes, say $(c_i, c_i')_{i=0,1,\dots}$ so that:

- the process $(c_i)_{i=0,1,\dots}$ (respectively, $(c_i')_{i=0,1,\dots}$) treated alone has the same transition probabilities as $(d_i)_{i=0,1,\dots}$,
- there are dependencies between the choices of the processes c_i and c_i'; due to these dependencies the processes c_i and c_i' should converge and reach the same state.

The Coupling Lemma says that if $\Pr(c_t \neq c_t') \leq p$, then the variation distance between the distribution of c_t and the stationary distribution of this process is at most $2p$. The Coupling Lemma is usually formulated and applied for discrete state Markov chains. However, we shall use it for non-discrete Markov chains (where normally it is impossible to construct a good coupling).

As in Sect. 2.1, in order to describe a step of the chain $(d_i)_{i=0,1,\dots}$, we consider 4 random variables: x, y, u, v, where x, y denote delays chosen for transmitting to U_{i+1} and u, v the delays for transmitting to D_{i+1}, while x, u denote the delays chosen by the station from layer i that is first ready for starting a transmission.

Let us consider two copies of the Markov process $(d_i)_{i=0,1,\dots,}$. Let the first process be in state d, the second one in state d'. Defining a coupling for these processes is surprisingly easy: for both process choose exactly the same values for x, y, u, v.

Lemma 4. *With probability at least $\frac{1}{4}$ after a coupling step both processes reach the same state.*

Proof. W.l.o.g. assume that $d \leq d'$. Let us consider the first process. With probability at least 0.5 we have $x = \min(x, y + d)$. Similarly, $u = \min(u, v + d)$ with probability at least 0.5. Since x, y are stochastically independent from u, v, with probability at least $\frac{1}{4}$ the layer delay for the first process after performing the step is $|x - u|$.

For the second process note that if $x = \min(x, y + d)$ and $u = \min(u, v + d)$, then $x = \min(x, y + d')$ and $u = \min(u, v + d')$ as well, as $d \leq d'$. It follows that in this case the second process reaches the layer delay $|x - u|$ as well.   □

By Lemma 4, after T steps the probability that the processes still differ is at most $(\frac{3}{4})^T$. Hence, by Coupling Lemma, for $t > \frac{(-\log \varepsilon + 1)}{2 - \log 3}$ the variation distance between the distribution of d_t and the stationary distribution μ is at most

$$2 \cdot \left(\frac{3}{4}\right)^t \leq 2 \cdot \frac{3}{4}^{\frac{(-\log \varepsilon + 1)}{2 - \log 3}} = 2 \cdot (2^{\log 3/4})^{\frac{(-\log \varepsilon + 1)}{\log(4/3)}} = 2 \cdot 2^{\log \varepsilon - 1} = \varepsilon$$

and Theorem 2 follows.   □

For comparison, let k_i denote the density of probability distribution of d_i. We have not derived a closed expression for k_i, however it is possible to estimate numerically the total variation distance between k_i and k_{i+1} separately for each single i. For instance, we get

$$\frac{1}{2} \int_0^1 |k_3(x) - k_2(x)| \, dx < 0.001$$

$$\frac{1}{2} \int_0^1 |k_4(x) - k_3(x)| \, dx < 0.0001$$

$$\frac{1}{2} \int_0^1 |k_5(x) - k_4(x)| \, dx < 0.00001$$

This confirms the general observation from Theorem 2.

3   Propagation Time

Our previous calculations were limited to the delays within a layer. Now, given probability distribution of these delays we aim to find the expected time of message delivery from layer i to layer $i + 1$.

Propagation Time for a Fixed Delay. For a moment let us fix the value of a layer delay to d. Let $\hat{T}_d$ denote the time that elapses between the moment when the first node at the current layer gets the message and the time when the first node of the next layer gets the message. According to the notation used in Sect. 2, $\hat{T}_d = \min\{X_d, Y_d\}$. Hence,

$$\hat{T}_d = \min \{ \min\{x, d + y\}, \min\{u, d + v\}\}.$$

In order to compute cumulative density function of the random variable $\hat{T}_d$ we use the similar argument as in Sect. 2. Notice that the condition $\hat{T}_d \geq z$ is equivalent to $(x \geq z) \wedge (y \geq z - d) \wedge (u \geq z) \wedge (v \geq z - d)$. Since x, y, u, v are stochastically independent and uniformly distributed over $[0, 1]$, we thereby get

$$\Pr(Z > z) = (1 - z)^2(1 - \max(0, z - d))^2$$

Hence,

$$\Pr(Z < z) = \begin{cases} 1 - (1 - z)^2 & \text{if } z \in [0, d], \\ 1 - (1 - z)^2\big(1 - (z - d)\big)^2 & \text{if } z \in [d, 1]. \end{cases}$$

By differentiating this cumulative density function we get the following formula for the density function $f_{\hat{T}_d}$:

$$f_{\hat{T}_d}(z) = \begin{cases} 2 - 2z & \text{if } z \in [0, d], \\ D(d, z) & \text{if } z \in [d, 1], \end{cases}$$

where $D(d, z) = 4 + 6d + 2d^2 - 12z - 12dz - 2d^2z + 12z^2 + 6dz^2 - 4z^3$. Let d be a fixed delay. We aim to compute the expected time

$$\mathbf{E}[\hat{T}_d] = \int_0^1 z f_{\hat{T}_d}(z)\, dz$$

After some straightforward operations we get $\mathbf{E}[\hat{T}_d] = W(d)$, where $W(d)$ is the following polynomial:

$$W(d) = \tfrac{1}{30}\left(6 + 15d - 20d^2 + 10d^3 - d^5\right) \tag{5}$$

Propagation Time for Stationary Distribution. As before, let $\mu(d)$ denote the stationary distribution for the layer delay. If the initial layer delay occurs according to the distribution $\mu(d)$, then the expected propagation time $\hat{T}$ for one layer can be computed as follows:

$$\mathbf{E}[\hat{T}] = \int_0^1 W(z)\mu(z)\, dz = \int_0^1 W(z)\left(\int_0^1 \mu(t)k(t, z)\, dt\right) dz = \int_0^1 \mu(t)\left(\int_0^1 W(z)k(t, z)\, dz\right) dt$$

$$= \int_0^1 \mu(t)\left(\int_0^1 \tfrac{1}{30}(6 + 15z - 20z^2 + 10z^3 - z^5)k(t, z)\, dz\right) dt$$

By Lemma 2 we obtain

$$\mathbf{E}[\hat{T}] = \tfrac{787}{2835} + \int_0^1 \mu(t)\left(\int_0^1 R(z)k(t, z)\, dz\right) dt$$

$$= \tfrac{787}{2835} + \tfrac{162949}{48648600} + \int_0^1 \mu(t)\left(\int_0^1 S(z)k(t, z)\, dz\right) dt$$

$$= \tfrac{787}{2835} + \tfrac{162949}{48648600} + \tfrac{315286}{1206079875} + \epsilon \approx 0.28$$

where $\epsilon \leq 0.00086$ and $R(z), S(z)$ are polynomials given by the formulas

$$R(z) = \tfrac{11z^2}{126} - \tfrac{26z^3}{135} + \tfrac{7z^4}{36} - \tfrac{z^5}{10} + \tfrac{z^6}{45} - \tfrac{2z^7}{315} + \tfrac{z^8}{180} - \tfrac{z^9}{1134}$$

$$S(z) = \tfrac{263z^2}{62370} - \tfrac{z^3}{1215} - \tfrac{11z^4}{756} + \tfrac{529z^5}{18900} - \tfrac{233z^6}{8100} + \tfrac{94z^7}{4725} -$$
$$\tfrac{3z^8}{280} + \tfrac{173z^9}{34020} - \tfrac{19z^{10}}{11340} + \tfrac{29z^{11}}{155925} + \tfrac{z^{12}}{28350} - \tfrac{z^{13}}{90090}$$

The integral $\int_0^1 \frac{1}{30}\left(6 + 15z - 20z^2 + 10z^3 - z^5\right)k(t, z)\, dz$ is computed separately on two intervals, where $k(t, z)$ is a polynomial. So we compute two integrals of polynomial functions. The result is an expression with the values of 2 polynomials at respectively, 2 points for each polynomial. In each case the result is a constant plus a polynomial on d without a free term. As the result, say $\mathcal{R}$, is used for computing the integral $\int_0^1 \mu(t)\mathcal{R}(t)\, dt$, the constant term can be placed in front of the integral. After that we perform the same procedure again and again. We retrieve the constants step by step to achieve a better precision and at each step we obtain an integral of the polynomial with a higher degree, so we only have to compute more and more moments, to achieve a better approximation. Using three iterations of that method we conclude that $\mathbf{E}\left[\hat{T}\right] = 0.281212 + \epsilon$, where $|\epsilon| < 10^{-3}$. Therefore, by linearity of expectation

$$\lim_{n \to \infty} \frac{\mathbf{E}\left[T_n\right]}{n} = \mathbf{E}\left[\hat{T}\right] \approx 0.28 \tag{6}$$

The method presented above may also be used for the computations of a precise approximation of the variance of T_n. We have shown earlier the convergence to stationary distribution $\mu(x)$ of the Markov Chain. If we assume that delays occur with $\mu(x)$, then the random variables for the time execution between two frames can be considered as independent. Having the expected value for the time of delivery in a single layer, what we need to obtain is the second moment of the random variable denoting the time delivery with an assumption that the delay d is fixed. Finally, we conclude that

$$\lim_{n \to \infty} \frac{\mathbf{Var}[T_N]}{n} \approx 0.035 \tag{7}$$

4   Final Remarks

The results obtained in this paper can be generalized for other propagation models. For instance, we may modify the braid chain protocol in the way that the delay to send a message to the next node in the chain is chosen according to the exponential distribution (which is a standard model in many areas of telecommunication). Using the proof techniques presented above we can show that there is $|\epsilon| \leq \frac{0.006}{\lambda^2}$ such that

$$\lim_{n \to \infty} \frac{\mathbf{E}\left[T_n\right]}{n} = \frac{313}{800} \cdot \frac{1}{\lambda} - \epsilon,$$

where $\mathbf{E}\left[T_n\right]$ is the expected value of time needed to pass through n levels.

The key technical point of our analysis is the technique used in Sect. 3. The approach based on polynomials and iterative integrating leads to surprisingly precise results in a conceptually easy way. We feel that this approach can be used in many different occasions, substantially simplifying the analysis.

References

1. Baquero, C., Almeida, P.S., Menezes, R.: Fast estimation of aggregates in unstructured networks. In: ICAS, pp. 88–93. IEEE Computer Society (2009)
2. Cai, Z., Lu, M., Wang, X.: Distributed initialization algorithms for single-hop ad hoc networks with minislotted carrier sensing. IEEE Trans. Parallel Distrib. Syst. **14**(5), 516–528 (2003)
3. Cichoń, J., Klonowski, M.: On flooding in the presence of random faults. Fundam. Inform. **123**(3), 273–287 (2013)
4. Cichoń, J., Lemiesz, J., Zawada, M.: On message complexity of extrema propagation techniques. In: Li, X.-Y., Papavassiliou, S., Ruehrup, S. (eds.) ADHOC-NOW 2012. LNCS, vol. 7363, pp. 1–13. Springer, Heidelberg (2012). https://doi.org/10.1007/978-3-642-31638-8_1
5. Cichoń, J., Gebala, M., Zawada, M.: Fault tolerant protocol for data collecting in wireless sensor networks. In: The 22nd IEEE Symposium on Computers and Communications, July 2017
6. Feller, W.: An Introduction to Probability Theory and Its Applications, vol. 2, 2nd edn. Wiley, New York (1971)
7. Flajolet, P., Sedgewick, R.: Analytic Combinatorics, 1st edn. Cambridge University Press, New York (2009)
8. Lindvall, T.: Lectures on the Coupling Method. A Wiley-Interscience Publication, New York (1992)
9. Mosk-Aoyama, D., Shah, D.: Computing separable functions via gossip. In: ACM PODC 2006, pp. 113–122. ACM (2006)

Author Index

Printed in the United States
By Bookmasters